The Chemistry and Testing Of Dairy Products

By
J. A. NEWLANDER, Ph.D.
Professor Emeritus of Dairy Industry

and

HENRY V. ATHERTON, Ph.D.
*Associate Professor of Dairy Industry
University of Vermont
College of Agriculture and Home Economics
Burlington, Vermont*

THIRD EDITION
REVISED 1964

OLSEN PUBLISHING COMPANY • MILWAUKEE, WISCONSIN

Copyright 1964
by Olsen Publishing Co.,
Milwaukee, Wis.

All rights reserved.
This book or any part thereof
must not be reproduced in any form
without written permission of the publisher.

Printed in the United States of America

637.127
N54t
1964

Preface

This book is a complete revision and expansion of the original "Testing and Chemistry of Dairy Products" by J. A. Newlander. Information on the composition and properties of milk has been brought up to date and related to modern methods for making the physical and chemical tests which are so important in quality control of milk and milk products.

It is the intent of the authors to prepare a somewhat detailed discussion of the chemistry of the dairy products and the testing procedures to give an insight into the composition of milk and its derivatives, and the reasons for the special conduct of the different tests.

The revisions and additions which have been made in this book recognize the many changes which have taken place in the field of quality control during the years since World War II. The literature has been carefully reviewed and those contributions are presented which have importance in assessing the compositional quality of milk.

Additional chapters have been added to summarize information in several areas not covered in present texts or else receiving only piece-meal treatment. Experience in teaching agricultural students, fieldmen, regulatory workers and laboratory personnel has demonstrated the need for an organized presentation of this material.

This text is designed to discuss the several aspects of dairy testing and quality control in such a manner that the beginning student can appreciate the varied physical and chemical properties of milk and their influence on consumer acceptance of the resulting dairy products. Suggested laboratory tests are presented to illustrate the principles discussed. At the same time, the more advanced students and those responsible for quality control in the dairy industry will find this a valuable reference to the methods of conducting the different tests in the dairy laboratory.

Table of Contents

Chapter I

                                                          Page

COMPOSITION OF MILK 1
    Definition and Composition of Milk.
    Chemical Composition and Properties of Fat, Compound Lipids, Derived Lipids, Casein, Albumin, Globulin, Lactose, and Ash.
    Initial and Final Solubility of Lactose.
    Energy Value of Milk in Relation to Fat.
    Vitamins in Milk — A, D, E, K. Thiamin, Riboflavin, Grass Juice Factor, Niacin, Pantothenic Acid, Pyridoxine, Biotin, Choline, Inositol, Folic Acid, P-amino Benzoic Acid, B_{12}. Ascorbic Acid (C). Chemical Composition and Properties.
    Enzymes and Their Reactions in Milk.
    Gases in Milk.

Chapter II

PHYSICAL PROPERTIES OF MILK 35
    Taste, Color, Freezing Point, Boiling Point, Specific Gravity. Specific Heat, Surface Tension, Bound Water, Foam, Fat Clumping, Electrical Conductivity, Effect of Heat.

Chapter III

SAMPLING METHODS 55
    Daily and Composite Samples.
    Definition of Composite Sample.
    Methods of Taking Composite Samples.
    Care and Testing of Composite Samples.
    Accuracy of Composite Samples.
    Sampling Other Dairy Products — Cream, Evaporated Milk, Sweetened Condensed Milk, Dried Milk, Malted Milk, Butter, Cheese, Ice Cream and Frozen Dairy Products.

Chapter IV

BABCOCK, GERBER, MOJONNIER TESTS FOR FAT 67
    The Babcock Test for Milk, Cream, Skimmilk, Whey, Cheese, Churned Milk, Frozen Milk.
    The Gerber Test for Whole Milk, Chocolate Milk, Chocolate Drink, Cream, Skimmilk, Buttermilk, Whey, Frozen Desserts.
    Comparing the Babcock and Gerber Tests.
    The Mojonnier Test for Fat in Milk, Skimmilk, Buttermilk, Whey, Cream, Ice Cream, Evaporated Milk, Sweetened Condensed Milk, Whole Milk Powder, Skimmilk Powder, Butter, Cheese, Cottage Cheese, Creamed Cottage Cheese.

Chapter V

SOME OTHER TESTS FOR FAT 115
    Modified Babcock Tests (Acid Reagents). American Association Test for Buttermilk, Whey, and Skimmilk. Pennsylvania Test for Ice Cream, Whole Milk, Whey, Cream, Chocolate Milk, Sweetened Condensed Milk, Evaporated Milk, Dried Whole Milk, Butter, Cheese. Acetic Acid and Sulfuric Acid Method for Ice Cream Mix.

Modified Babcock Tests (Alkaline Reagents). Minnesota Test for Ice Cream, Nebraska Test for Ice Cream, Garrett-Overman Ice Cream Test.
The Detergent Tests — DPS Detergent Method for Fat in Milk, Cream, and Ice Cream. TeSa Fat Test for Milk, Chocolate Milk, Low Fat Frozen Desserts, Cream, Ice Cream, High Fat Frozen Desserts.
Optical Methods.

Chapter VI
TESTING FOR SOLIDS IN DAIRY PRODUCTS 132
Official Method (AOAC) for Total Solids in Milk and Cream.
Mojonnier Test for Total Solids in Milk, Skimmilk, Buttermilk, Whey, Cream, Ice Cream, Evaporated Milk, Sweetened Condensed Milk, Whole Milk Powder, Skimmilk Powder, Butter, Cheese, Cottage Cheese.
Dietert Method for Total Solids in Dairy Products.
Cenco Moisture Balance.
Toluene Distillation Method for Moisture in Powdered Milk.
Other Methods for Determining Total Solids in Dairy Products.

Chapter VII
THE LACTOMETER AND ITS USES 146
Quevenne and New York State Board of Health Lactometer.
Specific Gravity Determinations of Milk — by Lactometer, Westpahl Balance and Pycnometer.
Formulae for Solids in Milk.
Relationship of Fat to Solids Not Fat in Milk.
Adulteration of Milk — Watering, Skimming, Watering and Skimming.

Chapter VIII
TESTS FOR MILK QUALITY 163
Phosphatase Tests for Pasteurization — Scharer Field Test. Cornell Phosphatase Test for Milk and Fluid Products and Cheese and Solid Dairy Products. Long Method, Short Method. Rapid Field Test.
Sediment Test of Milk.
Viscosity — Pipette Method and Borden Flow Meter.
Feathering and Oiling-Off of Cream in Coffee. Protein Stability, Keeping Quality, Rancidity, Homogenization Efficiency.

Chapter IX
THE ADULTERATION OF MILK AND MILK PRODUCTS 187
Determining Added Water — Immersion Refractometer and Thermister Cryoscope.
Substitution of Vegetable Fats for Milk Fat in Dairy Products.
Improving Quality or Appearance — Preservatives in Milk, Boiled Milk Test, Test to Determine Coloring Matter.
Accidental Adulteration — Residues in Milk. Testing for Quaternary Ammonium Compounds, Hypochlorites and Chloramins in Water and Milk. Rapid Disk Assay Method for Penicillin in Milk.

Chapter X
ANALYSIS OF BUTTER AND CHEESE 207
Butter
Composition of Butter.
Overrun in Butter — Factory, Churn and Composition.
Moisture Tests — Laboratory and Official Methods, Preparation of Samples — Official, Mechanical Stirrer and Factory Methods.
Salt Test — Cornell and Standard Solution Methods.
Fat Test — Kohman Method.

Sediment Tests for Butter, Cream, Salt.
Free Water in Butter.

Cheese

Composition of Cheese.
Sampling for Analysis.
Moisture Tests — Troy, Olive Oil, Karl Fischer Methods.
Sediment Test for Cheese — Spicer and Price, Orthophosphoric Methods.
Salt Tests — ADSA Test, Marquardt Test, and Mercurimetric Test.

Chapter XI

THE ACIDITY OF MILK AND ITS PRODUCTS 235

Apparent and Real Acidity.
Testing for Acidity — American Dairy Science Association Recommended Procedures for Milk, Skimmilk, Buttermilk, Whey, Sweet Cream, Sour Cream, Condensed Milk, Ice Cream Mix, Chocolate Milk, Butter, Cheese. Rapid Acid Test.
Factors Affecting Titratable Acidity — Action of Calcium Phosphate, Stage of Lactation, Mastitis, Enzyme Activity, Effect of Feed.
Hydrogen Ion Concentration — Determination by Colorimetric and Potentiometric Methods.
Buffer Systems in Milk.
Relationship of Acidity in Different Products.

Chapter XII

DAIRY CHEMISTRY PROBLEMS 255

Solutions.
Moisture and Solids.
Standardization.
Percentage.

Chapter XIII

BACTERIOLOGY OF MILK 263

Bacteria — Characteristics, Growth and Sources of Milk Bacteria.
Counting Bacteria in Milk — Plate Method, Direct Miscroscopic Method, Methylene Blue Reduction Test and Resazurin Test.
Economic Importance of Milk Bacteria.
Mastitis Tests — Microscopic Method, Brom Thymol Blue, Chloride, Hotis, Modified Whiteside.
Pathogenic Bacteria.
Yeasts and Molds.
Production of High Quality Milk.

Chapter XIV

NUTRITIVE VALUE OF MILK AND ITS PRODUCTS 294

Food Value of Milk, Butter, Cheese, Buttermilk, Skimmilk and Chocolate Milk.
Milk for Infants, Normal and Processed Milks, Soft Curd Milk.

Chapter XV

THE CHEMISTRY OF CLEANING 316

Milk as a Soil.
Sanitation Chemicals — Duties of Detergents.
The Water Supply — Temporary and Permanent Hardenss.
Chemical Sanitization — Chlorine Compounds, Iodine-bearing Compounds, Quanternary Ammonium Compounds, Chlorophenols, Acid Sanitizers.
Cleaner-Sanitizers.
Methods of Analysis — Water Hardness by Versenate Method, Quaternary Ammonium Compounds in Milk and Water.

APPENDIX 340
    USPHS Milk Code — Legal Standards for Dairy Products.

    Food and Drug Administration (U. S. Dept. of Health, Education, and Welfare) — Definitions and Standards of Certain Dairy Products.

    Grade Names Used in U. S. Standards for Farm Products — Dairy Products.

    International Atomic Weights.

    Average Composition of Milk and Its Products.

    Standard Units of Measure and Weight.

    Comparative Temperature Scales of Centrigrade, Reaumur, and Fahrenheit.

    Specific Gravity and Weight per Gallon of Water, Milk, and Cream.

    Composition of Milk of Different Species of Animals.

    Composition of Milk of Different Breeds of Cattle.

    Glossary of Terms.

    Federal and State Standards for the Composition of Milk Products.

List of Illustrations

	Page
Hortvet Cryoscope	37
Cenco-du Nuoy Tensiometer	44
Milk Samplers — Milk Thief, Scovell, McKay	58
Babcock Milk Test Bottle	70
Automatic Acidity Tester	72
Babcock Milk Tester	73
Dividers	74
Babcock Test Bottle Shaker	75
Babcock Tempering Bath	76
Method of Reading Whole Milk Tests	77
Butterfat Column Reader	78
Meniscus Reader	78
Acid Hydrometer and Cylinder	79
Earthen Jar for Waste Acid with a Convenient Cover for Draining the Bottles	84
Test Bottle Bath and Shaker	84
Babcock Bottle Washer	85
Babcock Cream Test Bottle	88
Cream Test Balances	89
Reading the Babcock Cream Test	90
Skimmilk Test Bottle	91
Skimmilk Test Bottle Showing Completed Test	91
Gerber Centrifuges and Speed Indicator	96
Gerber Lock Stopper. Key and Cross Section of Stopper	96
Automatic Feeder for Acid (Gerber)	97
24-Bottle Stainless Steel Rack and Cover (Gerber)	97
Gerber Test Reader	97
Automatic Syringe for Acid (Pump Type)	98
Fat Column in Milk Test Bottle	98
Automatic Acid Measure	99
Model D Mojonnier Milk Tester with Balance and Weights	104
Adding Reagents	105
Flask Hanger with Flask Suspended to Balance Arm	105
Method of Placing Dish Upon the Balance Pan	107
Correct Procedure When Pouring Ether-Fat Solution into Dish	107
Fat Extraction Flask	108
Correct Position of Flask When Shaking	109
Fat Dish	109
TeSa Apparatus	127
Sepascope	129

List of Illustrations (Continued)

	Page
Construction of Sepascope	129
Weighing the Solids Sample	133
Dish Contact Maker	133
Dietert-Detroit Solids Determinator	136
Cenco Moisture Balance	138
Toluene Distillation Apparatus	140
Brabender Semi-Automatic Rapid Moisture Tester	143
Quevenne and N. Y. State Board of Health Lactometers	147
Westphal Balance	150
Scharer Field Kit	166
Range of Sediment as Prepared for Official Grading	172
Milk Sediment Tester	173
Borden Flow Meter	174
Advanced-Fisk Milk Cryoscope	191
Disk Plate Technique for Detecting Antibiotics in Milk	203
Butter Trier	209
Moisture and Butterfat Balance	212
Desiccating Cabinet	214
Desiccator	214
Analytical Balance	215
Mettler Gram-atic Balance	215
Thelco Moisture Oven	216
Nafis Acid Test	238
Beckman Zeromatic Ph Meter	251
Milk Dilution Bottle	268
Showing Method of Making Dilutions in Plating Milk	268
Quebec Colony Counter	269
Photomicrographs of Milk Samples	271
Microscope	272
0.01 ml. Transfer Syringe	273
Breed Pipette	273
Losee Reduction Incubator	275
Mold Mycelia Filaments in Butter	289
Water Analysis Kit	334

List of Tables

	Page
1. Composition of Milk of Various Mammals	2
2. Grams of Protein and Lactose per 100 Grams of Fat in the Cows' Milk Received by Two Norwegian Dairies in 1948	3
3. Fatty Acids in Milk Fat	6
4. Iodine Number of Some Fats and Oils (Normal Range)	9
5. Seasonal Variation in Milk Fat Tests	11
6. Effect of Period of Lactation on Milk Fat Tests	11
7. Energy Value of Different Milks	13
8. Composition and Properties of the Components of Casein	15
9. Amino Acids in Milk Proteins	16
10. Protein Analyses of Skimmilk Comparing Newer and Older Nomenclature	21
11. Chemical Analyses of Casein, Albumin and Globulin	22
12. Initial and Final Solubilities of Lactose	24
13. Gas Content of Milk	33
14. Freezing Points of Milk, Cream, and Skimmilk	37
15. The Freezing Point of Some Authentic Milk Samples Produced in the United States as Determined by the Hortvet Cryoscope	38
16. Composition of the Solid and Liquid Portions of Frozen Milk	39
17. Effect of Freezing Milk at Varied Temperatures for Different Periods of Time on Cream Line	39
18. Specific Gravity of Milk Constituents	40
19. Relation of the Fat Test to the Specific Gravity of Normal Milk	42
20. Specific Heat of Milk and Its Products	43
21. The Bound Water Content of Some Milk and Milk Products	47
22. Effect of Temperature on Foaming Ability	47
23. Percentage Reduction of Some Properties and Constituents During Pasteurization of Goats' Milk	53
24. Effect of Temperature on Fat Test	62
25. Accuarcy of the Drip Sample	63
26. Speed of Centrifuge	82
27A. Comparison of Mojonnier, Gerber, and Babcock Fat Tests of Milk	101
27B. Relative Efficiency of Babcock, Gerber Weight and Gerber Measure Methods When Compared to Mojonnier Test	102
28. Summary of Methods for Making Fat and Total Solids Tests by Mojonnier Method	110-112

List of Tables (Continued)

Page

29. Reported Accuracy of Various Methods for Determining the Total Solids of Milk and Skimmilk 155
30. Relation of Fat to Other Solids in Milk and to the Specific Gravity 156
31. Feathering Values with Citric Acid 177
32. Feathering Values with Sodium Citrate 177
33. Interpretation of Feathering Value 177
34. Limitations of the Various Methods Suggested for Detecting Adulteration of Butterfat at 10% Level 194
35. Composition of Green Cheddar Cheese 225
36. Apparent Acidity of Milk from Different Breeds 235
37. Method of Making Acid Standards 243
38. Acidity and pH Values of Milk During First Month of Lactation 246
39. Relationship of Hydrogen Ion Concentration and pH Ions per Liter of Solution 248
40. Color Changes of Indicators Used in Determining pH 250
41. Relation of Titratable Acidity of Milk to the pH Value 253
42. Effect of Temperature of Milk on Bacterial Development 291
43. Amino Acid Analysis of Bovine Milk Proteins 295
44. Average Composition of Human and Cows' Milk 304
45. Vitamin A Potency of Creamery Butter 309
46. Composition and Energy Value of Various Cheeses 310
47. Summary of Food Nutrients in Dairy Products and Human Requirements (Average) 313
48. Extremes in Composition of Milkstone 319
49. Common Detergent Ingredients 323
50. Reactions of Temporary Hardness 326
51. Methods of Controlling and Removing Water Hardness 327
52. Quantity of Anionic Solution Required to Titrate Various Concentrations of Added QAC in Water and Milk 338

CHAPTER I

Composition of Milk

Definition. Milk is the whole, fresh, lacteal secretion obtained by the complete milking of one or more healthy cows, excluding that obtained within 15 days before and 5 days after calving, or such longer period as may be necessary to render the milk practically colostrum-free.

The main constituents of milk are water, fat, casein, lactose (milk sugar), albumin and ash. The constituents other than the water are known as total solids (T.S.) and the total solids minus the fat as solids not fat (S.N.F.). Other common terms are milk plasma which consists of all the substances except the fat, and milk serum which consists of all the constituents except the fat and casein. Thus, the former corresponds practically to skimmilk and the latter to whey, the watery product left in cheese making.

Composition. The composition of normal cows' milk varies to a great extent. There are any number of tables giving the average composition of milk and each one differs slightly, but for all practical purposes the composition of normal whole milk corresponds closely to the figures in Table 1.

Table 1 represents the average analyses of a large number of samples of normal milk. The composition of individual samples of cow's milk would vary greatly, especially the fat and water. In general, the water will range from 82 to 90 percent; the fat from 2.5 to 8.0 percent; the casein from 2.3 to 4.0 percent; the sugar from 3.5 to 6.0 percent; the albumin from 0.4 to 1.0 percent and the ash from 0.5 to 0.9 percent.

Table 1. Composition of Milk of Various Mammals

	H₂O %	Fat %	Sugar %	Protein %	Ash %
Cow	87.00	4.00	5.00	3.30	.70
Human	87.41	3.78	6.21	2.00	.30
Goat	85.71	4.78	4.46	4.29	.76
Sheep	83.00	5.30	4.60	6.30	.80
Mare	90.18	1.59	6.73	2.14	.42
Ass	91.23	1.15	6.00	1.50	.40
Rat	68.30	14.79	2.83	11.77	1.50
Monkey	...	3.93	5.89	2.09	.26
Buffalo	82.05	7.98	5.18	4.00	.79
Camel	87.61	5.38	3.26	2.98	.70
Reindeer	65.32	19.73	2.61	1.91	1.43
Llama	86.55	3.15	5.60	3.90	.80
Hog	81.82	6.85	5.00	6.19	.98
Dog	79.26	8.17	4.00	7.53	1.36
Cat	82.17	3.33	4.91	9.08	.51
Rabbit	69.50	10.45	1.95	15.54	2.56
Elephant	68.00	19.60	8.80	3.10	.50
Whale	70.10	19.60	...	9.50	1.00
Porpoise	41.11	45.80	1.33	11.19	.57
Fox	81.88	5.42	5.11	...	.88
Hippo	90.43	4.51	4.40	...	.11
Dolphin	48.76	43.71	...	...	.46
Porcupine	...	31.00	...	...	...

Historically, milk fat has served as the basis for most pricing systems for milk. Consequently, many studies have been made to evaluate the factors which affect the fat content of a milk supply. However, the value of the non-fat milk solids must not be overlooked. Nutritional studies showing the importance of these milk solids in the diet, coupled with increased awareness of the economic consequences of variations in the solids-not-fat (SNF) content of milk on the yield of the many manufactured dairy products, have resulted in widespread interest in the factors which influence the SNF content of milk.

Syrrist[1] recently reported that the seasonal variations in the amount of each milk constituent were less than variations in total milk produced. However, the relative composition of milk varied many times during the year while total milk received usually increased and decreased once each year. Table 2, taken from the Norway report, indicates the relationship of protein and lactose to fat during the year.

Wilcox and Pfau[2] investigated the hereditary and environmental factors influencing the SNF content of over 5000 samples of milk from two breeds during the 1953-1958 period. They observed definite and systematic effects of age, pregnancy, and stage of lactation. Seasonal variations were attributed to the combined effects of temperature and feeding since both high temperatures and poor quality feeds resulted in milk with low SNF content while good feed and low temperatures tended to raise the SNF concentration. SNF content was high at the start of the lactation period, then sharply decreased to the lactation low point at 40-60 days. It then rose gradually until the sixth

Table 2. Grams of Protein and Lactose per 100 Grams of Fat in the Cows' Milk Received by Two Norwegian Dairies in 1948.[1]

Grams per 100 grams fat	Gausdal		Misvaer	
	Protein	Lactose	Protein	Lactose
January	96	139	94	136
February	94	137	91	134
March	91	131	95	138
April	92	136	85	128
May	91	133	86	131
June	85	117	78	112
July	79	119	84	125
August	82	113	79	113
September	86	116	85	122
October	91	127	92	126
November	92	131	89	116
December	92	125	95	126
Average	89	127	88	125

month of lactation, followed by a sharp increase until the end of the lactation period. However, the latter rise was attributed to gestation effects since the SNF content of milk from animals which were not bred increased little, if any, during the final stages of lactation. Heredity was shown to have an important influence on SNF content.

Reports[3,4] from England agree that underfeeding and breeding are the common causes of low SNF content in milk. Freeman[5] also observed from Kentucky studies that the "seasonal" variations noted in the literature were more likely due to the general pattern of freshening and availability of suitable feeds within an area rather than to changes on the calendar. Missouri studies[6] on the effect of controlled diurnal environmental temperature cycles on milk composition showed that both fat and SNF content of milk increased during high (70-100°F or 60-110°F) and low (10-40°F) temperature cycles. Total nitrogen content of milk increased during the low temperature cycles. Low temperatures did not affect the chloride content but high temperature cycles gave increased chloride values. Changes in milk composition occurred very slowly following a change in environmental temperature.

Water. As seen from the composition, water is the constituent that gives bulk to milk. It holds the solids either in solution or in suspension.

Fat.

a. Composition of milk fat. The fat in milk is spoken of as milk fat, butterfat or simply fat. It is present in minute droplets or globules which vary in diameter from .001 to .01 mm. or 1/25,000 to 1/2500 of an inch depending on the breed, stage of lactation and the individual cow. The higher testing breeds have the largest fat globules. Campbell[6a] found that on an average the volume of the fat globule in Guernsey milk is 80 percent greater than one in Holstein milk. The fat globules are usually largest during the first two weeks of lactation with the most rapid rate of decline occurring during the next two months. Thereafter the rate of decline in size is slow, but usually continues to the end of the normal lactation period.

The globules float around in the milk in a true emulsion, oil-in-water, the fat being in the dispersed phase. The spherical shape is due to the force of surface tension, or the affinity one molecule has for the other. Hence, there is a tendency for the molecules to move toward the center one, which results in the milk fat globule assuming the smallest surface area possible in relation to its volume. This form must necessarily be spherical.

The fat globules are kept apart by the viscosity of the liquid, and by the layer of colloidal substance around the outside of the globule. This layer consists of both protein and phospholipids, and has to be removed or broken before two fat globules can unite. A negative electric charge is present in the globules and this also tends to keep them apart. Agitation as in churning breaks the adsorbed layer and the globules coalesce with a final result of butter.

Milk fat, like most other animal fats, is a triglyceride. That is, one molecule of glycerol is joined chemically to three molecules of fatty acids. However, milk fat differs from other animal fats in that it has many more kinds of fatty acids. The principal ones are butyric, capric, caproic, caprylic, lauric, myristic, oleic, palmitic, stearic, and linoleic. Thus, butterfat is a mixture of different fats, each of which is made up of a molecule of glycerol and three molecules of fatty acids. For example, glycerol plus three molecules of butyric acid gives the fat butyrin; glycerol plus oleic acid gives olein. In most cases, however, a molecule of glycerol, instead of being united to three molecules of the same acid, is joined to two or three different fatty acid molecules, such as stearic, oleic, and palmitic acids. The resulting fat in this case would be termed stearyl olyl palmitin. The following chemical equations illustrate the above examples:

1. $C_3H_5(OH)_3$ + $3C_3H_7COOH$ = $C_3H_5(C_3H_7COO)_3$ + $3H_2O$
   Glycerol   Butyric acid   Butyrin   Water

2. $C_3H_5(OH)_3$ + $3C_{17}H_{33}COOH$ = $C_3H_5(C_{17}H_{33}COO)_3$ + $3H_2O$
   Glycerol   Oleic acid   Olein   Water

3. $C_3H_5(OH)_3$ + $\begin{Bmatrix} C_{17}H_{35}COOH \\ C_{17}H_{33}COOH \\ C_{15}H_{31}COOH \end{Bmatrix}$ = $C_3H_5 \begin{Bmatrix} C_{17}H_{35}COO \\ C_{17}H_{33}COO \\ C_{15}H_{31}COO \end{Bmatrix}$ + $3H_2O$
   Glycerol   Stearic, oleic and palmitic acids   Stearyl olyl palmitin   Water

The structural formula of a fat is as follows:

$$\begin{array}{l} CH_2OH \\ | \\ CHOH \\ | \\ CH_2OH \end{array} + \begin{array}{l} HO-\overset{O}{\overset{\|}{C}}-C_3H_7 \\ HO-\overset{O}{\overset{\|}{C}}-C_3H_7 \\ HO-\overset{O}{\overset{\|}{C}}-C_3H_7 \end{array} = \begin{array}{l} CH_2O-\overset{O}{\overset{\|}{C}}-C_3H_7 \\ | \\ CHO-\overset{O}{\overset{\|}{C}}-C_3H_7 \\ | \\ CH_2O-\overset{O}{\overset{\|}{C}}-C_3H_7 \end{array} + 3H_2O$$

Glycerol + Butyric acid = Butyrin

New methods of analysis in recent years have enabled workers to provide a much more comprehensive picture of the composition of milk fat than was possible in early texts. Recent studies[7] have shown that, of the approximately 40 different fatty acids found in milk (as compared with 6-10 in most edible fats) about 18 are saturated and 22 unsaturated. The saturated acids account for about 60-65 percent of the total fatty acids and the unsaturated acids the remaining 35-40 percent.

Most of the fatty acids in milk fat have an even number of carbon atoms. The saturated acids range from butyric acid with 4 carbon atoms to lignoceric acid with 24 carbons. Palmitic acid, stearic acid and myristic acid (the C_{14}, C_{16}, and C_{18} saturated acids) comprise 72-78 percent of the total saturated acids and 45-50 percent of the total fatty acids present in milk fat. Table 3 gives the formulae, melting points, and percentage of the different acids in milk fat. This shows that the remaining saturated acids are present in only small amounts. The presence of butyric acid in milk fat serves as a distinguishing characteristic since no other fats contain this low molecular weight fatty acid. Branch-chain acids, straight-chain acids with an uneven number of carbon atoms, and straight-chain acids with 20-24 carbon atoms have been reported in trace amounts.

Table 3. Fatty Acids in Milk Fat

Acid	Carbon atoms	Formula	Melting point ° F.	Average[10] percentage
		Acetic Series $C_nH_{2n}O_2$		
Butyric	4	C_3H_7COOH	17.8	2.93
Caproic	6	$C_5H_{11}COOH$	14.9	1.90
Caprylic	8	$C_7H_{15}COOH$	60.8	0.79
Capric	10	$C_9H_{19}COOH$	87.8	1.57
Lauric	12	$C_{11}H_{23}COOH$	118.4	5.85
Myristic	14	$C_{13}H_{27}COOH$	136.4	19.78
Palmitic	16	$C_{15}H_{31}COOH$	147.2	15.17
Stearic	18	$C_{17}H_{35}COOH$	156.7	14.91
		Oleic Series $C_nH_{2n-2}O_2$		
Oleic	18	$C_{17}H_{33}COOH$	57.2	31.90
		Linoleic Series $C_nH_{2n-4}O_2$		
Linoleic	18	$C_{17}H_{31}COOH$	−0.05	4.50

The acids in the acetic series are saturated while those in the other series are unsaturated. A hydrocarbon which unites with oxygen or such monovalent elements as bromine, chlorine, iodine or hydrogen is spoken of as an unsaturated compound. This union is possible because of the double linkage between two adjacent carbon atoms. This is the unsaturation point and here an atom may be added and the compound thus pass into the single linkage class. The saturated hydrocarbons have single linkage and therefore do not have this particular property of combination.

$$H-\underset{|}{\overset{H}{C}}=\underset{|}{\overset{H}{C}}-H + I_2 \rightarrow H-\underset{\underset{I}{|}}{\overset{H}{\overset{|}{C}}}-\underset{\underset{I}{|}}{\overset{H}{\overset{|}{C}}}-H$$

Unsaturated Compound   Saturated Compound

The unsaturated fatty acids in milk present many problems for the chemist attempting to fractionate and identify them.

They range in chain length from 10 to 24 carbons. They exist in different geometrical configurations. The fatty acid may exist with isomers having the double bond in different positions in the carbon chain. In addition, there are small quantities of polyunsaturated fatty acids in milk fat, these having 2, 3, 4, or 5 double bonds.

The lower members of the monounsaturated fatty acids (C_{10}-C_{16}) have the double bond in the 9, 10 position. These are present only in limited concentration. The C_{18} monounsaturated acid (oleic acid) exists in a number of isomeric forms and is one of the major constituents of milk fat. Oleic acid accounts for approximately 30 percent of the total fatty acid content. It is apparent that most of the early analyses include the lower molecular weight unsaturated acids along with oleic acid and thus values for oleic acid appear a little high. The polyunsaturated acids account for approximately 3-5 percent of the fatty acid content in milk fat.

The major fatty acids of milk may be divided into two groups — volatile and non-volatile.

Volatile	Non-Volatile
1. Butyric	6. Myristic
2. Caproic	7. Palmitic
3. Caprylic	8. Stearic
4. Capric	9. Oleic
5. Lauric	10. Linoleic

The first four fatty acids are soluble and the rest insoluble. All are saturated except oleic and linoleic which are unsaturated. Fats from the volatile fatty acids are rather unstable compounds and are quite easily decomposed. They have considerable influence on the flavor (and off-flavors) in milk. The unpleasant odor from rancid butter is due to freed fatty acids, mainly butyric. Normal processing and storage methods employed in the dairy industry do not produce any marked chemical changes in the saturated fatty acids. The unsaturated fatty acids, on the other hand, are of particular significance in the oxidative deterioration of milk and its products. A recent contribution by Jack and Smith[8] brings together much of the current knowledge concerning the chemistry of milk fat. The reader is directed to this excellent review.

A soap is formed by the union of a non-volatile fatty acid usually palmitic or stearic with sodium (or potassium). Animal fats and vegetable oils are boiled with sodium hydroxide and a soap is formed according to the following reaction:

$$C_3H_5(C_{17}H_{35}COO)_3 + 3NaOH = 3C_{17}H_{35}COONa + C_3H_5(OH)_3$$
$$\text{Stearin} \quad \text{Sodium} \quad \text{Sodium stearate} \quad \text{Glycerol}$$
$$\text{hydroxide} \quad \text{(soap)}$$

A similar reaction will take place if a fat is heated with steam under pressure in which case the resulting products will be the free fatty acid and glycerol.

$$C_3H_5(C_{17}H_{35}COO)_3 + 3H_2O = 3C_{17}H_{35}COOH + C_3H_5(OH)_3$$
$$\text{Stearin} \quad \text{Water} \quad \text{Stearic acid} \quad \text{Glycerol}$$

b. Factors affecting composition of milk fat. Milk fat varies in composition, the principal changes being in the amounts of the butyric, palmitic, stearic and oleic acids. This variation is due mainly to breed of animal, kind of feed and season of year.

The fat from Jerseys and Guernseys is usually firmer than that from Holsteins and Ayrshires, as the former contains less of the volatile and unsaturated fatty acids than does the latter. The fatty acids have varied melting points which affect the firmness of the milk fat. An increase in the low melting point acids such as oleic and butyric will give a soft milk fat, while an increase in stearic acid produces a hard fat.

Such feeds as linseed oil meal and pasture grass increase the oleic and linoleic acids with the production of soft fat. Cottonseed meal produces a firm fat because it lowers the amount of butyric acid, which has a low melting point, and thus indirectly increases the percent of higher melting point acids. In general, feeds rich in carbohydrates and low in oils produce the firmer fats, while succulent roughages and oily concentrates give the softer fats.

Milk fat contains more of the volatile and of the unsaturated fatty acids during the spring and summer than in the fall and winter. This, however, is more a feed factor than a seasonal one.

c. Properties of milk fat. (1). *Melting point.* The melting point of milk fat varies normally between 89.5° F. and 96.5° F. with an average of 93° F. This wide variation in melting point is to be expected because of the change in proportions of the several fatty acids in milk fat due to feed and breed, etc. Because of the different melting points of the various fatty acids, butter softens slowly instead of melting suddenly. The melting point increases slightly (1-4° F.) toward the end of the lactation period. The greater increases apply especially to the Holstein breed.

(2). *Specific gravity.* The specific gravity of milk fat at 70° F. is 0.93, while at 135° F. to 140° F., the temperature at which fat tests are tempered, it is 0.892. However, a specific gravity of 0.9 is used in calculating what the capacity of the graduated portion of the Babcock test bottle should be. This will be explained later under the Babcock test.

(3). *Solubility.* Milk fat is readily dissolved by either ethyl ether, carbon tetrachloride, chloroform or benzene. It is moderately soluble in acetone, slightly so in alcohol and insoluble in water.

(4). *Iodine number.* The iodine number measures the amount of unsaturated glycerides in a fat. It is the grams of iodine combining with 100 grams of fat. Since unsaturated fatty acids have double linkages at certain points in the carbon chain, iodine will join at these points and the amount of the halogen absorbed will determine the degree of unsaturation. Table 4 gives some typical iodine numbers.[9]

Table 4. Iodine Number of Some Fats and Oils (Normal Range)

Butter 30-34	Olive oil 77- 91
Tallow 35-40	Cottonseed oil 104-116
Lard 48-64	Linseed oil 175-201

It will be noted that the oils have a greater degree of unsaturation than the fats, being higher in such fatty acids as oleic and linoleic. Butter contains less of the unsaturated fats than tallow or lard and therefore has a lower iodine number.

The iodine number of milk fat increases quite sharply during the pasture season and when oily feeds are fed in quantities. It also rises during the last few weeks of the lactation period, and to some extent during fasting. Thus, any factor that increases the proportion of unsaturated fatty acids, raises the iodine number.

(5). *Reichert-Meissl number.* Milk fat differs from other common fats in having a larger percentage of volatile fatty acids. Therefore, a method of measuring the amount of these acids gives a means of distinguishing milk fat from other fats. The Reichert-Meissl number is used for this purpose. Specifically, the Reichert-Meissl number is equivalent to the number of milliliters of N/10NaOH required to neutralize the volatile fatty acids in 5 grams of fat.

Most Reichert-Meissl values of fats are below 1. Tallow and lard are around 0.3 to 0.5. Butter, however, is high with an ordinary range of 27 to 31. This number is an important means of distinguishing milk fat from other fats.

It varies appreciably with the season of year being highest in March, dropping during the summer and reaching its lowest point in October. Various oils such as corn oil, linseed oil and cottonseed oil added to a dairy ration of hay, silage and grain will cause a lowering of the number. Cows that are underfed yield a milk fat with a low Reichert-Meissl number.

(6). *Saponification number.* The saponification number of a fat is the number of milligrams of potassium hydroxide (KOH) necessary to saponify one gram of fat. It is a measure of the number of chains of fatty acids present. Milk fat has more chains per unit of weight than most fats because it contains more fatty acids of the lower series, and since it requires as much KOH to saponify a high molecular weight fatty acid as a low one, milk fat would have a high saponification number. According to Holland[10] the saponification number of butterfat is 231, cocoanut fat 257, peanut oil 190, corn oil 191 and soybean oil 192. Other fats as oleo oil, lard and tallow range around 192 to 203.

(7). *Acid value.* The acid value of a fat indicates the amount of free fatty acids present, that is, those that are not joined to the glycerol radical. It is the number of milligrams of potassium hydroxide required to neutralize the free fatty acids in one gram of fat. In fresh butter the value according to Hunziker[11] will range between 0.40 to 0.56. Any higher figure would indicate deterioration leading to a rancid flavor.

d. **Factors affecting fat test of milk.** (1). *Breed of animal.* Milks from the different breeds have average fat tests approximately as follows: Holstein 3.4%, Ayrshire 4.0%, Guernsey 4.9% and Jersey 5.3%. Individual animals in each of the classes will of course vary more or less from the above figures. In general, the amount of milk yielded is in inverse proportion to the percentage of fat, and, thus, the total pounds of fat yielded by the different breeds will not differ greatly. (See appendix table for composition of milk of different breeds of cattle.)

(2). *Season of year.* The fat test of milk is lowest during the summer months and highest in the winter. It is a temperature effect, the test varying almost inversely with temperature. In Table 5 from a study by Becker and Arnold[12] are shown the fat tests by calendar months. The lowest test 4.76 percent occurs in August and the highest 5.46 in December.

Table 5. Seasonal Variation in Milk Fat Tests
(Effect of period of lactation eliminated)

Month	Fat %	Month	Fat %	Month	Fat %	Month	Fat %
January	5.33	April	5.03	July	4.81	October	5.06
February	5.27	May	4.80	August	4.76	November	5.44
March	5.17	June	4.83	September	4.92	December	5.46

(3). *Period of lactation.* After the first two or three months following calving, the fat test increases steadily each month to the end of the lactation period. Advancing lactation exerts a greater influence than the season of year. In Becker and Arnold's study (Table 6) the fat percentage dropped slightly the second month after calving, from 4.61 in the first month to 4.59 in the second, and then rose to 5.56 percent in the twelfth month.

Table 6. Effect of Period of Lactation on Milk Fat Tests
(Seasonal variation eliminated)

Month	Fat %	Month	Fat %	Month	Fat %	Month	Fat %
First	4.61	Fourth	4.83	Seventh	5.15	Tenth	5.38
Second	4.59	Fifth	5.02	Eighth	5.29	Eleventh	5.50
Third	4.68	Sixth	4.99	Ninth	5.29	Twelfth	5.56

(4). *Interval of milking.* In general the longer the interval between milkings the lower the test, and conversely, the shorter the interval the higher the test. If the intervals are equal the tests will not necessarily be the same. Experiments have shown that if the cows are milked at 6 a.m. and 6 p.m. the evening milking will test appreciably higher due to the greater amount of exercise on the part of the animal during the daytime. However, if the milkings occur at midday and midnight, the tests will be approximately the same.

Compound Lipids. In addition to the true fats there are also present in milk, phospholipids or mixed glycerides. In these lipids two of the OH groups of the glycerol are joined to fatty acids as in the true fats, but the third OH group of the glycerol is joined with phosphoric acid and a choline radicle. Lecithin is the main phospholipid of milk. Traces of cephalin and sphingomyelin are also present. Lecithin is found on the surface of the fat globules. It goes with the fat in skimming but a good share of it remains in the buttermilk upon churning cream, being detached from the globules when these unite in the churning process. Horrall[13] found from 0.027 to 0.044 percent lecithin in whole milk, 0.013 to 0.035 in skimmilk, 0.14 to 0.16 in buttermilk from sweet cream and 0.10 to 0.17 in buttermilk from soured cream. Thurston and Peterson[14] estimated that buttermilk of average fat content contained 0.39 percent lecithin. Crane and Horrall[15] found that the amount of phospholipids in butter averaged about 0.15 percent of the total fat. These workers reported that the ratio of lecithin to total phospholipids averaged 46 percent for whole milk fat, 54 percent for fat from cream, 48 percent for butter lipids and 34 percent for buttermilk lipids.

Lecithin is soluble in fats and fat solvents except acetone. At the same time, it has a strong attraction for water (the phosphoric acid and choline portion are water soluble), giving it considerable importance as a natural emulsifying agent. Since lecithin is soluble in ether and alcohol, it will cause slightly higher readings in fat tests employing these reagents. This effect applies mainly to buttermilk and will be discussed under buttermilk testing. The reader is directed to the recent work of King[16] which presents a comprehensive discussion of the physical and chemical properties of the fat globule membrane.

The following structural formula of a lecithin shows how the acids are united with glycerol.

$$\text{Glycerol} \begin{cases} CH_2O - \overset{\overset{O}{\|}}{C} - C_{17}H_{35} & \ldots\ldots\ldots\ldots\text{Stearic acid} \\ CHO - \overset{\overset{O}{\|}}{C} - C_{17}H_{33} & \ldots\ldots\ldots\ldots\text{Oleic acid} \\ CH_2O - \overset{\overset{O}{\|}}{P} - OH & \ldots\ldots\ldots\ldots\text{Phosphoric acid} \\ \quad\quad\quad\;\; | \\ \quad\quad\quad\;\; O \\ \quad\quad\quad\;\; | \\ \quad\quad\quad\; CH_2 \\ \quad\quad\quad\;\; | \\ \quad\quad\quad\; CH_2 & \ldots\ldots\ldots\ldots\text{Choline} \\ \quad\quad\quad\;\; | \\ \quad\quad\; N(CH_3)_3\, OH \end{cases}$$

a Lecithin

Early workers noted that trimethyl amine ($[CH_3]_3N$), a breakdown product of choline, has a strong fishy odor. They assumed the presence of a fishy flavor in butter was indicative of the breakdown of this constituent part of lecithin. This assumption has been questioned for some time and recent information by European workers[17] clearly refutes this explanation of the origin of fishy flavors. Although trimethyl amine may be present in fishy butter, the addition of this chemical to good butter does not produce the off-flavor. Finnish workers[18] believe that tests for trimethyl amine in butter are unreliable and that the oily and fishy flavor of sour, stored butter results from the decomposition products of fat autoxidation. The oxidation products of linoleic acid appear to be of prime importance in the development of these flavor defects in butter.

Derived Lipids. Under the class of derived lipids are the sterols. They are alcohols of high molecular weight and are in the unsaponifiable portion of fats. Cholesterol ($C_{27}H_{45}OH$) is a sterol in milk and its content is around .015 percent. One of its forms, 7-dehydro cholesterol is a precursor of vitamin D, being activated by irradiation with ultra violet rays.

Energy Value of Milk in Relation to Fat. A milk testing 6.0 percent fat will not have twice the energy value of a milk testing 3.0 percent because the other solids in the milk do not vary in the same proportion. A simple formula that will give the energy values of different milks in relation to their fat contents is that of Sharp's[19]: $E = 12(F + 2)$. E represents the calories in 100 grams milk and F the percent fat. For example, the calories per 100 grams and per quart of milk would be for 3, 4, and 6 percent fat tests as shown in Table 7. This simple method of computing the energy value of milk gives relative values of different testing milks which agree closely with those obtained by the Gaine's formula — lbs. fat $\times$ 15 + lbs. milk $\times$ 0.4 = lbs. of 4 percent milk.

Table 7. Energy Value of Different Milks

Fat	In 100 Grams	In One Quart
%	Calories	Calories
3.0	60	585
4.0	72	702
6.0	96	936

Casein. Casein is the principal protein compound of milk. It contains the elements carbon, hydrogen, oxygen, nitrogen, sulphur and phosphorus. It exists in milk in combination with calcium and consequently is often termed calcium caseinate. Casein is not in solution in milk, but in a very fine suspension, called the *colloidal state*. The particles have a diameter around 80 millimicrons or 1/300,000 of an inch.

Casein is precipitated by acids, alcohol, rennet and heavy metals. The precipitate by acid is casein itself as the acid frees it from its union with calcium. Heat and acid are interrelated in precipitating casein. The higher the temperature the less acid it requires for coagulation, and conversely, the more acid the lower will be the heat of coagulation. The alcohol acts as a dehydrating agent, it removes the water from the casein and this results in precipitating the casein. The precipitate is calcium caseinate the same form as exists in milk. The rennet curd is a new compound, paracasein or calcium paracaseinate. Chemical analyses of casein and paracasein show no composition difference in these two products. However, paracasein has never been reversed to casein which is a good evidence that a chemical change of some kind has taken place.

Rennet extract is a liquid containing the enzyme rennin which is obtained from the stomach lining of young calves. Pepsin, which is secured from the stomachs of young pigs, will also act similarly to rennet but not to the same degree. Its curdling power is approximately but one-fourth that of rennin.

Heavy metals, such as mercury, silver, lead, copper, zinc, aluminum and iron, will precipitate casein. Casein has a specific gravity of 1.25 to 1.31 and a molecular weight of at least 12,800 and may be as high as 375,000. Like other proteins it is *amphoteric* in reaction, that is, it partakes of the nature of both a base and an acid. This dual role results from the fact that the protein molecule has both amino (NH_2) and carboxyl (COOH) terminals, the former acting as a base and the latter as an acid.

Proteins, being amphoteric and containing the amino and carboxyl groups, dissociate either into acids or bases. This dissociation leads to positive and negative charges on the protein particles, being positive on the acid side and negative on the basic. At a certain hydrogen-ion concentration the two dissociations will be equal and thus the positive and negative charges will be similar. This condition is known as the "isoelectric point" of a substance. Casein has an isoelectric point of pH 4.7. At this hydrogen-ion concentration, casein can be most readily precipitated from milk. Therefore, if fresh milk with a pH of 6.6 is allowed to sour to a pH of 4.7, the electric charge will become zero and the milk will begin to show evidence of curdling.

Until recent years, casein was considered to be a pure, homogeneous protein. Then, various workers were able to make incomplete fractionations of casein. Mellander[20] published results of electrophoretic studies in 1939 which definitely showed the existence of at least three separate casein fractions and he designated them α-casein, β-casein, and γ-casein. Now there is evidence that these fractions are not strictly homogenous. Hipp and co-workers[21] observed that the electrophoretic pattern of unfractionated casein indicated the three components were present to the extent of 75 percent α-casein, 22 percent β-casein, and 3 percent γ-casein. Table 8, derived from data by Hipp, et al.,[21] shows certain differences in the three forms of casein. The action of rennet on milk changes the α-casein to α_1 paracaseinate and α_2 paracaseinate while the β-casein and γ-casein are co-precipitated but unchanged. Pyne's[22] recent review of the Chemistry of Casein presents most of our present knowledge on this subject and the interested reader is directed to this comprehensive report.

Table 8. Composition and Properties of the Components of Casein[21]

	α	β	γ
Percent Nitrogen	15.58	15.53	15.40
Percent Phosphorus	0.99	0.55	0.11
Percent Sulfur	0.72	0.86	1.03
Isoelectric point (pH)	4.7	4.9	5.8
Mobility (μ)	−6.75	−3.05	−2.01
Specific rotation $[\alpha]_D^{25}$	−90.5	−125.2	−131.9

Casein is an important food constituent, both in milk and in its products. The various kinds of cheese contain around one-third casein and, consequently, have a high protein food value. Casein is a high quality protein as it contains all the essential amino acids. *Essential* amino acids are those needed by the body but cannot be synthesized by it. Therefore, they must be supplied in the food. *Non-essential* amino acids can be synthesized by the body. Table 9, divided into two groups, essential and non-essential amino acids, gives the percentage of each in casein and albumin.

Table 9. Amino Acids in Milk Proteins

Essential Amino Acids[22a]	Protein		Non-essential Amino Acids	Protein	
	Casein	Albumin		Casein	Albumin
	%	%		%	%
Phenylalanine	3.90	1.30	Glycine	0.45	0.37
Isoleucine	6.90		Alanine	1.85	2.41
Leucine	9.70	14.03	Serine	0.40	1.76
Lysine	6.25	8.10	Aspartic acid	4.10	9.30
Histidine	1.83	1.52	Citrulline	...	...
Arginine	3.72	3.00	Tyrosine	7.49	1.95
Tryptophan	1.40	2.69	Proline	8.70	3.76
Threonine	3.50	...	Hydroxyproline	10.50	...
Methionine	3.50	2.62	Cystine	0.34	3.79
Valine	7.90	3.30	Glutamic acid	21.77	12.89

The following essential amino acids have been found in beta-lacto-globulin*: Lysine 9.75 percent, histidine 1.54 percent, arginine 2.89 percent, tryptophan 1.94 percent and methionine 3.22 percent.
* Chibnall, A. C.—Proceedings of the Royal Society, B. 131 (1942).

Commercially, casein has a wide variety of uses, the greatest being in the paper industry. Here, casein serves as a water repellent, as a binder and to give a smooth surface to the paper. In the plastics industry large quantities of casein are used in the manufacture of belt buckles and buttons and fair amounts in making fountain pen barrels, combs and steering wheels. Casein has found wide application in the manufacture of paints. These are cheaper than oil paints and dry much sooner. However, not being water resistant, they must be used only for inside work. Excellent glues are made from casein which are used extensively in the manufacture of furniture and plywoods. "Casein wool" was first made in Italy in 1936 under the name of Lanital and in this country in 1941, using the term Aralac. The first important use of Aralac was in the felt used for men's hats, and later as a cloth for suits and dresses.

1. **Glycine, $C_2H_5NO_2$**
   or Amino-acetic acid

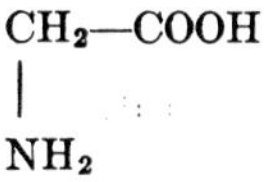

   CH_2-COOH
   $|$
   NH_2

2. **Alanine, $C_3H_7NO_2$**
   or α-Amino-propionic acid

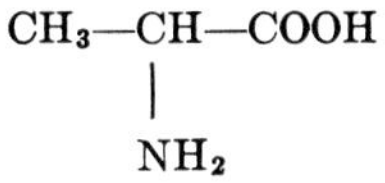

   $CH_3-CH-COOH$
   $\qquad |$
   $\qquad NH_2$

3. **Valine, $C_5H_{11}NO_2$**
   or α-Amino-β-methyl butyric acid

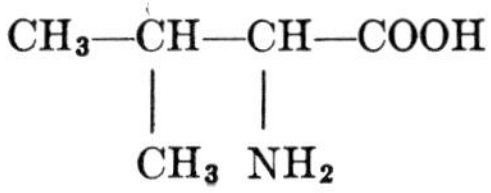

   $CH_3-CH-CH-COOH$
   $\qquad | \quad\ \ |$
   $\qquad CH_3\ NH_2$

4. **Threonine, $C_4H_9NO_3$**
   or α-Amino-β-hydroxy butyric acid

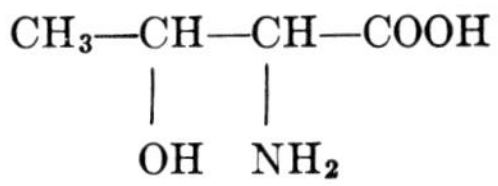

   $CH_3-CH-CH-COOH$
   $\qquad |\quad\ \ |$
   $\qquad OH\ NH_2$

5. **Leucine, $C_6H_{13}NO_2$**
   or α-Amino-γ-methyl valeric acid

   $CH_3-CH-CH_2-CH-COOH$
   $\qquad |\qquad\qquad |$
   $\qquad CH_3\qquad\ NH_2$

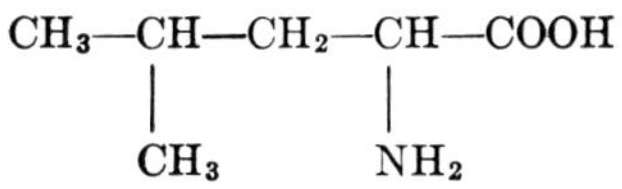

6. **Isoleucine, $C_6H_{13}NO_2$**
   or α-Amino-β-methyl valeric acid

   $CH_3-CH_2-CH-CH-COOH$
   $\qquad\qquad\ \ |\ \ \ \ |$
   $\qquad\qquad\ CH_3\ NH_2$

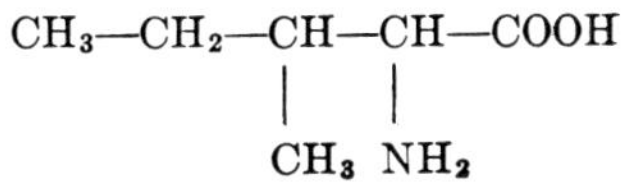

7. **Serine, $C_3H_7NO_3$**
   or α-Amino-β-hydroxy propionic acid

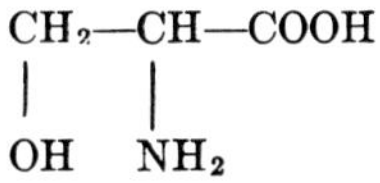

   $CH_2-CH-COOH$
   $|\qquad |$
   $OH\quad NH_2$

8. **Aspartic acid, $C_4H_7NO_4$**
   or Amino-succinic acid

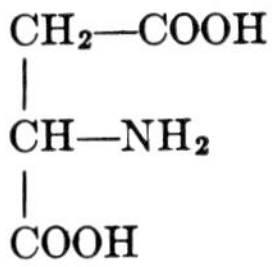

   CH_2-COOH
   $|$
   $CH-NH_2$
   $|$
   $COOH$

9. **Glutamic acid, $C_5H_9NO_4$**
   or α-Amino-glutaric acid

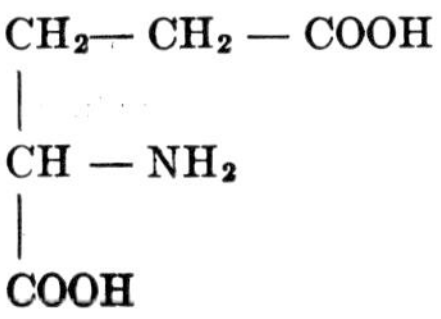

   CH_2-CH_2-COOH
   $|$
   $CH-NH_2$
   $|$
   $COOH$

10. **Glutamine, $C_5H_{10}N_2O_3$**
    or Glutamic acid 5-amine

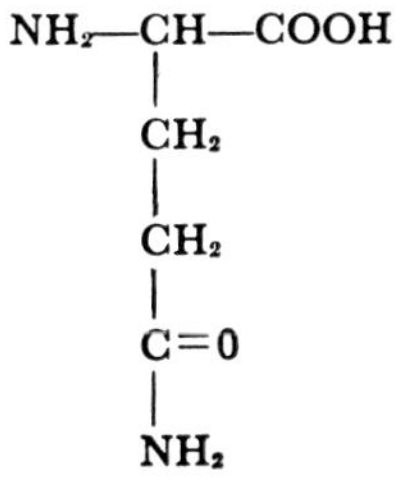

    $NH_2-CH-COOH$
    $\qquad\ |$
    $\qquad CH_2$
    $\qquad\ |$
    $\qquad CH_2$
    $\qquad\ |$
    $\qquad C=O$
    $\qquad\ |$
    $\qquad NH_2$

18   CHEMISTRY AND TESTING OF DAIRY PRODUCTS

11. Proline, $C_5H_9NO_2$
or Pyrrolidine-2-carboxylic acid

```
CH2—CH2
 |    |
CH2  CH—COOH
  \  /
   NH
```

12. Hydroxy proline, $C_5H_9NO_3$
or 4-Hydroxy pyrrolidine-2-carboxylic acid

```
HO—CH—CH2
    |    |
   CH2  CH—COOH
     \  /
      NH
```

13. Phenylalanine, $C_9H_{11}NO_2$
or α-Amino-β-phenyl-propionic acid

```
       H    H
       C == C
      /      \
    HC        C—CH2—CH—COOH
      \\    //        |
       C — C         NH2
       H   H
```

14. Tyrosine, $C_9H_{11}NO_3$
or α-Amino-β-parahydroxy-phenyl propionic acid

```
        H    H
        C == C
       /      \
    HO-C       C—CH2—CH—COOH
       \\    //        |
        C — C         NH2
        H   H
```

15. Cystine, $C_6H_{12}S_2N_2O_4$
or Di (α-Amino-β-Thio-propionic acid)

```
CH2 — S — S — CH2
 |             |
CH — NH2      CH — NH2
 |             |
COOH          COOH
```

16. Methionine, $C_5H_{11}SNO_2$
or α-Amino-γ-methylthiol-butyric acid

```
CH3—S—CH2—CH2—CH—COOH
                |
               NH2
```

COMPOSITION OF MILK

17. **Cysteine,** $C_3H_7O_2NS$
    β-thiol-α-aminopropionic acid

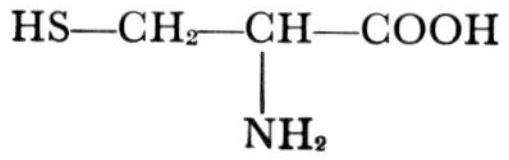

    $$HS-CH_2-CH-COOH$$
    $$|$$
    $$NH_2$$

18. **Arginine,** $C_6H_{14}N_4O_2$

    or α-Amino-δ-guanadine-valeric acid

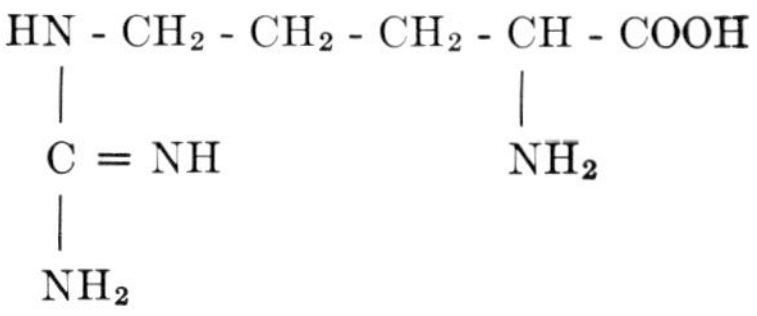

    $$HN-CH_2-CH_2-CH_2-CH-COOH$$
    $$|\qquad\qquad\qquad\qquad\quad |$$
    $$C=NH\qquad\qquad\quad NH_2$$
    $$|$$
    $$NH_2$$

19. **Lysine,** $C_6H_{14}N_2O_2$

    or α-ε-Diamino-caproic acid

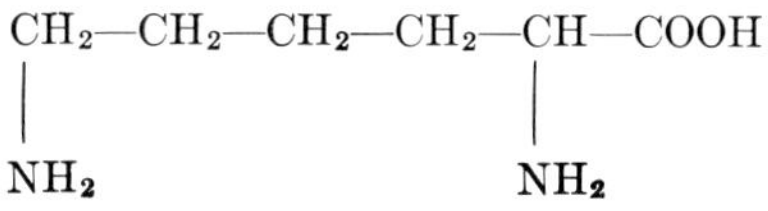

    $$CH_2-CH_2-CH_2-CH_2-CH-COOH$$
    $$|\qquad\qquad\qquad\qquad\quad |$$
    $$NH_2\qquad\qquad\qquad\quad NH_2$$

20. **Histidine,** $C_6H_9N_3O_2$

    or α-Amino-β-imadazole-propionic acid

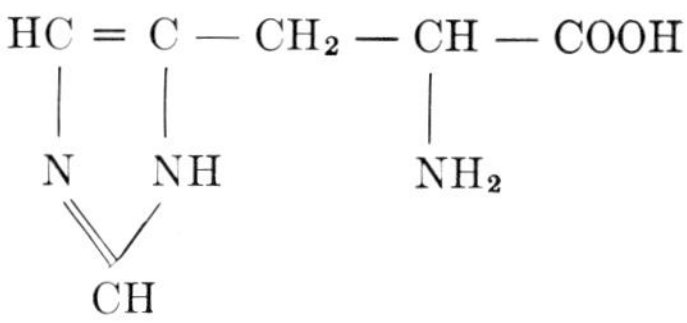

    $$HC=C-CH_2-CH-COOH$$
    $$|\quad\ |\qquad\qquad\quad |$$
    $$N\quad NH\qquad\quad\ NH_2$$
    $$\diagdown\!\!\diagup$$
    $$CH$$

21. **Tryptophan,** $C_{11}H_{12}N_2O_2$ or α-Amino-β-indole propionic acid

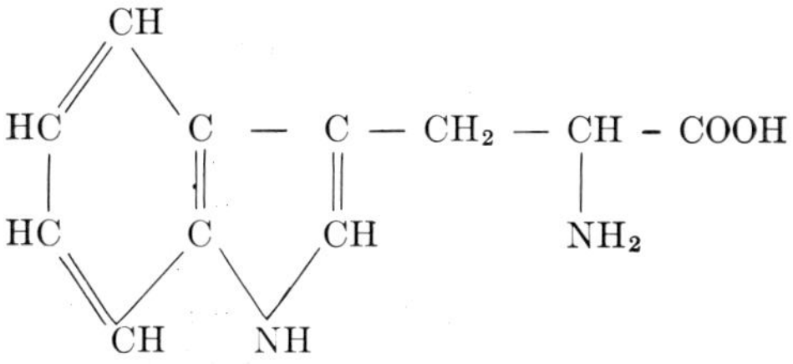

    $$\begin{array}{c} CH \\ HC \diagup \quad \diagdown C - C - CH_2 - CH - COOH \\ |\qquad \| \quad \|\qquad\qquad\quad | \\ HC\qquad C\quad CH\qquad\quad NH_2 \\ \diagdown\qquad\diagup \\ CH\quad NH \end{array}$$

22. **Citrulline,** $C_6H_{13}O_3N_3$
or α-Amino-δ-ureidovaleric acid

$$\begin{array}{c} NH_2 \\ | \\ C=O \\ | \\ HN-CH_2-CH_2-CH_2-CH-COOH \\ | \\ NH_2 \end{array}$$

23. **Thyroxine,** $C_{15}H_{11}I_4NO_4$ or (3, 5, 3', 5'—Tetraiodo—4'—hydroxy-diphenyl ether) alanine

Albumin. Albumin or lactalbumin like casein belongs to the protein group. It is generally considered to be in true solution in milk, but some investigators claim it to be in a very fine colloidal state. It is not acted upon by rennet nor coagulated by acids at ordinary temperatures, but is precipitated by heat especially at the proper pH value, which is around 4.5. It contains the elements carbon, hydrogen, oxygen, nitrogen and sulphur, but no phosphorus as in the case of casein. It is usually listed as a high quality protein but all the essential amino acids have not as yet been reported. The molecular weight is uncertain, ranging from 1000 to 25,000, considerably lower than casein.

Just as modern methods of analyses have shown casein to be a mixture of several fractions, the older use of the term lactalbumin no longer indicates the true nature of this group of milk proteins. A recently published nomenclature[23] of the proteins in milk defines albumin or lactalbumin as "the portion of the milk serum proteins that is soluble in neutral one-half saturated ammonium sulfate or saturated magnesium sulfate." The fraction of milk serum proteins (skim milk proteins other than casein) that is insoluble in the above reagents is called the globulin or lactoglobulin fraction. Commercial lactalbumin, prepared by heating properly acidified whey includes both the lactalbumin and lactoglobulin. Table 10 compares the newer and older nomenclature of the proteins in skimmilk.[23]

Table 10. Protein Analyses of Skimmilk Comparing Newer and Older Nomenclature

Electrophoretic Analysis		Rowland Procedure	
(Protein)	(g/100 ml.)	(Classical fraction)	(g/100 ml.)
α-Casein	1.4-2.3		
β-Casein	0.5-1.0	Casein	2.2-3.5
γ-Casein	0.06-0.22		
		Total serum proteins	0.4-0.8
β-Lactoglobulin	0.20-0.42		
α-Lactalbumin	0.07-0.15	Lactalbumin fraction	0.15-0.40
"Blood" serum albumin	0.02-0.05		
Immune globulins	0.05-0.11	Lactoglobulin fraction	0.07-0.20
Other components	0.06-0.17	Proteose-peptone fraction	0.06-0.25

The chief value of albumin is its food value in milk. However, a nutritious but not very palatable albumin cheese may be made from whey, the watery by-product left in the manufacture of American Cheddar cheese. The whey is heated to 165° F. to 175° F., and then upon adding a little acid to bring the hydrogen-ion concentration to a pH around 4.5, the albumin will be precipitated and during the heating process will come to the surface from where it may be dipped off into molds.

Globulin. A third protein in milk is lactoglobulin. It is usually not listed in the composition of milk but is included in the percentage allowed for the albumin because the amount is not large in normal milk. However, in colostrum it is the chief protein, the amount being higher than that of casein. It is important in providing immunity to certain infections in the new born. It contains the same elements as casein — carbon, hydrogen, oxygen, nitrogen, sulphur and phosphorus. It can be separated from albumin by precipitating it with an excess of salts such as magnesium sulfate. Rennin and dilute acids do not affect it, but high temperatures will cause it to precipitate along with the albumin.

The classical lactoglobulin has been broken down into euglobulin and pseudoglobulin. The former is insoluble in water and 0.06 percent NaCl but soluble in 0.6 percent NaCl. Pseudoglobulin is soluble in water but precipitated when alcohol is added to its aqueous solution.

Table 11. Chemical Analyses of Casein, Albumin and Globulin

	Casein	Albumin	Globulin
	%	%	%
Carbon	53.50	52.51	51.86
Hydrogen	7.13	7.10	6.96
Oxygen	22.14	23.04	24.64
Nitrogen	15.80	15.43	15.44
Sulphur	0.72	1.92	0.86
Phosphorus	0.71	trace	0.24

Other Proteins. In addition to the casein, albumin, and globulin portion of milk, several other proteins have been identified. These are rather insignificant insofar as total composition is concerned but are of considerable importance in the behavior and nutritive value of milk and milk products. Riboflavin is present as a protein-phosphoric acid complex on the fat globule surface. Several flavoproteins are thought to exist on the fat globule surface where they assume importance in oxidation-reduction actions and are probably associated with the oxidation of fat. Some proteose-peptone constituents are present in milk, either naturally occurring or resulting from degradation of natural proteins. Although only two enzymes have been isolated from milk, it is known that there are 8 or 10 which are native to milk. These will be discussed later.

Whitney[24] has summarized the state of our present knowledge concerning the minor proteins in cows' milk. He observed that xanthine oxidase, lactoperoxidase, alkaline phosphatase, and trypsin inhibitor have been conclusively shown to be native to milk and purified to the extent that knowledge is available concerning their composition and behavior. Proteose, amylases, lipases, and lactenins I and II are native to milk and some of their properties are known. He lists γ-casein and a "red protein" as known to be present in milk and have been purified but these appear to be heterogeneous and their origin not firmly established. Lactose-destroying enzymes, aldolase, catalase, copper protein, δ-proteose, and the sunlight-flavor precursor have been shown to be present in milk but their origin and properties are still uncertain.

Sugar. Lactose is the chemical term for the sugar in milk. It is a di-saccharide, being a union of the two hexose sugars, glucose and galactose. It has the same formula as sucrose, or cane sugar ($C_{12}H_{22}O_{11}$), but differs in molecular constitution. Lactose is in solution in milk. It is less soluble than sucrose and therefore less sweet. The following figures show the relative sweetness of some common sugars to that of sucrose:

Sucrose,	$C_{12}H_{22}O_{11}$	100	Maltose,	$C_{12}H_{22}O_{11}$	32.5
Fructose,	$C_6H_{12}O_6$	173	Galactose,	$C_6H_{12}O_6$	32
Invert sugar,		130	Lactose,	$C_{12}H_{22}O_{11}$	16
Glucose,	$C_6H_{12}O_6$	74			

Whittier[25] thoroughly described the structure of lactose, its chemistry and physiochemical properties as well as its manufacture and fermentation in a review of over 300 references up to 1944. The practical aspects of lactose production were well presented by Fisher.[26] Several excellent papers on lactose were presented at the 1958 Annual Meeting of The American Dairy Science Association.[27]

Lactose is an important food constituent of milk, especially for babies. It gives a firmer flesh than other sugars, aids in calcium and phosphorus assimilation thus forming better bones and teeth, and furthermore lessens the requirement of vitamin D needed. The pharmaceutical industry uses considerable quantities of lactose as a filler in medicine and as a coating for pills. The value of lactose as a medium for the growth of the mold, Penicillium notatum, the producer of the drug, penicillin, is largely responsible for the increase in lactose production from less than 8 million pounds in 1943 to 21 million pounds in 1947. A series of patents issued after World War II for newer methods of producing edible lactose at relatively low cost stimulated anew the interest of food manufacturers in the use of lactose to improve food products and lower ingredient cost. Lactic acid is produced in milk by the action of bacteria upon the milk sugar. This action may be represented by the equation:

$$C_{12}H_{22}O_{11} + H_2O = 4C_3H_6O_3$$
$$\text{Lactose} \quad\quad \text{Water} \quad\quad \text{Lactic Acid}$$

Lactose fermenting yeasts, such as Torula cremoris and Torula sphaerica are capable of fermenting milk sugar to alcohol and carbon dioxide. Foamy cream in hot weather is a result of this reaction.

Two forms of lactose are known — alpha lactose and beta lactose. The former exists both as a hydrate and as an anhydride but the latter only in the anhydride state. The ordinary lactose of commerce is the alpha hydrate, that is, it contains water of crystallization.

Initial and Final Solubility

When either alpha or beta lactose is dissolved in water, the one will change to the other form until an equilibrium is reached. For example, if a sufficient quantity of α-lactose is placed in 100 ml. of water at 32° F., immediately 5.0 grams of it will dissolve. This is the *initial solubility* of lactose at that temperature. Then as the mixture stands the α-form gradually changes over to the β-form because β-lactose is more soluble than the α-lactose and as the change-over continues more α-lactose dissolves until both forms are saturated with respect to each other. At this equilibrium point there will be 5.0 grams alpha and 6.9 grams beta lactose or a total of 11.9 grams. This is the *final solubility* of lactose at 32° F. Table 12 shows the initial and final solubilities of lactose at different temperatures.

Table 12. Initial and Final Solubilities of Lactose

Temperature	Alpha	Beta	Total
°F	g	g	g
32	5.0	6.9	11.9
59	7.3	9.6	16.9
77	8.9	12.7	21.6
102	12.4	19.1	31.5
120	17.6	24.8	42.4
147	26.6	39.2	65.8
165	35.1	51.1	86.2
192	58.7	80.5	139.2

The steps in the manufacture of commercial lactose are briefly as follows: Heat whey to nearly boiling, add lime to precipitate the albumin and phosphates and then evaporate the solution to about 30 percent solids. Next filter the concentrated solution and again evaporate to about 80 percent solids at which point the sugar crystals will form. Cool, centrifuge out the crystals, and refine with carbon black.

Beta lactose is much more soluble than the alpha form and therefore much sweeter. It is nearly as sweet as cane sugar. In

1931 Sharp[19] patented his method of obtaining this sugar commercially. Briefly the method consists of adding alpha lactose to a saturated solution of lactose held above 93.5° C. At this temperature the sugar solution is saturated with the beta lactose and unsaturated as regards the alpha form, and thus the added alpha lactose will dissolve and beta lactose to an equivalent amount will necessarily be forced to crystallize out. It is then put through a process of filtration in a heated centrifuge and recrystallized. The resulting sugar is practically in the beta form entirely.

Lactose is a reducing sugar, that, is, it acts on a copper salt and reduces it to cuprous oxide, a brick red compound. A non-reducing sugar produces cupric oxide, which gives a black color. Fehling's solution is used for this test and is made up as follows:

Solution A. Copper sulphate, $CuSO_4$, 34.65 grams in 500 ml. of water.

Solution B. Potassium hydroxide, KOH, 125 grams and Rochelle salt, $NaKC_4H_4O_6 \cdot 4H_2O$, 173 grams in 500 ml. of water.

These solutions are made and kept separately in order to avoid reactions that would render them inoperative.

The test is made by measuring 2 ml. each of solutions A and B into a test tube and adding around 1 ml. of the sugar solution. Heat the mixture over a Bunsen flame and then note if a red precipitate or black color forms. Since Fehling's solution is essentially copper hydroxide in solution, the following reactions illustrate what takes place according to whether the sugar is a reducing or non-reducing sugar:

Reducing sugar

$$2Cu(OH)_2 + heat + sugar\ solution = Cu_2O + 2H_2O + O$$
Cuprous oxide (brick red precipitate)

Non-reducing sugar

$$Cu(OH)_2 + heat + sugar\ solution = CuO + H_2O$$
Cupric oxide (black coloration)

A non-reducing sugar consists of an aldose sugar and a ketose sugar wherein, the aldehyde group (CHO) of the former is joined with the ketone group, C = O, of the latter and thus no reduction of the Fehling's solution is possible. Sucrose, a combination of glucose, an aldose sugar, and fructose, a ketose sugar, belongs to this class. If, however, sucrose is hydrolyzed by a dilute acid to glucose and fructose, reduction will take place, since these hexose sugars separately have reducing properties. The following structural formulae will show the difference between the two di-saccharides:

| Glucose (1–5) | + Galactose (1–5) | = | Lactose (4–1) |

| Glucose (1–5) | + Fructose (2–5) | = | Sucrose (1–2) |

It will be noted that in the case of lactose the fourth carbon atom of the glucose molecule is united with the first carbon of the galactose, leaving the aldehydic group (CHO), of the first carbon of the glucose molecule free. Therefore this sugar has reducing properties and is a *reducing sugar*. On the other hand in sucrose, the aldehydic group (CHO) of the glucose is joined to the ketonic group (C = O) of the fructose, and with no free aldehydic or ketonic groups, reduction cannot take place. Thus sucrose is a *non-reducing sugar*.

Ash. If milk is heated sufficiently to just evaporate the water, the residue will be the solids in milk. If, however, the solids are heated to a dull red heat, all that part which can be converted to volatile substances by oxidation will be driven off and the non-volatile part will be left as a white residue. This is the *ash*. It contains the minerals or inorganic constituents of milk.

The elements in comparatively large quantities in milk are: calcium, 0.112%; phosphorus, 0.095%; potassium, 0.138%; magnesium, 0.013%; sodium, 0.059%; chlorine, 0.109% and sulphur, 0.01%. In small amounts are: iron, 3.0 p.p.m.; zinc, 3.0 p.p.m.; silicon, 2.0 p.p.m.; copper, 0.3 p.p.m. and fluorine, 0.15 p.p.m. In traces are aluminum, manganese, iodine, boron, titanium, vanadium, rubidium, lithium and strontium. The ash content of milk varies little during the year.[1]

The percentage of minerals in the ash will be less than it was in the milk, because of the alterations due to oxidation occurring during the incineration. Some of the chlorine may be volatilized at too high heating, and the citric acid is destroyed in ashing. According to Van Slyke[28] the percentage of salts drops from around 0.9 in the milk to 0.7 in the ash.

In milk the potassium, sodium and chlorine are in solution; parts of the calcium and phosphorus are combined with the proteins while the remaining portions of the calcium and phosphorus together with the magnesium are partly in suspension and partly in solution. Sulphur is present in the amino acids, cystine and methionine of the casein and albumin.

Minerals are essential for life processes. Calcium and phosphorus are necessary for bone formation and good teeth. Iron is required in the formation of hemoglobin of the red corpuscles. Copper acts in conjunction with iron in this formation, as the iron cannot be used for this purpose unless copper is present. Iodine is needed in the production of the hormone, thyroxin, which prevents goiter. See Chapter XIV for nutritive values of dairy products.

Minerals also enter into the heat stability of milk, especially in the manufacture of evaporated milk. A certain balance is necessary between the calcium and magnesium on the one hand and the phosphates and citrates on the other. The first two minerals tend to cause coagulation while the latter two compounds prevent it. Hence, if a batch of milk contains an excess of calcium, the correct balance could be adjusted by the addition of sodium phosphate. Citric acid aids in maintaining salt balances by forming salts with calcium and magnesium.

Vitamins. Milk contains the fat-soluble vitamins A, D, E and K; the B-complex consisting of thiamin, riboflavin, a grass juice factor, niacin, pantothenic acid, pyridoxine, biotin, choline, inositol, folic acid, P-amino benzoic acid, and B_{12}; and vitamin C or ascorbic acid. Their chemical and physical properties will be considered here. For nutritional values see Chapter XIV.

a. *Vitamin A* is a colorless compound, having the formula, $C_{20}H_{30}O$. It does not exist as such in plants but as the yellow colored pigment, carotene, $C_{40}H_{56}$, which must be converted to vitamin A before it can function. There are four forms of carotene, alpha, beta, gamma and hydroxy-β-carotene. The beta-carotene is the most active, and when taken into the animal body is converted into vitamin A. Therefore carotene is often called provitamin A. Both carotene and vitamin A may be present in the animal body. The yellow color of milk fat and of animal fat is due to the presence of carotenoids. The fat in Guernsey milk has a much more golden color than has that of Holstein milk because it has a higher content of carotene. However, the fat in Holstein milk contains more preformed vitamin A and since carotene must be changed to vitamin A before it can be used, the two milks will finally have approximately the same amount of vitamin A per unit of fat. Thus the color of milk fat as a guide in comparing the vitamin A potency of different milks is not reliable.

Vitamin A is soluble in oils and fats and practically insoluble in water. Both the carotene and the vitamin are stable to heat, acids and alkalis. Large losses may occur through oxidation and exposure to light.

b. *Vitamin D* is now recognized in two forms known as D_2 and D_3. Ergosterol, a sterol in plants such as yeasts, when irradiated will produce calciferol, which is the active form, vitamin D_2, $C_{28}H_{44}O$. 7-dehydro cholesterol $C_{27}H_{44}O$, a sterol in animal bodies especially present in the skin, when exposed directly to sunlight or ultra-violet light, will be transformed to the active form, vitamin D_3. Vitamin D is stable to heat, alkalis, acids and oxidation. It is soluble in oils and fats and insoluble in water. It is available commercially as viosterol.

c. *Vitamin E* is one of the higher alcohols ($C_{29}H_{50}O_2$). It is insoluble in water, soluble in oils and fats, stable to heat, light, alkalis and acids, but is destroyed when fats turn rancid. It is called the anti-sterility vitamin. Its active form is alpha tocopherol.

d. *Vitamin K* was recognized in 1934 and synthesized in 1939. It is present in milk only in trace quantities. K_1, a yellow oil, has the formula $C_{31}H_{46}O_2$ and K_2, a yellow crystalline substance, has the formula $C_{41}H_{56}O_2$. The compound has a quinoid ring which is responsible for its physiological activity, demonstrated by the fact that the much simpler synthetic compound "menadione" (2 methyl-1, 4-naphthoquinone or $C_{11}H_8O_2$) has equivalent activity with the natural vitamins. These compounds are soluble in organic solvents and are stable to heat but are destroyed by light and oxidizing agents.

e. *Thiamin or vitamin B_1* has been isolated chemically in crystalline form as thiamin hydrochloride, $C_{12}H_{18}ON_4SCl_2$. It is soluble in water but insoluble in fats and oils. Pasteurization by the holding method destroys around 9.0 percent of the thiamin according to Holmes et al.[29] but practically no loss occurs by the high-temperature short-time process. Autoclaving milk will destroy all the vitamin but it is quite stable to dry heat. Milk is not a rich source of this vitamin.

f. *Riboflavin, Vitamin B_2 or G,* consists of a ribose sugar attached to a colored compound called flavin. Its chemical formula is $C_{17}H_{20}N_4O_6$. Whey, the watery product in cheese making, is a rich source. Its greenish yellow color is due to the presence of this vitamin. The vitamin has been crystallized, is soluble in water, stable to heat, fairly stable to oxidation, acids and alkalis, but sensitive to light. It takes up sunlight readily and uses it to oxidize ascorbic acid, thus destroying this vitamin. It is therefore important to keep milk out of sunlight. The riboflavin content of milk shows very little response to summer pastures as compared to barn feeding. The different breeds of dairy cows yield in their milk approximately the same amount of this vitamin when considering the total yield for a period.

g. *A grass juice factor* has in recent years been shown to be present in milk as indicated by the better growth of rats fed on pasture milk. It has not as yet been chemically identified.

h. *Niacin, or nicotinic acid* the pellagra prevention factor, is present in milk to a small extent, 1.46 milligrams per liter.[30]

Its formula is $C_6H_5O_2N$. It is soluble in hot water and alcohol. It is obtainable commercially.

i. *Pantothenic acid* is a member of the vitamin B complex and is found in skimmilk and whey. It is a white powder, not readily water soluble. It is unstable to alkalis, acids and prolonged heat. It is stable to light and to oxidizing and reducing agents. Its exact functions are not as yet clear. Its formula is $C_9H_{17}O_5N$.

j. *Pyridoxine, B_6,* is also a member of the vitamin B complex. It is a white crystalline substance, soluble in water and alcohol, unstable in neutral or alkaline solutions, and relatively stable to acids and heat. Light causes a rapid destruction. It appears to bear some relationship to the metabolism of both fat and protein. Milk is a good source. Its formula is $C_8H_{11}O_3N$.

k. *Biotin* also belongs to the B complex vitamins. It was discovered in 1936. It is a crystalline compound, an acid, soluble in alcohol and water and very stable chemically. Its formula[31] is $C_{10}H_{16}O_3N_2S$. Milk is a good source of this vitamin.

l. *Choline* is a constituent of the phospholipid lecithin but it is also classed with the vitamins. Its formula is $C_5H_{15}O_2N$. It is a colorless viscid fluid, unstable to alkalies but stable to acids.

m. *Inositol* has the formula $C_6H_{12}O_6$. It is a white crystalline compound and quite stable to both alkalies and acids. Inositol is soluble in water and insoluble in alcohol.

n. *Folic acid* has the formula $C_{19}H_{19}O_6N_7$. It is a yellow crystalline solid, slightly soluble in water but unstable in acid solution. At pH 1.0 autoclaving destroys most of its activity.

o. *Vitamin B_{12}* was isolated in 1948. It is a red crystalline compound. Its formula is $C_{63}H_{90}O_{14}N_{10}P$ Co. The presence of cobalt is one of the distinguishing differences between this vitamin and others in the B complex.

p. *Ascorbic acid or vitamin C* is water soluble and its chemical formula is $C_6H_8O_6$. It has also been called hexuronic acid and cevitamic acid. It is lost rather easily through oxidation. Heat treatments of milk, especially in copper containers, activate the speed of oxidation. Vat pasteurization destroys around 18.0 percent[29] but according to King[32] high temperature pasteurization causes no loss. However, most milks do not have a sufficient amount of vitamin C to fill the need of children. Fruit juices are the main source. The vitamin is very sensitive to alkalies, but fairly stable in weak acid solutions. It is a strong reducing agent and its content in a substance can be determined by titrating with the dye, 2, 6-dichloro-phenol-indophenol. Ascorbic acid is available commercially.

q. *P-amino benzoic acid or vitamin B_x* ($C_7H_7O_2N$) was recognized as a vitamin in 1940. It is a white, crystalline substance, slightly soluble in water. It melts at 186-187° C. Milk contains approximately 0.1 mg. of this vitamin per liter of milk.

Enzymes. Enzymes are organic compounds secreted by living cells and act as catalysts, in that they aid a chemical reaction without themselves being changed. A number of enzymes are present in milk among which are lipase, protease, catalase, peroxidase, reductase, phosphatase, diastase and lactase.

a. *Lipase.* This enzyme acts on fats, freeing the fatty acids from their union with glycerol. It is the cause of rancidity in dairy products, the flavor being due mainly to the highly volatile butyric acid that is liberated. It may be activated in milk by first cooling the milk to 40° F., then warming it to 86° F. and finally cooling to below 50° F. Homogenizing raw milk will cause it to turn rancid within a few minutes. The enzyme is inactive at 104° F. and destroyed at 131° F. Milk from cows in late lactation and from certain individual cows at any time will often have a rancid flavor. It occurs more often during barn feeding as green feeds tend to prevent it. Rancidity may be detected by taste and also by measuring the surface tension[33] of milk, as it drops from a normal of 49-51 to 39-40 dynes per centimeter in rancid milk. Lactic acid bacteria use up the free fatty acids and then the surface tension is restored to normal. Herrington[34] published an extensive review of the literature on lipase in 1954.

b. *Protease or galactase.* Proteins are split by this enzyme to peptones and amino acids and thereby it aids in the ripening of certain cheeses. It is inactivated by heat at 165° F. to 175° F.

c. *Catalase.* In mastitis the bacteria and leucocytes increase the catalase content of milk. If hydrogen peroxide is added to milk in a test tube, the enzyme will liberate molecular oxygen and the amount of gas produced will indicate the quality of the milk. This method has been used to some extent in detecting mastitis but has now been superseded by other tests. A temperature of 150° F. to 160° F. for 30 minutes inactivates this enzyme. Its presence does not affect the quality of milk.

d. *Peroxidase.* This is another enzyme in milk that acts on hydrogen peroxide but it differs from catalase in liberating the oxygen in the active or nascent form. It is inactivated by a temperature of 175° F. and this fact is made use of in the "boiled milk" test. If active as in raw milk it will liberate nascent oxygen from hydrogen peroxide which in turn will react with an added compound paraphenylene hydrogen chloride and turn it blue. In milk heated above 175° F. the enzyme is destroyed and no action will take place when these compounds are added.

e. *Reductase or Schardinger's enzyme.* The quality of milk is determined by means of this enzyme in the methylene blue reduction test. It reduces the methylene blue in the presence of an aldehyde. A temperature of at least 176° F. is required to inactivate it.

f. *Phosphatase.* This enzyme is essential to the life of the bacteria producing it. Tests for proper pasteurization are based on the activity of phosphatase. It has the property of liberating phenol from alcoholic esters of phosphoric acid and the measurement of this free phenol indicates whether or not the milk has been properly pasteurized. Regular pasteurization temperature of 143° F. for 30 minutes will cause inactivation.

g. *Diastase.* Starch is split by this enzyme. No starch is present in milk but the enzyme gains entrance to some milks with the blood as in colostrum milk and in diseased udders.

h. *Lactase.* Some lactose is broken down to the simple sugars by lactase and then changed to alcohol. However, little of this activity occurs in milk.

Gases. Milk contains the gases — oxygen, carbon dioxide and nitrogen. The percentages of each change immediately upon contact with the air, as in the milking process. Table 13 gives the amounts as found in the original state and later in ordinary commercial mixed milk. A large part of the carbon dioxide is lost, while the oxygen and nitrogen through absorption are increased. The oxygen content of milk is of economic importance in its relation to the undesirable oxidized flavor of milk.

Table 13. Gas Content of Milk

Investigator		Oxygen Vol. %	Carbon dioxide Vol. %	Nitrogen Vol. %
Frayer[35]	First drawn (anaerobically)	0.092	6.58	1.18
Frayer[35]	Raw commercial mixed milk	0.43	3.63	1.37
Noll and Supplee[36]		0.47	4.45	1.29

BIBLIOGRAPHY

1. Syrrist, G. Economic Effects of the Seasonal Variations in the Delivery of Milk to Dairies in Norway. Sci. Rept., Agric. Coll. of Norway, 37, 11 (1958).
2. Wilcox, C. J., and Pfau, K. O. Study Causes of Variations in SNF Percentage of Milk. N. J. Agric. (May-June, 1958).
3. Provan, A. L. Solids-Not-Fat in Milk. J. Soc. D. Tech. 8, 2 (1955).
4. Rowland, S. J. Low Solids-Not-Fat — The Factors Involved. The Agr. Rev. (July, 1957).
5. Freeman, T. R. Effect of Breed, Season, and Stage of Lactation on Certain Constituents and Properties of Milk. Ken. Bull. 667 (1959).
6. Merilan, C. P., and Bower, K. W. Influence of Daily Environmental Temperature Cycles on Composition of Cows' Milk. Mo. Bull. 687 (1959).
6a. Campbell, M. H. A Study of the Fat Globule Size in Milk. Vt. Bull. 341 (1932).
7. Riemenschneider, R. W. Fatty Acids of Milk. Proc. Conf. on Milk Conc. USDA, Philadelphia, Pa. (1957).
8. Jack, E. L., and Smith, L. M. Chemistry of Milk Fat: A Review. J. D. Sci. 39, 1:1-25 (1956).
9. Conant, J. B. Organic Chemistry. The MacMillan Co. N. Y. p. 89 (1937).
10. Holland, et al. J. Ag. Rsh. 24, p. 365 (1923).
11. Hunziker, O. F. The Butter Industry, 3rd edit. (1940).
12. Becker, R. B., and Arnold, P. T. Influence of Season and Advancing Lactation on Butterfat Content of Jersey Milk. J. D. Sci. 18:397 (1935).
13. Horrall, B. E. A Study of the Lecithin Content of Milk and Its Products. Ind. (Purdue) Bul. 401 (1935).
14. Thurston, L. M., and Petersen, W. E. Estimation of Fat in Buttermilk. J. D. Sci. 11:278 (1928).
15. Crane, J. C., and Horrall, B. E. Phospholipids in Dairy Products. II. Determination of Phospholipids and Lecithin in Lipids Extracted from Dairy Products. J. D. Sci. 26:935-942 (1943).
16. King, N. The Milk Fat Globule Membrane. Commonwealth Agr. Bur. Farnham Royal, Bucks, England (1955).
17. Van Der Waarden, M. Flavour Defects in Butter Stored at Low Temperature. D. Sci. Abst. 7,3:214-215 (1945).
18. Storgards, T., and Hietaranta, M. On the Formation of Oily and Fishy Flavors in Butter. Proc. XII Int. Dairy Congress, 2:389-391 (1949).
19. Sharp, P. F. Dairy Chemistry, Cornell University.
20. Mellander, O. Elektrophoretische Untersuchung von Casein. Biochem Z. 300:240-245 (1939).
21. Hipp, N. J. et al. Separation of α, β, and γ Casein. J. D. Sci. 35:272-281 (1952).
22. Pyne, G. T. The Chemistry of Casein. D. Sci. Abst. 17,7:531-554 (1955).
22a Vickery, H. B. Private Communication. Conn. Ag. Exp. Sta. (1942).
23. Jenness, R. et al. Nomenclature of the Proteins of Bovine Milk. J. D. Sci. 39:536-541 (1956).
24. Whitney, R. McL. The Minor Proteins of Bovine Milk. J. D. Sci. 41,10:1303-1323 (1958).

25. Whittier, E. O. Lactose and Its Utilization: A Review. J. D. Sci. 27:505-537 (1944).
26. Fisher, C. H. The Sugar We Get from Milk. USDA Yearbook of Agr. pp. 322-328 (1950-51).
27. Choi, R. P. et al. Lactose Symposium. J. D. Sci. 41,2:319-334 (1958).
28. Van Slyke. L. L. Modern Methods of Testing Milk and Milk Products. Orange Judd Pub. Co. New York, p. 15, 3rd edit. (1927).
29. Holmes, A. D. et al. Effect of High-Temperature Short-Time Pasteurization on the Ascorbic Acid, Riboflavin, and Thiamin Content of Milk. J. D. Sci. 28:29-34 (1945).
30. Bailey, Jr., E. A. et al. A Method for the Estimation of Nicotinic Acid in Milk. J. D. Sci. 24:A269 (1941).
31. Hofmann, K. et al. Characterization of the Functional Groups of Biotin. J. Biol. Chem. 141:207-214 (1941).
32. King, C. G., and Waugh, W. A. The Effect of Pasteurization Upon the Vitamin C Content of Milk. J. D. Sci. 17:489-495 (1934).
33. Tarassuk, N. P., and Smith, F. R. Relation of Surface Tension of Rancid Milk on the Growth of Strep. lactis. J. D. Sci. 23:163-171 (1940).
34. Herrington, B. L. Lipase: A Review. J. D. Sci. 37,7:775-789 (1954).
35. Frayer, J. M. The Dissolved Gases in Milk and Dye Reduction. Vt. Bul. 461 (1940).
36. Noll, C. I., and Supplee, G. C. Factors Affecting the Gas Content of Milk. J. D. Sci. 24:993-1015 (1941).

REVIEW QUESTIONS

1. Distinguish between milk plasma and milk serum.
2. Define a fat.
3. Why does butter soften gradually rather than suddenly when placed in a warm room?
4. What is meant by an unsaturated fat?
5. What is the difference between a true fat and a phospholipid?
6. If Holstein milk testing 3.5 percent fat sells for 26 cents a quart, what should Jersey milk testing 5.5 percent sell for on a comparable energy value basis?
7. Give three ways in which casein differs from albumin.
8. Define an amphoteric reaction. Isoelectric point.
9. Write the structural formula of an amino acid.
10. What are the sulphur containing amino acids?
11. Distinguish between a reducing and a non-reducing sugar.
12. What are some of the most important minerals in milk, for the human body?
13. Why does Guernsey milk have a more golden color than Holstein milk?
14. Is the color of milk fat a reliable guide for comparing the vitamin A potency of different milks? Explain.
15. How may the enzyme lipase be activated?

CHAPTER II

Physical Properties of Milk

Taste. Fresh milk that has been produced under ideal conditions will have no pronounced flavor, but will have a slightly sweet and pleasant taste. This is primarily due to the relationship of the lactose and chloride contents.[1] If this relationship is disturbed so that the chloride becomes relatively greater, as in late lactation or in mastitic conditions, the flavor is adversely affected. The fat and protein contents, while not primarily concerned with the taste of milk, give body to the flavor. Milks with low fat tests and correspondingly low solids not fat, will taste somewhat flat while the milks with higher percentage of these constituents will have a fuller flavor.

Any pronounced flavors in freshly drawn milk are usually due to certain types of feed. Pasture grass, fresh clover and alfalfa, and silages will give to milk characteristic flavors which are rather objectionable but can be controlled by proper methods and schedules of feeding. Certain weeds such as garlic or wild onion will taint the flavor of milk very soon after ingestion by the animal and the objectionable flavor will follow into the cream and butter. In general, dry feeds produce no marked off flavors. If the feeds giving the strong flavors are fed to the cows after milking, the barns kept well ventilated and hygienic conditions of milking used, the milk obtained should be of fine flavor and quality. Occasionally, fresh milk will give a "cowy" or "unclean" flavor as the result of ketosis or an udder abnormality.

Color. The white color of milk results from the dispersion of reflected light by the fat globules and the colloidal particles of casein and calcium phosphate. The yellow color is due to the pigment carotene which is fat soluble. The deeper color of Guernsey and Jersey milk as compared to that of the lower testing breeds is due to the greater amount of carotene present in the fat. All milks have a deeper color during pasture feeding than when barn fed, as grass is high in carotene, and this pigment is readily transferred from the feed to the fat in the milk or to the body fat of the animal. Feeding carrots will have the same effect, as these roots are very high in carotene content. Xanthophylls also give some yellow color. Skimmilk has a bluish tinge since the fat is removed and therefore less particles are left to disperse the reflected light.

Riboflavin is another pigment in milk, but its color does not appear until the fat and casein have been removed as in cheese making. It gives the remaining product, whey, a yellowish green appearance. This pigment is water soluble.

Certain organisms will affect the color of milk, but this would apply only to abnormal conditions such as the production of red milk by *Bacillus prodigiosus* and blue milk by *Pseudomonas cyanogenes*.

Freezing Point. Milk has a freezing point which is quite constant and is generally given as $-0.55°$ C. or $31.01°$ F. It depends on the soluble constituents, mainly lactose and chlorides. These compounds have a close inverse relation, that is, when one rises in amount the other lowers and vice versa, which fact accounts in large measure for this constancy. Doan[2] gives figures in Table 14 as the freezing points of whole milk, cream, and skimmilk in trials with ten samples of milk that were centrifugally separated.

The figures (Table 14) show a very close agreement for whole milk, cream, and skimmilk irrespective of their differences in composition. This illustrates the fact that the freezing point depends, not on the composition of the sample, but on the concentration of the dissolved substances, mainly, lactose and salts. The fat existing in milk as coarse globules and the casein as colloids have no effect on the freezing point.

A nationwide survey[3] conducted in 1953 showed statistically significant differences in the freezing point of milk of certain of the eight cities included in the survey. Dahlberg[4] has pointed out in a recent report that freezing points of milk may vary daily, seasonally, as well as according to breed, feed, and many other

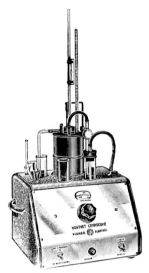

Hortvet Cryoscope
Courtesy Fisher Scientific Co.

Table 14. Freezing Points of Milk, Cream and Skimmilk

Sample No.	Whole milk			Cream			Skimmilk		
	Fat	S.N.F.	Freezing point	Fat	S.N.F.	Freezing point	Fat	S.N.F.	Freezing point
	%	%	°C.	%	%	°C.	%	%	°C.
1			−0.558			−0.559			−0.558
2			−0.537			−0.540			−0.538
3			−0.546			−0.546			−0.548
4			−0.544			−0.545			−0.545
5			−0.549			−0.550			−0.548
6	4.30	8.47	−0.545	22.0	6.85	−0.545	0.03	8.91	−0.546
7	3.95	8.51	−0.538	19.5	7.91	−0.539	0.04	9.05	−0.536
8	4.05	8.53	−0.540	27.5	6.32	−0.538	0.04	9.02	−0.537
9	4.00	8.53	−0.543	24.0	7.02	−0.545	0.03	9.41	−0.542
10	4.05	8.36	−0.541	20.0	7.25	−0.540	0.03	8.80	−0.540
Avge.			−0.544			−0.545			−0.544

factors. He noted that the freezing point of milk is related to that of the blood of the cow, requiring two or three hours to approach equilibrium.

After reviewing over one hundred references on the freezing point of milk, Shipe[5] has recommended that the present official standard be re-examined. The data of Doan,[2] Denis-Nathan,[6] and others (See Table 15) indicate that the value of −.550° C., which is presently accepted by AOAC as the official freezing point of milk, is too low. Although the difference does not appear to be great, it may cause some problems when freezing point data are used to determine if a sample of milk contains added water. This will be further discussed in the section on milk adulteration.

Table 15. The Freezing Point of Some Authentic Milk Samples Produced in the United States as Determined by the Hortvet Cryoscope[4]

Author	Year	Number of Samples	Freezing Point, ° C. Highest	Lowest	Average
Authentic Herd Samples					
Hortvet	1921	15	−0.545	−0.562	−0.551
Bailey	1922	45	−0.530	−0.560	−0.543
Lampert	1939	24	−0.529	−0.547	−0.536
Krienke, Arrington	1949	13 Jerseys —	—	—	−0.542
		6 Holst'ns —			−0.535
Sato, Hankinson, Gould, Armstrong	1957	15	−0.540	−0.555	−0.546
Robertson	1957	1627	−0.517	−0.577	−0.543
Average	1921-1922	60			−0.545
Average	1921-1957	1745			−0.543
Authentic Samples of Individual Milkings					
Hortvet	1921	60	−0.534	−0.562	−0.547
Bailey	1922	179	−0.530	−0.566	−0.545
Krienke, Arrington	1949	29 Jerseys —	—	—	−0.549
		13 Holst'ns —			−0.539
Sato, Hankinson, Gould, Armstrong	1957	44	−0.533	−0.560	−0.546
Kleyn, Shipe	1957	541 a.m.—	−0.513	−0.565	−0.537
		536 p.m.—	−0.515	−0.571	−0.544
Average	1921-1957	1417			−0.542

Denis-Nathan[6] using the Hortvet cryoscope found the freezing point range of milk to be −0.528° C. to −0.561° C. with an average of −0.541° C.

As milk freezes it becomes very uneven in composition. The frozen portion is low in fat and other solids, while the liquid portion becomes concentrated with the milk solids, so much so, that milk never freezes solid in its entirety.

In Table 16 Richmond[7] gives the compositions of frozen and liquid portions of milk, respectively, when 10 percent of the milk was frozen.

Table 16. Composition of the Solid and Liquid Portions of Frozen Milk

	Melted Ice %	Liquid Portion %
Water	96.23	85.62
Fat	1.23	4.73
Proteids	0.91	3.90
Sugar	1.42	4.95
Ash	0.21	0.80
Specific gravity	1.0090	1.0345

Table 16 shows that the frozen portion is high in water and low in solids, but that the ratio of the solids to each other is not altered. This fact shows that only the water separates out upon freezing.

Freezing alters the physical condition of milk and it never returns to its original state. This is of utmost importance in the bulk milk handling system since milk may freeze during the first milking, melt during the succeeding additions of warm milk and thus appear normal at the time of pickup. Reid[8] found freezing causes the fat globules to lose their complete emulsion, to clump and become distorted and irregular in shape and size. The cream line was considerably lessened in milk that had been allowed to cream, then frozen and later thawed at 60° F.

Table 17. Effect of Freezing Milk at Varied Temperatures for Different Periods of Time on Cream Line[8]

Temperature Frozen in Degrees	Frozen 3 hours Percentage loss after thawing			Frozen 5 hours Percentage loss after thawing		
	1 hr.	3 hrs.	24 hrs.	1 hr.	3 hrs.	24 hrs.
0° F.	39.00	42.44	45.90	40.75	42.65	45.90
6° F.	38.84	41.82	43.16	40.12	42.47	45.40
10° F.	31.68	33.33	38.34	39.83	41.24	43.40
15° F.	24.90	29.70	33.20	28.70	33.80	38.88
20° F.	15.78	24.65	28.67	18.00	26.19	32.70
25° F.	8.31	13.98	25.57	15.90	27.90	30.10
30° F.	0.58	4.25	12.19	2.21	5.89	13.04

It will be noted in Table 17 that the lower the temperature of freezing and the longer the time, the greater was the loss of cream line.

The casein is also affected by freezing. It is partly broken from its existence in milk as calcium caseinate and precipitated as flakes. This condition together with some free fat particles give to thawed-out milk an unnatural appearance. The flavor is also affected, being rather watery to the taste. Thus, the avoidance of freezing is an important factor as regards market milk.

Boiling Point. Since a dissolved substance in water will lower the freezing point of the mixture below that of the water itself, it will also act conversely and increase the boiling point. Therefore milk with its dissolved substances will have a higher boiling point than pure water. The average temperature of boiling milk according to Eckles[9] is 100.17° C. or 212.3° F. It may go slightly higher with milk containing larger percentages of solids as in case of Jersey milk. Judkins and Mack[10] give the boiling range 212° F. to 214° F. Davies[11] cites the boiling point as 100.55° C. (213° F.). Bowen[12] gives the boiling point as 1° F. greater than that of water or 213° F.

Specific Gravity. The specific gravity of a substance is the ratio of its weight to the weight of an equal volume of another substance taken as a standard. Water is the standard for solids and liquids, while hydrogen is the standard for gases. The average specific gravity of normal whole milk is 1.032 at 60° F. (15.5° C.). Milk is heavier than water since all the solid constituents except the fat have a greater specific gravity than water. In Table 18 Sharp and Hart[13] and Richmond[7] give specific gravity figures of the different milk solids.

Table 18. Specific Gravity of Milk Constituents

	Sharp and Hart at 30° C.	Richmond at 15° C.
Fat	0.913	0.930
Plasma solids	1.592	1.616
Lactose	1.630	1.666
Citric acid	1.680	(as lactose)
Protein	1.350	1.346
Ash	5.500	5.500

It will be noted that Sharp and Hart used a temperature of 30° C. (87° F.) for their specific gravity readings. They claim that a temperature of 60° F. as specified by health and food officials is the most unsatisfactory temperature that could have been chosen, as the fat is not in the proper physical condition for specific gravity determinations. Solid fat has a greater specific gravity than liquid fat at the same temperature, and therefore when milk has been held at a low temperature for some time and then warmed to 60° F. it will have a higher specific gravity than if kept warm for a period before cooling to 60° F. In the first case the fat would still be more or less solid and in the second instances more or less liquid. They recommend that milk be warmed for one-half minute at 45° C. and then cooled to 30° C. before taking a specific gravity reading. This procedure will insure the fat being in the proper liquid state. The temperature at which the specific gravity determination is made should be stated[14] since water and milk expand approximately 1 part in 1000 parts for each 5° C. (9° F.) change in temperature.

Normal milk may range in specific gravity from 1.029 to 1.035 with the majority of samples between 1.030 and 1.033. A milk low in fat will have the lower specific gravity and conversely one rich in fat will have the higher specific gravity. At first hand it would appear that the reverse would be true because of the low specific gravity of fat, but in normal milk as the fat increases the solids not fat also increase and their combined specific gravities more than offset the effect of the fat. Thus, Jersey milk will be slightly heavier than Holstein milk. Table 19 based on the average of around 250,000 analyses of milk shows this point clearly.

A number of workers have observed that, when milk is allowed to stand quietly for several hours following milking, the specific gravity will tend to increase rapidly for a short time and then more slowly until it finally becomes stable. The rapid rise in specific gravity immediately after milking has been attributed to the escape of gases from the milk. The slower later change is now generally attributed to the slow solidification of fat. The phenomenon is referred to as the "Recknagel effect," after the man who first described it and results in an average change of 0.001 in specific gravity. In some cases the specific gravity of milk has steadily increased for 22 hours after milking,[15] a fact which would appear to refute some earlier claims that the action was solely the result of the escape of air bubbles from the milk.

Specific Heat. The specific heat of a substance is the ratio between the amount of heat required to raise its temperature one degree and the amount to raise the temperature of an equal mass of water one degree. The unit of heat is the "calorie." It is the amount of heat required to raise one gram of water one degree centigrade, 3.5° C. to 4.5° C. Since water is the standard for fluids, specific heat may also be defined as the number of calories required to raise the temperature of one gram of the substance 1°C. A larger unit is the British thermal unit (B.T.U.).

Table 19. Relation of the Fat Test to the Specific Gravity of Normal Milk

Fat* %	Solids Not Fat* %	Total Solids* %	Specific Gravity†
3.0	8.33	11.33	1.03092
3.1	8.40	11.50	1.03112
3.2	8.46	11.66	1.03128
3.3	8.52	11.82	1.03144
3.4	8.55	11.95	1.03148
3.5	8.60	12.10	1.03160
3.6	8.65	12.25	1.03172
3.7	8.69	12.39	1.03196
3.8	8.72	12.52	1.03184
3.9	8.76	12.66	1.03192
4.0	8.79	12.79	1.03196
4.1	8.82	12.92	1.03200
4.2	8.86	13.06	1.03208
4.3	8.89	13.19	1.03212
4.4	8.92	13.32	1.03216
4.5	8.95	13.45	1.03220
4.6	8.98	13.58	1.03224
4.7	9.01	13.71	1.03228
4.8	9.04	13.84	1.03232
4.9	9.07	13.97	1.03236
5.0	9.10	14.10	1.03240

† The specific gravity was calculated by the author using the formula $\frac{1}{4} L + 0.2F = \%$ S.N.F., where L is the Quevenne lactometer reading and F is the percent fat. The specific gravity is obtained by prefixing 1.0 to the lactometer reading. Other formulae may be applied and the results will be relatively the same.

*Reprinted by permission from Market Milk, second edition, by E. Kelly and C. E. Clement, published by John Wiley & Sons, Inc.

It is the amount of heat required to raise the temperature of 1 pound of water 1° F. at its maximum density, 39° F.

By knowing the specific heat of a substance, the amount of heat or refrigeration, that is required to raise or lower it to a certain temperature, may be calculated. For example, (1) how much heat would be required to raise the temperature of 100 grams of water 5° C., and how much for the same amount of milk, and (2) how much heat would be required to raise 100 pounds of water or 100 pounds of milk 5° F.? The solutions are obtained by using the formula — Weight (mass) × Specific heat × Change of temperature = Units of heat. Thus:

PHYSICAL PROPERTIES OF MILK

		Weight	Specific heat	Temperature change	Units of heat required
(1)	Water	100 grams	× 1.0	× 5° C. =	500 calories
	Milk	100 grams	× 0.94	× 5° C. =	470 calories
(2)	Water	100 lbs.	× 1.0	× 5° F. =	500 B. T. U.
	Milk	100 lbs.	× 0.94	× 5° F. =	470 B. T. U.

To convert B. T. U. to calories multiply by 252, since one pound equals 453.6 grams and one degree Fahrenheit equals 5/9° C., or $453.6 \times 5/9 = 252$.

Table 20 gives specific heat of some dairy products at different temperatures.[16]

Table 20. Specific Heat of Milk and Its Products

Product	Specific heat at various temperatures			
	0° C. (32° F.)	15° C. (59° F.)	40° C. (104° F.)	60° C. (140° F.)
Whole milk ...	0.920	0.938	0.930	0.918
Skimmilk	0.940	0.943	0.952	0.963
Whey	0.978	0.976	0.974	0.972
Cream 15% ...	0.750	0.923	0.899	0.900
Cream 30% ...	0.673	0.983	0.852	0.860
Cream 45% ...	0.606	1.016	0.787	0.793
Butter			0.556	0.580
Butterfat			0.500	0.530

Specific heat of milk and cream varies widely depending on the fat content and the temperature at which the determination is made. The greatest specific heat for milk and cream[12] is from 50° F. to 75° F., which is partly within the range of optimum growth of some kinds of bacteria. Thus, any cooling apparatus for dairy products should provide ample cooling surface to lower the temperature promptly through this critical stage.

Surface Tension. The molecules in a liquid are attracted to each other, and this creates a pull from the surface. Beneath the surface the molecules are surrounded by other molecules and the attraction is equalized, but at the surface the balance is broken and a tension results. Surface tension may thus be defined as a state of stress at the surface of a liquid due to the attraction of the molecules for each other. It is expressed in dynes. A *dyne* is the force that, acting on a mass of one gram, gives it an acceleration of one centimeter per second.

Surface tension may be measured in several ways, one of the simplest being the "falling drop" method. A tube of uniform bore is used and the number of drops falling per unit of time are counted and compared to that of water. Thus a liquid showing 200 drops as compared to water with 100 drops would have a surface tension one-half as great as that of water. A more accurate method is the "platinum ring" procedure. The ring is connected to a delicate balance and the pull required to draw it out of a liquid is measured by the balance.

Cenco-du Nuoy Tensiometer
Courtesy Central Scientific Company

The surface tension of water is 72-73 dynes. Dahlberg and Hening[17] found the surface tension of skimmilk, whole milk and 30 to 35 percent cream to be, respectively, 57.4, 55.3 and 49.6 dynes.

An increase in fat and protein content lowers the surface tension of milk. Aging lowers the surface tension of raw milk slightly, but has a tendency to raise that of pasteurized milk. Pasteurization increases the surface tension to a small extent while homogenization lowers it. An increase in temperature of milk lowers its surface tension markedly. Coagulation of normal milk will cause an increase of approximately 10 dynes in the surface tension, but after standing several days no appreciable difference in the normal and coagulated samples of milk appears.

A thorough review of the surface tension of milk (79 references) was presented by Whitnah[18] who concluded that the free fatty acids and proteins were surface-active constituents which cause the surface tension of milk or milk products to be lower than that of water.

Viscosity. Viscosity of a substance may be defined as its resistance to flow, due to internal friction between molecules as they shear each other. The unit of viscosity is the *poise* named after Poiseuille, one of the pioneers in this line of investigation. The poise is the force (dynes) required to move for a distance of one centimeter in one second of time a plane one square centimeter in area past another plane of the same area, which are one centimeter apart and the space filled with the liquid. In other words it is a unit pressure per unit area per unit of time per unit of distance. A smaller unit is the *centipoise* which is one-hundredth of a poise. The viscosity of water at 20° C. is slightly more than a centipoise, but is usually considered as one centipoise. Whole milk averaging 4.32 percent fat had in a trial by Bateman and Sharp[19] an average viscosity of 1.6314 centipoises and skimmilk an average of 1.404 centipoises.

Milk is more viscous than water due to the casein, fat and albumin, listed in descending order of greatest effect. The viscosity of milk may be measured in several ways. A common method is the use of a pipette with a uniform bore and comparing the rate of flow of a given volume of milk with the rate of flow of the same amount of water. One objection to the use of pipettes is the effect of turbulence. The liquids may run too fast and set up eddies, and hence the liquid in the central part may move too fast. The more accurate method is to employ a viscosimeter. This is an instrument that measures the viscosity by measuring the resistance of a solid when rotated while suspended in the liquid.

Many factors affect the viscosity of liquids, which would include milk. Any factor that induces clumping of the fat globules will increase the viscosity. Low temperatures and aging have this effect. Mechanical agitation of whole milk decreases the viscosity because the fat globule clumps are partially broken up, while in case of skimmilk it has no effect due to the small amount of fat present. Homogenized milk would not be affected either as the globules are already broken up. Homogenization increases the viscosity of whole milk, but slightly decreases that of skimmilk. This process breaks up the fat globules into much smaller ones and thereby provides a larger surface area than previously. A film of protein is adsorbed by the surface of the globules, and this surface being much greater than in case of the non-homogenized milk, a much greater adsorption takes place and this causes a greater viscosity. In regard to the skimmilk some of the protein particles would be broken and therefore the viscosity would be reduced. Pasteurization temperature slightly lowers the viscosity through breaking of the clumps of fat globules, but when subjected to high heat as under pressure the viscosity is increased due to the proteins being affected. Dilution of milk with water lowers the viscosity as there is almost a direct relationship of total solids in milk to viscosity. Viscosity also increases with development of high acid, and with increased fat content.

Viscosity of milk is relatively unimportant, but as regards cream it is very important. Freshly separated cream even though high in fat is rather fluid and low in viscosity, so much so, that the housewife suspects it of being low in fat. Thus, it is necessary to age the cream to increase its viscosity so it will appear rich and whip well.

Bound Water. Milk, cream, and other liquid dairy products have been shown to contain bound water in appreciable quantities.[20] Normal whole milk samples contain 2 to 3.5 percent bound water according to this report. The bound water contents of milk and several dairy products are shown in Table 21. The authors observed that casein binds approximately half of the bound water; albumin accounts for 30 percent, the fat globule membrane about 15 percent, and the other milk solids about 4 percent. However, Ling[15] considered these values misleading and recomputed the values according to the concentration of each milk constituent. On this basis, phospholipids bound 6 grams of water per gram of phospholipid; albumin and globulin bound 1.3 grams per gram, and casein bound 0.7 gram per gram.

Pyenson and Dahle[20] observed that rennet casein appeared to bind more water than acid casein, thus possibly accounting for some of the differences in the two types of casein. They also believed that their observations on the binding of water by milk proteins explained at least a part of the Recknagel phenomenon. It was also noted that pasteurization lowered the bound water content of milk while aging generally increased it regardless of the treatment which was given to the sample.

Table 21. The Bound Water Content of Some Milk and Milk Products[20]

Sample	Hours aged at 40° F.	Percent solids	Percent bound water
1. Raw milk	24	13.25	3.18
2. Skimmilk	24	9.44	2.13
3. Cream	8	29.08	2.50
4. Cream	8	43.20	3.42
5. Buttermilk	0	8.25	1.75
6. Cond. skimmilk	24	25.33	11.62
7. Colostrum	24	19.17	4.65

Foam. The ability of milk and its products, skimmilk and cream, to foam is a common observation. In case of whipping cream it is desirable to have a stable foam, but at other times such as filling milk cans, bottles, and separating milk it is undesirable. Proteins are the chief cause of foam. They are adsorbed by the thin film surrounding an air bubble and this gives stability to the entrapped air. The protein in this film is similar to that of the milk proper. Sharp, Myers and Guthrie[21] report the protein content of the foam in skimmilk to be from 0.12 to 0.63 percent higher than in the skimmilk beneath.

Sanmann and Ruehe[22] studied the factors that affect the foaming of milk. Table 22 shows the amounts of foam they

Table 22. Effect of Temperature on Foaming Ability

Temperature ° F.	Percentage increase in volume due to foam		
	Whole milk %	Skimmilk %	Cream, 20% %
35	88	130	49
40	73	122	40
50	30	84	36
60	11	28	26
70	8	18	20
80	8	17	15
90	16	21	8
100	29	26	31
110	32	29	35
120	31	28	37
130	30	27	37
140	28	25	35
150	26	24	31
160	25	24	29
170	23	23	27
180	22	21	25

obtained by agitating 250 ml. samples with an electrical stirring device. Low temperatures, 35° F. to 40° F., favored the formation of the largest volume of foam, while temperatures around 60° F. to 90° F. gave the lowest. The volume of foam then increased with higher temperatures, but not to the maximum obtained at the low temperatures.

They also found that pasteurization had no appreciable effect on foaming ability but homogenization increased it when measured at 40° F. and 80° F., and decreased the amount of foam at 140° F. As regards the effect of the solids in milk on foaming, the fat generally has a depressing effect and the solids not fat an augmenting result. The phospholipid, lecithin, probably acts as a foam breaking substance. In the churning process of cream the agitation liberates a large part of the lecithin, which then goes into the buttermilk and consequently very little foam is present there. The mass of air bubbles in the cream, when being churned, flattens out as soon as the butter comes.

California studies[23] have demonstrated the presence of two types of foam in milk, cream, and buttermilk which may appear separately or simultaneously. One foam appeared to be a protein type, the other a phospholipid-protein type. In studying the foaming of these dairy products at temperatures between 5° C. (41° F.) and 50° C. (122° F.), each of the three products exhibited minimum foaming at 30-35° C. (86-95° F.). The protein-type foam predominated at higher temperatures. When skimmilk is caused to foam at 35° C., increases in fat content cause a decrease in foam volume and stability up to a level of approximately 5 percent fat. Further increases in fat content to 20 percent fat produce a steady increase in both foam volume and stability and then both remain constant with additional fat added. When foaming was done at 6° C. (42.8° F.), foam volume remained unchanged even though fat content was increased. Foam stability reached a minimum at about 5 percent fat and then increased rapidly as fat was increased to 10 percent, with highly stable cream-type foams forming when the fat content was increased to above the 10 percent level. Artificial milks and creams resembled the natural products only if phospholipids were combined with the fat prior to emulsification.

In the separation of milk, the slime that is deposited in the separator bowl results from coagulated protein material in the air cells which are formed by the foaming of the milk during separation. The walls of these cells finally collapse and the coagulated protein together with the cell walls constitute the slime.

Fat Clumping. Cream forms at the surface of milk because milk fat is lighter than the other solids. In freshly drawn milk the fat globules are separate individuals and thus will not rise rapidly, for the rate of rise according to Stoke's law is in proportion to the square of the radius. However, clumps or clusters of fat globules soon form and as they have a greater radii, a more speedy rise will occur. These clusters are not clear fat as they entrap more or less of the milk plasma and thereby do not rise as rapidly as if they were clear fat. Nevertheless, the clumps rise much faster than individual globules. Troy and Sharp[24] state that the fat globules rise so slowly as independent individuals, that it would require many times the normal creaming time of milk for them to reach the cream layer, whereas the clusters rise fast enough to correspond with normal creaming time.

It is now generally recognized that clumping or clustering of fat globules is the principal factor in gravity cream rising, but it is not yet clear as to why the globules clump. Sharp and Krukovsky[25] conclude it is due to an agglutinating substance which is adsorbed on the surface of the solid or solidifying fat globules and not on the liquid globules. Therefore, greater clumping occurs at lower temperatures. In general practice, milk is cooled to 40° F. to 45° F. to obtain good gravity creaming as in bottled milk. Dahlberg and Marquardt[26] set forth an electrical theory of fat clustering. The presence of a large quantity of calcium ions imparts some positive charges to the weak, negatively charged fat globules, and creates an electrical affinity between the globules.

Dunkley and Sommer[27] found that globulins or the fraction euglobulin were essential to fat clustering, as they cause the globules to adhere to each other. Also an optimum concentration of salts is necessary to favor the clustering of fat by euglobulin.

Certain practices accelerate or inhibit fat clumping. Heating milk to high temperatures destroys clumping. If milk is pasteurized at not higher than 143° F., as in the *holding* process, the clumping is not appreciably affected but at higher temperatures the cream line will show a considerable decrease, because[27] temperatures higher than those used for proper pasteurization cause a partial denaturation of the globulin fractions and less fat clumping occurs.

Homogenization breaks up the fat globules to about one-fourth their original diameter and no creaming results due either to these small globules or as Dunkley and Sommer[27] suggest to the denaturing effect of the homogenization on the globulin fractions. It does, however, induce clumping especially with the higher fat contents and increased pressures. These clusters apparently are not similar to those in untreated milk as very little upward movement occurs. If, however, homogenized milk or cream is mixed with skimmilk or normal whole milk, creaming takes place very readily and a striking cream line is formed.

Milk should not be agitated too much at temperatures below 50° F. as the fat clumps will be broken and the cream line diminished. Pasteurized milk should be bottled before it has time to clump, which occurs within 5 to 30 minutes after cooling, so that the clustering may take place in the bottle and thus provide a better cream line.

In general, good clumping means a good recovery of the fat and a deep cream layer with low fat in the skimmilk portion.

Electrical Conductivity. The soluble salts, especially the chlorides, in milk dissociate to some extent into ions which are electrically charged. These ions move about and thereby conduct electrical currents. Their movement is impeded somewhat by the fat globules and consequently whole milk has a lower conductivity than skimmilk. Since the lactose and chlorides vary inversely with each other and the lactose content of milk is rather variable, the conductivity of several different samples of milk may therefore be quite different. In case of mastitis the chloride content increases and the conductivity is augmented.

Since milk has conductivity, it also has resistance. The ohm is the unit of resistance, while the mho, the reciprocal of the ohm, is the unit of conductivity. The lower the conductivity the greater the resistance and vice versa. The property of electrical resistance of milk is applied to electrical pasteurization. A film of milk is passed between two electrodes and the resistance of the milk causes heat to be liberated and the milk is thereby subjected to high-temperature short-time pasteurization which is at least 160° F. for at least 15 seconds.

Horrall[28] determined the conductivity of the milk from 18 cows with normal udders and found a range of 46.1 to 49.2 $\times$ 10^{-4} mho. It was much higher in the milk from cows having udder infections, the range being 45.8 to 83.0×10^{-4}. Fresh skimmilk and buttermilk have conductivities around 50×10^{-4} mhos, but when allowed to sour, the conductivity increases usually about 5×10^{-4} mhos for each 0.1 percent rise in acidity.

Effect of Heat. Heating milk to near the boiling point will cause a film or skin to form on the surface. This skin is due mainly to the calcium caseinate but the other constituents in milk are also present to some extent.

Autoclaving milk, wherein temperatures around 245° F. or higher result, will cause browning. The brown color is due to the heat effecting a change between the casein or amino acids and the sugar. The browning, however, is dependent on the presence of both the casein and lactose, as the heating of either constituent separately will not give the color. Patton[29] has presented a comprehensive review of this important aspect of milk behavior.

Heating milk to high temperatures causes a cooked flavor to appear. In the holding method of pasteurization, 143° F. for 30 minutes or the high-temperature short-time method, 160° F. for 15 seconds, very little heated taste is noticeable but at higher temperatures or longer periods of heating, the cooked flavor becomes very apparent. Gould[30] found the flavor to appear at 158-162° F. for 30 minutes holding and at 169-172° F. for momentary heating.

This cooked taste has been shown to be due to the production of sulfhydryls by high temperatures. These compounds come from the proteins in milk, the most likely source[31] being the β-lactoglobulin with the protein of the fat globule membrane as a possible additional source. Sulfhydryl compounds appear to be wholly responsible for the cooked flavor of heated milk and milk products. Gould added to milk a sulphite salt, such as ammonium or sodium, and artificially produced a cooked flavor. The same effect was obtained by adding the compound glutathione a tripeptid consisting of cysteine, glycine and glutamic acid, which contains the sulfhydryl ($-SH$) linkage.

Sulfhydryl compounds are active antioxidants and prevent oxidized flavors in milk heated to high temperatures. Oxidized flavors in milk do not usually appear until the sulfhydryls are oxidized and the cooked flavor has disappeared. The addition of copper to milk delays the production of sulfhydryls through oxidation. Consequently, higher temperatures are required to produce a cooked flavor when copper is present. One part per million of this metal raises the temperature by approximately 15° F. for momentary heating to produce a cooked flavor.

High temperatures are required to coagulate fresh milk of low acidity. Sommer and Hart[32] heated fresh milk to 137° C. and it required from three to 20 minutes for it to curdle. The development of acid, however, lowers the temperature requirements for coagulation. McInerney[33] found that milk containing 0.57 percent acid will, on the average, precipitate at a temperature between 60° F. and 65° F., that containing 0.50 percent at 75° F. to 80° F., 0.40 percent at 100° F. to 110° F., 0.35 percent at about 150° F. and 0.25 percent acid in milk will not cause coagulation until heated to 180° F.

A mineral balance is connected with heat coagulation. Calcium and magnesium unite with the citrates and phosphates, and if an excess of the first is present, coagulation will occur at lower temperatures. Conversely, the addition of citrates or phosphates would absorb the excess of calcium and magnesium and thus induce stability. Calcium and magnesium on one hand and the citrates and phosphates on the other have opposite effects on the heat coagulation. A proper balance of these two classes of milk constituents will prevent coagulation.

Ling[15] has noted that a concentration of protein occurs at the milk-air surface and this, following steady evaporation of water from the surface, gives rise to the formation of the familiar skin. He also observed that a similar reaction occurring at the gas-liquid interface of foam or bubbles may result in the formation of milkstone, a problem of deep concern in the sanitary care of milk handling equipment.

Haller and his associates observed a lower percentage loss of several constituents and properties of goats' milk when the milk was pasteurized by the holder method than occurred when the HTST system was used. Table 23, prepared from data of Haller, et al.[34] indicates this difference. Aschaffenburg and co-workers[35] in England have shown that Ultra-High Temperature pasteurization of milk (275° F.) did not affect the freezing point, fat, or solids-not-fat content of raw milk.

Table 23. Percentage Reduction of Some Properties and Constituents
During Pasteurization of Goats' Milk[34]

	HTST %	Holder %
Soluble Calcium	3.6	7.5
Soluble Phosphorus	3.2	4.7
Soluble Proteins	7.4	4.8
Curd Tension (Method 1)	5.1	38.4
Curd Tension (Method 2)	6.9	48.4
Ascorbic Acid	0	39.5

BIBLIOGRAPHY

1. Roadhouse, C. L. and Koestler, G. A. Contribution to the Knowledge of the Taste of Milk. J. D. Sci. 12, p. 421 (1939).
2. Doan, F. J. Some Observations on the Freezing Point of Cream and Its Use in Detecting Added Water. J. D. Sci. 10, p. 353 (1927).
3. Dahlberg, A. C., et al. Sanitary Milk Control and Its Relation to the Sanitary, Nutritive, and Other Qualities of Milk. Nat. Res. Council Pub. 250 (1953).
4. Dahlberg, A. C. Does the Freezing Point Determination Tell What We Need to Know About Added Water in Milk? Proc. Milk Ind. Foundation Lab. Sect. 29-39 (1958).
5. Shipe, W. F. The Freezing Point of Milk: A Review. J. D. Sci. 42,11:1745-1762 (1959).
6. Denis-Nathan, L. The Cryoscopy of South African Milk. Union South Africa Dept. Agr. Sci. Bul. 119 (1933).
7. Richmond, H. D. Dairy Chemistry, C. Griffin and Co., London, p. 155 (1899).
8. Reid, W. H. E. The Deleterious Effect of Freezing on Several of the Physical Properties of Milk. Mo. Res. Bul. 100 (1927).
9. Eckles, Combs and Macy. Milk and Milk Products. McGraw-Hill Book Co., New York (1936).
10. Judkins, H. F. and Mack, M. J. The Principles of Dairying, John Wiley and Sons, New York (1941).
11. Davies, W. L. The Chemistry of Milk. D. Van Nostrand Co., New York (1936).
12. Bowen, J. T. Refrigeration in the Handling, Processing and Storing of Milk and Milk Products. U.S.D.A. Misc. Pub. 138 (1932).
13. Sharp, P. F. and Hart, R. G. The Influence of the Physical State of the Fat on the Calculation of Solids From the Specific Gravity of Milk. J. D. Sci. 19, p. 683 (1936).
14. Davis, J. G. A Dictionary of Dairying. 2nd Ed. Leonard Hill, Ltd. London (1955) p. 962.
15. Ling, E. R. A Textbook of Dairy Chemistry, Vol. I, 3rd Ed. Revised. Champan and Hall, Ltd. London (1956).
16. Hammer, B. W. and Johnson, A. R. The Specific Heat of Milk and Milk Derivations. Ia. Res. Bul. 14 (1913).
17. Dahlberg, A. C. and Hening, J. C. Viscosity, Surface Tension and Whipping Properties of Milk and Cream. N. Y. (Geneva) Tech. Bul. 113 (1925).
18. Whitnah, C. H. The Surface Tension of Milk: A Review. J. D. Sci. 42,9:1437-1449 (1959).
19. Bateman, G. M. and Sharp, P. F. A Study of the Apparent Viscosity of Milk as Influenced by Some Physical Factors. J. Agr. Res. Vol. 36, No. 7, p. 647 (1928).
20. Pyenson, H., and Dahle, C. D. Bound Water and Its Relation to Some Dairy Products. J. D. Sci. 21:169-185, 601-614 (1938).
21. Sharp, P. F., Myers, R. P. and Guthrie, E. S. Accumulation of Protein in the Foam of Skimmilk. J. D. Sci. 19, pp. 655-662 (1936).
22. Sanmann, F. P. and Ruehe, H. A. Some Factors Influencing the Volume of Foam on Milk. J. D. Sci. 13, pp. 48-63 (1930).
23. Richardson, G. A., and El-Rafey, M. S. The Role of Surface-Active Constituents Involved in the Foaming of Milk and Certain Milk Products. III. Milk Lipids, Including Phospholipids. J. D. Sci. 31:223-239 (1948).
24. Troy, H. C. and Sharp, P. F. Physical Factors Influencing the Formation and Fat Content of Gravity Cream. J. D. Sci. 11, pp. 189-229 (1928).
25. Sharp, P. F. and Krukovsky, V. N. Differences in Adsorption of Solid and Liquid Fat Globules as Influencing the Surface Tension and Creaming of Milk. J. D. Sci. 22, p. 750 (1939).
26. Dahlberg, A. C. and Marquardt, J. C. The Creaming of Raw and Pasteurized Milk. N. Y. (Geneva) Tech. Bul. 157 (1929).
27. Dunkley, W. L. and Sommer, H. H. The Creaming of Milk. Wis. Res. Bul. 151 (1944).
28. Horrall, B. E. A Study of the Lecithin Content of Milk and Its Products. Purdue (Ind.) Bul. 401 (1935).

29. Patton, Stuart. Browning and Associated Changes in Milk and Its Products: A Review. J. D. Sci. 38,5:457-478 (1955).
30. Gould, I. A. Cooked Flavor in Milk, a Study of Its Cause and Prevention. J. D. Sci. 24, p. A 301 (1941).
31. Josephson, D. V. and Doan, F. J. Observations on Cooked Flavor in Milk. The Milk Dealer 29: 2, pp. 35-62 (1939).
32. Sommer, H. H. and Hart, E. B. Heat Coagulation of Evaporated Milk. Wis. Res. Bul. 67 (1926).
33. McInerney, T. J. Temperatures at Which Milk Will Coagulate. J. D. Sci. 3, p. 224 (1920).
34. Haller, H. S., et al. The Effect of Pasteurization on Some Constituents and Properties of Goats' Milk. D. Sci. Abst. 4,1:46-47 (1942).
35. Aschaffenburg, R., et al. An Investigation of the Effect of Ultra-High-Temperature Treatment on the Freezing Point and Composition of Milk. J. Soc. Dairy Tech. 11,2:93-95 (1958).

REVIEW QUESTIONS

1. Why should such feeds as silages and fresh alfalfa be fed to cows after milking?
2. To what factors is the white color of milk due?
3. Define specific gravity. Density.
4. Which is heavier, Jersey or Holstein milk? Why?
5. Why does milk have a lower freezing point than water?
6. In partly frozen milk, which portion has the higher fat test, the frozen or the liquid portion?
7. What are some of the factors that affect the surface tension of milk?
8. Why does freshly separated cream, even though high in fat, seem to be of low test to the housewife?
9. What constituents in milk are the chief cause of foam?
10. Why is buttermilk rather high in lecithin content?
11. Would cream form at the surface of milk if the fat globules did not cluster?
12. What is the difference between an ohm and a mho?
13. What causes milk to brown when subjected to high temperatures?
14. To what compounds in milk is cooked flavor due?
15. What is meant by heat stability in milk?
16. What are some of the factors that affect the freezing point of milk?

CHAPTER III

Sampling Methods

1. Daily or Individual Samples

The proper sampling of any dairy product is the first step in obtaining an accurate chemical analysis of its constituents. This applies especially to the fat because of its low specific gravity. As soon as milk is drawn, the fat globules will begin to move upward, slowly at first, but as clustering increases, the speed of upward motion accelerates. Therefore, it is necessary to redistribute these fat globules before taking a sample.

To obtain a representative sample at the barn the most effective mixing is to pour the milk from one pail to another at least three times and then immediately withdraw the sample. This method of mixing would also apply in regard to a bottle of market milk, except more transfers from one bottle to another would be necessary if the cream line was well established.

At the weigh stand the method of taking a sample may have a marked effect on the fat test. Patrons expect uniformity in test from day to day and the operators emphasize accuracy. Daily tests, however, may not be uniform due to a number of factors. Cold weather increases the fat test but lowers the milk yield, and the opposite effect occurs in the warm months. Cows, just freshened, may give a higher fat test for a short time followed by a decrease for a period, and then gradually increase to the end of the lactation period. Underfeeding increases the test but at the expense of the milk yield. Individual cows, even of the same breed, vary in fat test and so if they are removed from the herd or have recently freshened or become dry, the test of the herd will be altered. Due to these variations it is well for the patron to check the total fat as well as the fat test itself.

The usual procedure at the weigh stand is to dump the milk through a strainer into the weigh can, note the weight and then dip out a sample. The milk is seldom stirred before taking the sample as it is considered that the pouring gives sufficient mixing. Several investigators have checked on this point and found considerable variations in test. Tracy and Tuckey[1] compared the fat tests from samples taken at the front and rear of the weigh tank. The average test of 72 samples was 4.40 at the front, 4.60 at the rear and after thoroughly mixing, 4.51 percent. Bailey[2] also found lower tests at the dump or front end. These results applied especially to cans of milk that had been cooled and allowed to cream for some time. The milk in the upper part of the can being richer would flow in first and move toward the rear of the tank while the last milk, which in some cases was nearly skimmilk, remained at the front and consequently lowered the test of the sample taken at that point. The size of the mesh of the strainer screen, and its location if extended below the surface of the milk, also had some effect. Rich viscous cream on some milks would not readily go through the screen and consequently some fat was held back in the strainer box until the milk was allowed to flow from the weigh can. This retention obviously affected the homogeneity of the batch of milk, the samples taken outside of the box testing too low and those inside too high.

Marquardt and Durham,[3] on the other hand, found no appreciable difference in the tests from the milk in the weigh can whether or not agitation occurred before sampling. Vermont studies[4] have shown that the design of the weigh tank is very important in obtaining accurate samples. Although agitators appeared to be valuable additions to some weigh tanks, the fact that they may be shut off, discarded, or become mechanically inoperative was considered a possible disadvantage. This work showed that about half of the tanks studied under actual operating conditions were yielding incorrect samples and, of these errors, nearly 60 percent resulted in fat tests which were too high. Weigh tanks can be checked for sampling accuracy by hand agitating the milk in the tank for 60 seconds prior to sampling. The recommended procedure[5] for checking mixing efficiency of weigh tanks, by comparing tests on samples of the same lot of milk taken at several locations within the tank, did not always give a true evaluation of the situation.

The conversion to milk cooling in bulk tanks on farms has introduced new sampling problems. Care must be taken that the milk in bulk tanks is not subjected to conditions which are conducive to churning or freezing during cooling and storage. There is general agreement that cold milk in bulk tanks should be agitated for at least 3-5 minutes prior to sampling. However, it has been shown[6] that these minimum requirements are seldom met because of the pressures placed on bulk collectors to speed up collection time or because of the driver's inability to estimate the proper agitation period prior to sampling. Inexpensive timing devices are available to minimize the latter problem. This work has also shown that the milk in a bulk tank can be thoroughly mixed and yet have visible cream streaks on the milk surface. Samples may be collected for testing on each day of pickup or combined into a single sample covering a longer collection period. In some cases, fresh samples are taken at the farm and combined into composites prepared and held at the milk plant; in other cases, the composite samples are carried on the truck and additions are made at the farm. Samples may be taken by a dipper which has been properly rinsed and sanitized immediately before it is dipped into the farm supply, or sterile, paper single-service milk sampling tubes may be used. The latter are especially desirable for collecting samples intended for bacteriological analyses. Persons taking samples of bulk milk should be cautioned never to hold glass bottles directly over the milk because of the danger of breaking or accidental spilling of preserved milk into the tank supply.

Obtaining samples for testing from bulk collecting trucks, over-the-road tankers or large storage tanks present additional problems. It is generally assumed that the pumping of milk at each farm and the truck movements on the road during bulk collection will keep the milk sufficiently agitated for direct sampling on return to the plant. Where milk has remained motionless in a tank and creaming has begun, it is necessary to provide some means of agitation prior to sampling. The operator must allow sufficient agitation time to render the milk homogeneous which may require 20-30 minutes or more. Efficiency of mixing can be checked by determining if fat tests of samples taken at the manhole and the outlet are identical.

2. Composite Samples

Composite Sample Defined. A composite sample is the quantity of milk or cream secured by mixing together proportional parts of different lots of the product. For example, if a cow gave 25 pounds of milk in the evening and 40 pounds in the morning, and one milliliter was taken for each pound, the sample would consist of a mixture of 25 ml. of the evening's and 40 ml. of the morning's milk. This would be a true composite sample and a test of this mixture would give the same result as if each sample were tested and the percent fat computed.

Composite samples are taken in order to determine in one test the amount of fat in the milk or cream for a definite period and thereby save considerable expense and time.

Methods of Taking Composite Samples. Composite samples may be secured by the use of a small dipper holding half an ounce, by graduated pipettes, by paper or metal sampling tubes, or by

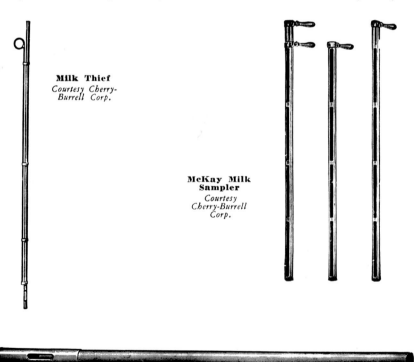

Milk Thief
Courtesy Cherry-Burrell Corp.

McKay Milk Sampler
Courtesy Cherry-Burrell Corp.

Scovell Milk Sampler
Courtesy Cherry-Burrell Corp.

automatic vacuum samplers. The latter system is widely used in large milk receiving operations because of its efficiency and ease of operation. Smaller receiving operations commonly use the dipper method for taking samples for the milk fat composite. The milk from a patron is first poured into a weigh can, weighed, and then a sample taken by the dipper or vacuum sampler and put in a composite sample jar containing a preservative. Small amounts of each delivery are thus added until the end of the period when the mixture is tested. At least 10 ml. of milk should be taken for each addition to the composite sample. At least 100 ml. of sample should be available at the time of testing. This test of the composite sample multiplied by the total weight of the milk delivered for the period will give the pounds of fat credited to the patron for that time.

Although the dipper method is generally used, it will not give a true composite sample because exact proportional amounts cannot be measured in this way. For all practical purposes, however, it gives fairly close results because a patron delivers nearly the same amounts of milk each day for the period.

Pipettes, graduated to tenths of a milliliter, are useful in obtaining a true composite sample of milk from individual cows. One milliliter for each pound of milk is a fair amount to take. This method will give a true sample as it is possible to measure the amount accurately. Obviously, this method of sampling is not practical for everyday use at the weigh stand as it is a time consuming procedure.

The sampling tube offers the best method of obtaining an aliquot sample quickly and accurately provided the containers have straight sides and are of uniform dimensions. The tubes remove a column of milk the entire depth of the mass, and this quantity is a certain proportional part of the whole, irrespective of the amounts in the containers.

Two common sampling tubes are the Scovell and McKay samplers. The Scovell sampler consists of a tube one-half to one inch in diameter, open at both ends, the lower end fitting closely into a cap which has several openings for admittance of the milk. As the tube is lowered into the milk it becomes filled to the same level as that in the container, and when the cap touches the bottom of the can the apertures are closed by pressing the cap tight against the end of the tube. In this manner a column of milk is retained which is an aliquot part of the total. The milk is emptied into the sample jar by inverting the tube.

The McKay sampler consists of two tubes, one inside the other with slots extending through both in such a way that they are open when the two handles are together and closed when the handles are at right angles. When inserting the sampler, the slots are left open so the tube will fill with milk, and then just before removing the instrument the slots are closed.

In all cases of sampling, the milk should be thoroughly mixed before taking the sample. Although it is possible by the use of tubes to secure a representative sample of milk without its being stirred, yet it is advisable to mix it so as to avoid the possibility of some of the cream sticking to the tube.

Care of Composite Samples. Composite samples should be kept in air tight containers to prevent loss of moisture which would cause the cream to toughen and tend to increase the fat test. Glass bottles with rubber stoppers are the most satisfactory containers for composite samples but any rubber stoppers that have become hardened should be replaced. Cork stoppers should not be used as they absorb some fat. Glass stoppers are also employed and would be satisfactory if each stopper could be kept attached to the bottle for which it was ground, but if mixed the stoppers will not give an air tight seal. The bottles must be plainly marked either with metal tags or with numbers painted on the bottles. The samples should be kept in a cool place and away from strong light in order to prevent toughening of the cream layer, and to prevent a lowering of the fat test. Sufficient preservative should be added to prevent the sample from souring, and each time a new amount of milk is added it should be mixed with the previous portions so all the milk will come in contact with the preservative. Molds should be guarded against by thorough cleaning of the bottles at the end of each period and disinfecting the stoppers with formalin or with a chlorine solution.

Composite samples maintained on the bulk pickup truck should be carried in an insulated case and iced in such a manner that the sample temperature never exceeds 40° F. Normally, this requires that the sample bottles be sitting in ice water. Higher temperatures frequently result in samples so badly churned by the constant agitation on the road that no test can be made. Uninsulated boxes may result in frozen composite samples during the winter in the northern states. The insulated case should be held firmly in place to prevent jarring and the cover should rest on top of individual bottles, held firmly in a rack, to prevent tipping or breaking. The bottle caps should be tight fitting and designed to protect the lip of the bottle.

Corrosive sublimate or mercuric chloride ($HgCl_2$) is the preservative generally used at milk plants. It is a violent poison and therefore as a precaution the tablets are colored pink or green and this color is imparted to the milk to indicate its nature. It is an effective preservative and one or two tablets, depending on their size, are sufficient to keep a half pint sample in good condition for a period of two weeks. At present, two sizes of tablets are packaged, the latter weighing close to 1.0 gram and the smaller 0.5 gram, containing respectively on an average 0.45 and 0.225 grams of pure mercuric chloride. Mercuric chloride should not be used in metal containers as it soon corrodes them.

In experimental trials where the milk samples are analyzed for solids and ash, it is necessary to use a liquid preservative. Formalin (CH_2O) is satisfactory for this purpose. One milliliter is sufficient to preserve the ordinary sized sample for two weeks. It has a very toughening effect on the casein, and extra agitation of the milk and acid is necessary to dissolve it. Formalin is a 40 percent solution of formaldehyde, made by mixing the formaldehyde, which is a gas at ordinary temperatures, in distilled water.

A less used preservative is potassium dichromate ($K_2Cr_2O_7$). Its effectiveness is lessened by light, and thus in general practice the mercuric chloride tablets are preferred.

Samples intended for official testing[7] may contain preservatives if inclusion of such preservative is not detrimental to the running of any other physical or chemical tests which are to be made in addition to the fat determination. Tablets containing $HgCl_2$ or $K_2Cr_2O_7$ (or other approved preservatives) which weigh no more than 0.5 gram per 8 fluid ounces of sample, or 0.1 ml. of a 36 percent solution of formaldehyde per fluid ounce of sample may be used under these conditions. However, the only preservative permitted for samples to be tested for phosphatase activity is chloroform ($CHCl_3$) and it is also specified[7] that only phenol-free stoppers can be used on the sample bottles containing milk for the phosphatase test.

Testing Composite Samples. By the time the composite sample is ready to be tested, a leathery cream layer may have formed and more or less fat will have dried on the inside of the jar. It is necessary to soften this fat and reincorporate it with the rest of the sample before a representative charge can be taken. To obtain a homogeneous mixture place the sample in water at 110° F. and warm the milk or cream to 95° F. to 105° F. Higher temperatures should be avoided as oiling-off may occur. In many cases the lower temperature is sufficient, but for older or leathery samples it is necessary to warm them to the higher limit to eliminate all solid particles of fat floating around in the liquid. Use a brush or spatula to remove any dried fat adhering to the stopper or the inside of the sample jar. Shake the bottles gently by a rotary motion, never vigorously, to mix in the melted fat. Continue this procedure until the sample is uniform and then give a final mixing by pouring from one vessel to another at least three times. Pipette immediately even though the temperature is above 70° F. The weight of the charge of milk may be slightly less but this error is offset by better drainage from the pipette and the elimination of the danger of the fat partially solidifying again if cooled. Cool the pipetted samples and then complete the test as for fresh milk or cream as given in Chapter IV.

Accuracy of Composite Samples. Theoretically a composite sample should test the same as the weighted average of daily tests of the same milk for the same period. In most cases, however, the composite samples show a lower test.

Lucas[8] found in a plant of 21 producers that the composite tests averaged 0.11 percent lower than the daily tests of the samples of milk. One factor that might cause this difference is the temperature to which the composite sample is heated in its preparation. A high temperature expands the milk and a pipetteful would not weigh as much as at a lower temperature. Figures given in Table 24 show the effect of temperature on the fat test.

Table 24. Effect of Temperature on Fat Test

Temperature	Specific volume	Percent fat
56° F.	0.9980	3.507
70° F.	1.0000	3.500
84° F.	1.0025	3.491
96° F.	1.0055	3.481
116° F.	1.0100	3.465
140° F.	1.0750	3.256

No appreciable differences in the tests occurred between the temperatures 56° F. to 96° F., but at the higher temperatures they were measurably lower. It was recommended that the samples be measured out for the Babcock test within the temperature range of 56° F. and 96° F.

Monroe[9] tested 290 seven-day composite samples of milk and reported the average test as 5.13 as compared to 5.22 percent for the daily tests, a difference of 0.09 percent.

Holland[10] in a study of composite milk samples found that molds may cause a decided decrease in the fat test. They hydrolyze the fat and the freed fatty acids are not recovered in the fat tests. Manus and Bendixen[11] state that mercuric chloride activates the enzyme lipase and this breaks up some of the fat into its component parts similar to the mold action.

A reliable method of sampling a day's run is by means of the drip sample. A valve may be placed in the milk line somewhere between the weigh tank and the cooler and allowed to drip any desired speed so as to obtain a fair amount of sample. Newlander[12] conducted such a trial at a milk plant having 61 patrons. Each patron's milk was tested individually, together with the following composites: A true composite at the weigh stand of all the milk received, two drip samples at different points in the line, and one true composite from the cans as ready for shipment. Table 25 gives the results.

Composite cream samples have usually been condemned because the large amount of fat in the cream and its viscous character precludes the securing of a sample that is as accurate as a daily test. However, trials have shown that if true composites are taken the tests will agree sufficiently close to the daily test to more than offset the cost of the extra labor in testing each day's delivery.

Table 25. Accuracy of the Drip Sample

Days	Individual tests of all patrons' milk %	Composite of milk as received %	Low end drip sample %	High end drip sample %	Composite from cans as shipped %
1	3.68	3.75	3.65	3.65	3.60
2	3.57	3.65	3.70	3.70	3.70
3	3.80	3.65	3.80	4.00	3.80
4	3.75	3.70	3.75	3.60	3.60
5	3.68	3.65	3.65	3.70	3.65
6	3.60	3.55	3.55	3.50	3.55
Average	3.68	3.66	3.68	3.69	3.65

A study of the accuracy of composite cream samples was made by Combs et al.[13] He compared the daily sample, the true composite and the dipper method of compositing. The composites covered a period of two weeks. The true composite consisted of one milliliter of cream for each pound delivered, and the dipper composite was made in the usual manner of taking a small dipperful of cream at each delivery irrespective of the amount. The fat tests obtained were: Daily 29.35, true composite 29.16 and dipper sample 29.08 percent.

3. Sampling Other Dairy Products

Cream. Cream samples are taken in the same manner as milk. Highly viscous heavy cream may require sampling tubes of wider base (⅜ or ½ inch) than is needed for milk or cream of lesser viscosity. Proper mixing is of utmost importance in preparing cream for sampling. Cream samples are not commonly combined into composites. It is recommended[7] that samples be tested daily and always within three days after collection since enzymatic deterioration of the cream will cause low tests.

Evaporated Milk.[7] Unopened cans of evaporated milk should be tempered in a water bath for two hours at a temperature of about 140° F. During this time, the can should be removed from the water bath at 15 minute intervals and shaken vigorously, then removed and cooled to room temperature. The lid of the can should then be taken off and the contents thoroughly mixed with a spoon or spatula. The sample may be tested in this condition or 40 grams of the prepared sample may be combined with 60 ml. of water and this mixture tested, the results obtained being corrected for the dilution.

Sweetened Condensed Milk.[7] The unopened can should be tempered at 90-95° F. When warm, the can should be opened, the contents scraped out including that from the sides of the can, and transferred to a dish of sufficient size to permit thorough mixing. Then mix until the mass is of uniform consistency. Combine 100 g of the thoroughly mixed product with enough water to make up to 500 ml. and mix thoroughly. Any results obtained should be corrected for the dilution.

Dried Milk or Malted Milk.[7] Sampling should be avoided under conditions of high humidity (rainy days, etc.). Samples should be taken at six or more locations on the top of the surface by means of a tubular trier of sufficient length to extend to the bottom of the barrel. Remove these cores and transfer to a clean, dry, container and seal immediately. Make the sample homogeneous by rolling and inverting the container. Then sift the sample through a No. 20 sieve onto a large sheet of paper, grinding the residue in a mortar and passing this through the sieve. Particles of wood or other materials which cannot be ground should be discarded. The sample should be sifted two or more times and then tested immediately or preserved in an air tight container until tested.

Butter.[7] Samples of bulk butter should be taken by means of a trier and combined in a glass jar which will prevent the entrance or escape of water. Print butter may be sampled by quartering the print and combining two opposite quarters. The samples may then be softened by warming in a water bath to a temperature of 95-100° F. Bath temperature should not exceed 105° F. The sample is frequently shaken. When the emulsion remains intact but is fluid enough to show the level of the sample immediately, the bottle is removed from the water bath and shaken at regular intervals until the butter is of a thick and creamy consistency. The proper amount of sample should be weighed immediately for testing.

Cheese.[7] Small quantities of cheese should be taken from several places by means of a trier or spoon which will reach to the center of the cheese. Samples of hard cheese should be passed through a food chopper or shredded finely and thoroughly mixed. Soft cheeses can be mixed best in a high-speed blender at room temperature providing final temperature does not exceed 75° F. Samples should be weighed and tested immediately after mixing is completed.

Ice Cream and Frozen Dairy Products.[7] Ice cream mix can be prepared in the same manner as cream. Frozen ice cream should be allowed to soften at room temperature and then mixed thoroughly with a spoon or beater. Ice cream containing fruits, nut meats, etc., should be softened and then mixed in a high-speed blender until homogeneous.

BIBLIOGRAPHY

1. Tracy, P. H. and Tuckey, S. L. Accuracy of Methods of Sampling Milk Deliveries at Milk Plants. Ill. Bul. 459 (1939).
2. Bailey, D. H. Methods of Sampling Milk. Pa. Bul. 310 (1934).
3. Marquardt, J. C. and Durham, H. L. Sampling Milk for Fat Test at Milk Plants. N. Y. (Geneva) Bul. 605 (1932).
4. Bradfield, Alec. Factors That Influence the Accuracy of Weigh Tank Sampling. Vt. Ag. Exp. Sta. Bul. 603 (1957).
5. Heinemann, B. Chairman, et al. Procedures for Sampling and Testing Milk by the Babcock Method. Committee Report Approved by the American Dairy Science Association (1953). J. Milk and Food Tech. 17:120-121, 125 (1954).
6. Dimick, Paul S. M.S. Thesis. The University of Vermont (1960).
7. Methods of Analysis A.O.A.C. Published by the Assn. of Off. Agr. Chem., Washington, D. C. Eighth Ed. (1955).
8. Lucas, P. S. Accuracy of Composite Samples. Mich. Tech. Bul. 158 (1938).
9. Monroe, C. F. Accuracy of Composite Milk Samples. Ohio Bul. 446 (1930).
10. Holland, R. F. A Study of Composite Milk Samples. Thesis for M.S. Degree at Cornell University (1938).
11. Manus, L. J. and Bendixen, H. A. Effects of Lipolytic Activity and of Mercuric Chloride on the Babcock Test for Fat in Composite Milk Samples. J. D. Sci. 39:508-513 (1956).
12. Newlander, J. A. Accuracy of the Drip Sample. The Creamery and Milk Plant Monthly, p. 30, July, 1928.
13. Combs, W. B., Thurston, L. M., Groth, A. E. and Coulter, S. T. The Accuracy of Composite Cream Samples. Minn. Bul. 243 (1927).

REVIEW QUESTIONS

1. What is the best method of mixing a sample of milk?
2. What effect does underfeeding have on the test of milk? Weather conditions?
3. What precautions should be taken when collecting milk samples from farm bulk milk cooling tanks?
4. How should composite samples be carried on farm bulk milk collection routes?
5. What is meant by a composite sample?
6. Give two methods of obtaining true composite samples of milk.
7. What is one of the best preservatives for composite samples?
8. What effect do molds have on the fat test of milk?
9. Why do composite samples usually test lower than daily samples?
10. Is it advisable to take composite cream samples?
11. How would you care for composite samples?
12. What is the difference between formaldehyde and formalin?

CHAPTER IV

Babcock, Gerber, Mojonnier Tests for Fat

I. The Babcock Test

Introduction. In 1890 Babcock gave to the dairy industry his invention for determining the fat content of milk and cream. It has been one of the most outstanding contributions to the industry. During its operation these past 70 years, very little change has been made in technique and procedure. The test was readily accepted by this country because of its simplicity and relatively low expense of operation and its close agreement with the standard gravimetric methods of analysis.

Previous to Babcock's test various volumetric methods were advocated both in Europe and in this country. Herreid[1] has reviewed the literature leading up to the Babcock test, and a few abstracts of these reviews will be given here to acquaint the reader with some of the attempts in perfecting a fat test for dairy products.

These brief accounts of the various methods proposed for determining the fat content of milk and its products will give some idea of the experimentation and thought given in devising some satisfactory method for testing milk before Babcock made his contribution to dairy science.

In 1854 Marchand presented what was probably the first volumetric method of estimating fat in milk. His test bottle was called a lactobutyrometer and the neck was graduated in tenths which were referred to as degrees. He used 10 cc. of milk, 1 or 2 drops of sodium hydroxide, 10 cc. of ether and 10 cc. of alcohol, and these contents were carefully mixed. The tube was then placed in water at 40° C. and allowed to remain until it had cooled to 30° C. When the ether-fat layer had fully separated the degrees of fat were read and converted to weight of fat in grams.

Schmidt (1878) modified Marchand's method by enlarging the lactobutyrometer so as to facilitate mixing of the milk and reagents. He also used 3 to 5 drops of five percent acetic acid instead of sodium hydroxide, stronger alcohol and read the tests at 20° C. instead of 30° C. The results were lower than gravimetric tests and a formula was devised to provide closer agreements.

Lieberman (1883) contributed a combination of volumetric and gravimetric procedures. Fifty cubic centimeters of milk were placed in a large cylinder and 5 cc. of sodium hydroxide and 50 cc. of ether were added. The contents were mixed and allowed to settle for 10 to 15 minutes, then a 20 cc. aliquot was evaporated on a water bath at 40° C. to 50° C. and dried for 15 minutes at 100° C. to 105° C. The air-free fat was then measured in a carefully calibrated flask and converted to percent by weight.

Schmid, W. (1888) introduced a new reagent, hydrochloric acid. Ten cc. of milk or 5 cc. of cream were measured into a 50 cc. tube, graduated into tenths, then 10 cc. of concentrated hydrochloric acid added, and the contents mixed and heated. Next the mixture was cooled and 30 cc. of ether added and mixed. Ten cc. of the clear ether layer were taken and evaporated over a bath at 100° C. The procedure required 15 minutes and checked within 0.1 percent of the gravimetric method.

DeLaval (1885) patented a method called a lactocrite test for estimating the fat content of milk. He was the first to use a centrifuge, substituting a disc for the bowl in his centrifugal separator. The reagent was a mixture of 20 parts concentrated acetic and one part of sulphuric acid. Ten cc. of milk and 10 cc. of the acid mixture were mixed and whirled at 6000 r.p.m. for 3 to 5 minutes. Tests could be read to 0.02 percent accuracy.

Short's method (1888) consisted of saponification of the fat in 20 cc. of milk with sodium and potassium hydroxide. The insoluble fatty acids were then obtained by boiling one hour with 10 cc. of a mixture of equal parts of sulphuric and acetic acids. The tubes containing the mixture were then allowed to stand in hot water and the fat column estimated in millimeters and the percent fat computed.

Parson (1888) criticized Short's method for not being accurate for skimmilk and buttermilk. His reagents consisted of 10 cc. of 50 percent sodium hydroxide, 5 cc. of 95 percent alcohol containing one ounce of castile soap per gallon and 50 cc. of gasoline. These compounds were mixed with 10 cc. of milk or cream in a flask and the fat-gasoline layer allowed to separate. Twenty-five cc. of this layer were evaporated, the fat collected in a calibrated tube, allowed to nearly solidify and then read to the uppermost meniscus. The reading was then converted to percent fat from a table. This method gave close agreements with gravimetric procedures, even in unskilled hands.

Failyer and Willard (1888) used a test bottle 10 inches long with a graduated portion 3.5 inches long and 4 to 5 millimeters in diameter. The reagents were hydrochloric acid and gasoline. Ten cc. of milk were added to the bottle and heated with 8 cc. of acid for about 5 minutes and then cooled. The gasoline, 15 cc. to 20 cc., was next added to dissolve the fat, and the bottles then placed in an inclined position in boiling water and air blown into the tube to evaporate the gasoline. Water was added to the base of the neck and the bottle whirled (time not stated) a few times. Boiling water was added to bring the fat into the graduated neck and a reading taken. The results agreed closely with a gravimetric method used by Babcock.

Leffman and Beam (1889-1892) submitted directions for a test that they claimed was devised in 1889. Test bottles with calibrated necks divided into 0.1 percent fat were used. Fifteen cc. of milk and 3 cc. of a mixture of equal parts amyl alcohol and strong hydrochloric acid were first mixed together. Then the bottle was filled nearly to the neck with concentrated sulphuric acid and mixed again. The bottles were finally filled with diluted sulphuric acid and whirled in a centrifuge for 1 to 2 minutes. The tests were read with dividers. Results agreed within 0.1 percent of the Adam's gravimetric method.

The Beimling test (1890) employed a centrifuge patented by H. F. Beimling, an Austrian chemist living in Philadelphia. This test was described by the Vermont Agricultural Experiment Station in bulletin 21, 1890. The same kinds of reagents were used as in Leffman and Beam's test and that of Gerber, but the glassware in the Gerber method was different from the other two. The results agreed closely with the Adam's procedure.

Gerber method (1892-1895) was formulated by Dr. N. Gerber, a Swiss chemist. This test employs special bottles and two reagents, sulphuric acid and amyl alcohol. Since this procedure is so widely used, specific directions for analyzing different dairy products will be given under "The Gerber Test."

Testing Whole Milk

Apparatus. In testing whole milk it is necessary to have milk test bottles, a pipette, acid measure, sulphuric acid, thermometer, centrifuge, speed indicator, reading dividers and tempering bath.

a. Milk test bottle. The standard bottle for testing whole milk is the 18 gram, 8 percent bottle graduated to read to 0.1

Babcock Milk Test Bottle
Courtesy Kimball Glass Co.

percent. The detailed specifications as officially[2] adopted are as follows:

Eight percent, 18 gram, 6 inch milk test bottle. The total height of the bottle shall be 150 to 165 mm. (5.9 to 6.5 inches). The bottom of the bottle shall be flat, and the axis of the neck shall be vertical when the bottle stands on a level surface. The charge of milk for the bottle shall be 18 grams.

(1) *Bulb.* The capacity of the bulb to the junction with the neck shall be not less than 45 ml. The shape of the bulb shall be either cylindrical or conical. If cylindrical, the outside diameter shall be between 34 and 36 mm.; if conical, the outside diameter of the base shall be between 31 and 33 mm., and the maximum diameter between 35 and 37 mm.

(2) *Neck.* The neck shall be cylindrical and of uniform diameter from at least 5 mm. below the lowest graduation mark to

at least 5 mm. above the highest. The top of the neck shall be flared to a diameter of not less than 10 mm. The graduated portion of the neck shall have a length of not less than 63.5 mm. The total percent graduation shall be 8. The graduations shall represent whole percent, five-tenths percent, and one-tenth percent, respectively, from 0.0 to 8.0 percent. The tenths percent graduations shall not be less than 3 mm. in length, and the five-tenths percent graduations not less than 4 mm. in length and shall project 1 mm. to the left; the whole percent graduations shall extend at least half-way around the neck to the right and shall project at least 2 mm. to the left of the tenths percent graduations. Each whole percent graduation shall be numbered, the number being placed to the left of the scale. The capacity of the neck for each whole percent on the scale shall be 0.20 ml. The maximum error of the total graduation or any part thereof shall not exceed the volume of the smallest unit of the graduation. Each bottle must be constructed so as to withstand stress to which it will be subjected in the centrifuge.

(3) *Testing for Accuracy.* Mercury and cork, alcohol and burette, and alcohol and brass plunger methods may be used for rapid testing of bottles, but accuracy of any questionable bottle must be determined by calibration with mercury (13.5471 grams of clean, dry mercury at 20° C. to be equal to 5 percent on the scale of the 18-gram milk test bottle and 10 percent on the scale of the 9-gram cream test bottle), bottle having been previously filled to 0 with mercury.

In calculating the volume of the graduated portion of the neck of the bottle, the specific gravity of butterfat is considered to be 0.9 at 140° F. Since 18 grams of milk are used the neck of the bottle would hold between the zero and 8 percent marks, $18 \times .08$ or 1.44 grams fat or a volume of $1.44 \div .9$ or 1.6 ml. Similarly the neck of a 10 percent bottle holds 2 ml.

b. Pipette. The milk is measured for testing by means of the pipette which holds 17.6 ml. up to the mark on the stem. Since milk is slightly viscous, the pipette will deliver only 17.44 ml. which weighs 18 grams considering the average specific gravity as 1.032. The specifications of the pipette follow:

Total length not more than 330 mm.
Outside diameter of suction tube: 6 to 8 mm.
Length of suction tube: 130 mm.
Outside diameter of delivery tube: 4.5 to 5.5 mm.

Length of delivery tube: 100 to 120 mm.
Distance of graduation mark above bulb: 15 to 45 mm.
Nozzle: Parallel with axis of pipette, but slightly constricted so as to discharge in 5-8 seconds when filled with water.
Graduation: to contain 17.6 ml. of water at 20° C. when the bottom of the meniscus coincides with the mark on the suction tube.
Delivery: 5 to 8 seconds.
The maximum error in the graduation shall not exceed 0.05 ml.
The pipette is to be marked "Contains 17.6 ml."
The pipette shall be tested by measuring from a burette the volume of water (at 20° C.) which it holds up to the graduation mark.

c. Acid Measure. Device used to measure H_2SO_4, whether graduated cylinder or pipette attached to Swedish acid bottle, must be graduated to deliver 17.5 ml.

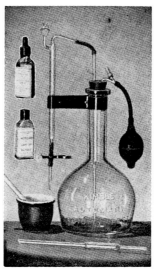

Automatic Acidity Tester
Courtesy Macalaster Bicknell Co.

d. Sulphuric acid. The sulphuric acid used in this test should have a specific gravity of not less than 1.82 and not more than 1.83. Halfway between 1.825 is preferable. This acid is a commercial product and is sometimes referred to as oil of vitriol. Its purpose is to dissolve all the solids not fat, and thus leaving the fat free so it can be quickly brought to the surface by centrifugal force.

e. Centrifuge or "tester." Standard centrifuge, however driven, must be constructed throughout and so mounted as to be capable, when filled to capacity, of rotating at necessary speed with minimum vibration and without liability of causing injury or accident. It must be heated, electrically or otherwise, to a temperature of at least 55° C. during centrifuging. It must be provided with a speed indicator, permanently attached, if possible. Proper rate of rotation may be ascertained by reference to Table 26. By "diameter of wheel" is meant the distance between the inside bottoms of opposite cups measured through the centrifuge wheel while cups are horizontally extended.

Babcock Milk Tester
Courtesy Will Corp.

f. Dividers or calipers. For measuring fat column.

g. Water bath for test bottles. Provided with thermometer and device which will maintain temperature at 55-60° C.

Procedure in Testing Whole Milk. The following are the steps in testing whole milk for fat:

(1). Temper the milk to about 100° F. The sample should not be shaken or mixed until it has reached a temperature of 90° F.[3]

(2). Mix the sample thoroughly. The most effective way of distributing the fat evenly throughout the milk is to pour the sample from one container to another several times, being careful to have the milk flow down the sides of the vessel to avoid raising air bubbles.

Dividers
Courtesy Will Corp.

(3). Immediately after mixing the sample, measure out a pipetteful of milk. This is done by sucking up the milk into a pipette until it is well over the mark and then quickly placing the forefinger (not thumb) over the end of the pipette to retain the milk. Then by releasing the pressure of the finger a little, some of the milk is allowed to escape until the bottom of the meniscus reaches the 17.6 ml. mark. During this operation the pipette should be held vertical, with the mark on a level with the eye. The forefinger must be dry, or otherwise the flow of the milk cannot be controlled. Blow out the last drop of milk in pipette about 10 seconds after free flow has ceased. The use of two pipettes will speed up the operation.

The milk may be readily transferred to the test bottle by placing the stem of the pipette into the neck of the bottle and allowing the bulb to hold it in position while the charge is flowing into the bottle. The earlier pipettes were not constructed with stems sufficiently slender to enter the necks of the bottles, and therefore it was necessary to hold both bottle and pipette in an awkward position when transferring the milk.

(4). After measuring the milk into the bottle, 17.5 ml. of sulphuric acid, having a specific gravity of 1.82-1.83 at 20° C. and tempered to 60-70° F., is added. Then the milk and acid are thoroughly mixed. The temperature of the milk at the time acid is added should not exceed 80° F.[3] The purpose of the acid **is** to burn up all the solids not fat, thus leaving the fat free to rise. The acid should be added slowly by letting it flow down the side of the neck and revolving the bottle gently until the neck is cleared of all adhering milk. The bottle should be held on a slant so that the acid will flow down the side of the neck and body and get under the milk, and thereby not char the fat, which it would do if allowed to fall through the milk.

It is best not to add acid to more than two bottles at a time before mixing, as there is danger of charring the fat if they are allowed to stand too long before shaking. In mixing, use a rotary motion to keep the curd from going into the neck. In order to get a clear test, it is necessary to shake the bottles

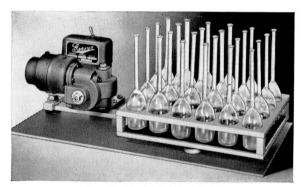

Babcock Test Bottle Shaker
Courtesy Will Corp.

thoroughly until all the curd is completely dissolved. A mechanical shaker is recommended because it will do a much better job of mixing than is done by hand and saves considerable time when making a number of tests. If the acid is the right strength, the mixture will turn brown at first and then gradually become black. If it turns black at the start, the acid is too strong, and a charred test will result. On the other hand, if the curd dissolves slowly and the mixture scarcely turns brown, the acid is weak and a curdy test will follow.

Much heat is generated upon mixing the milk and acid, a temperature around 225° F. is usually produced.

(5). Next place the bottles in the centrifuge, being careful that the machine is balanced, that is, each bottle must be counterbalanced by another on the opposite side in the corresponding cup.

In case of a hand machine, which is now seldom used, put a

dipper of hot water in the bottom of the tester to keep the samples warm. Change this water at the end of the first run. The temperature in steam machines is kept sufficiently high by the exhaust steam, and electric machines are equipped with heating elements.

(6). Centrifuge at full speed for five minutes. The proper speed is generally indicated on the machine, if not, the speed may be calculated as explained on page 82. At the end of the five minutes, stop the centrifuge and add hot water, around 140° F., to the test bottles, bringing the mixture up to the base of the neck. The water should be soft and must be free from oil. Centrifuge again for two minutes and then add hot water as before until the fat column is within the graduations. Give a final whirling of one minute.

(7). Transfer the bottles to a water bath, at a temperature of 135° F. to 140° F. The water should completely cover the fat column. Hold in the bath at least three minutes. It is neces-

Babcock Tempering Bath
Courtesy Will Corp.

sary to temper the fat column before reading in order to get uniform results. The fat will be contracted if too cold and thus read too low, or expanded if too hot and consequently read too high. (See page 83.)

(8). Read the test from the lowest point of the fat column to the top of the meniscus. The meniscus is the curve in the

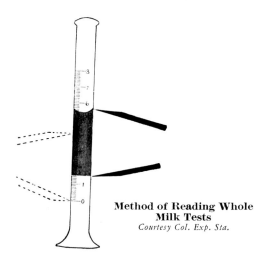

Method of Reading Whole Milk Tests
Courtesy Col. Exp. Sta.

upper end of the fat column. Reading the extreme points makes up for the small amount of fat left in the body of the test bottle. When reading take only one bottle at a time from the bath, as the fat columns cool rapidly. Greater accuracy in reading is secured by the use of dividers. In using the dividers, place one pointer at the bottom of the fat column and then spread the other until it reaches the top of the meniscus. Next place one of the pointers on the zero mark and read the division on which the other rests. The dividers should be made to move steadily with no slipping. Several mechanical devices and special illuminating lamps for greater ease in reading tests are available. These tend to increase the precision and repeatability of results.

It is advisable to run all tests in duplicate, as a check, in which case the average of the two readings is used. The duplicate should not vary more than one-tenth of one percent from the original. If the difference is greater than this, or if the test is imperfect otherwise, the sample should be retested.

It is generally agreed that the results of the Babcock test are somewhat lower than those obtained in the ether extraction (Mojonnier) test. In order to obtain agreement between the two tests, a recommendation has been made[4] that the capacity of the Babcock pipette be increased to 18.05 ml. and, if this is

Butterfat Column Reader
Courtesy H. C. Goslee

Meniscus Reader
Courtesy Macalaster Bicknell Co.

done, eliminate the upper meniscus by the addition of a colored mineral oil (glymol) as is done in the Babcock test for cream. However, other studies have indicated that the officially approved procedures would give satisfactory results if properly carried out. At the present time this change has not been accepted by AOAC.

Appearance of a Perfect Test. The following five points describe the appearance of a perfect test:

(1). The color of the fat column should be a straw yellow.

(2). The ends of the fat column should be clearly and sharply defined.

(3). The fat should be free from specks and sediment.

(4). The water just below the fat column should be perfectly clear.

(5). The fat should all be within the graduations.

Curdy Tests. If the fat column is too light in color, or curdy, it is due to one or more of the following factors:

(1). Temperature of the milk or acid, or both, too low.

(2). Acid too weak.

(3). Not enough acid used.

(4). Milk and acid not mixed thoroughly.

When the acid is a little weak, good results may still be obtained by using more acid and higher temperatures.

Charred Tests. If the fat column is darkened and contains black specks, or a black substance at the base, it is due to one or more of the following conditions:

(1). Temperature of the milk or acid, or both, too high.

(2). Acid too strong.

(3). Milk and acid mixed too slowly.

(4). Acid dropped through the milk.

(5). Too much acid used.

When the acid is too strong, clear tests may be secured by using less acid and lower temperatures.

Care of the Sulphuric Acid. Commercial sulphuric acid is usually clear and almost colorless, but when foreign particles drop into the acid it becomes darkened. This does not spoil its usefulness as long as most of the particles are dissolved; some undissolved material will do no harm unless it gets into the fat column.

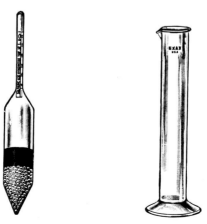

Acid Hydrometer and Cylinder
Courtesy Macalaster Bicknell Co.

When left in open containers, sulphuric acid will absorb moisture from the air and gradually become too weak for use in this test. For this reason the acid must be kept in tightly stoppered containers which may be either glass bottles or earthenware jugs. The stoppers should be of glass or rubber, since these materials resist the action of the acid. Cork or wooden plugs are not acid resistant and consequently should not be used. They not only become loose fitting but charred particles, that are insoluble, fall into the acid and this may render it unfit for testing purposes.

If it is desired to weaken acid that is too strong, a portion may be taken from the carboy and left exposed to the air until enough moisture has been absorbed by the acid to reduce it to the proper strength. Care should be taken that it does not absorb too much moisture. The time required will be from several hours to several days, depending on the amount of moisture in the room and strength of acid.

The acid may also be weakened by mixing it with a small amount of water. It will require approximately 30 ml. of water to lower the specific gravity of 1000 ml. of sulphuric acid from 1.833 to the desired 1.825. The acid must be added to the water, and not the reverse, or spattering will result from the sudden boiling of the water. Small amounts of acid should be added at first, so that too much heat will not be produced and thereby break the container. It will be necessary to cool the diluted acid to the proper temperature before using it for testing.

Advantages of Sulphuric Acid. Sulphuric acid has certain advantages over other acids for use in the Babcock test:

(1). It quickly dissolves the solids not fat and destroys the viscosity, thus liberating the fat.

(2). It generates much heat and so keeps the fat in a liquid condition, which is necessary for complete separation.

(3). It increases the specific gravity of the mixture surrounding the fat thereby aiding the separation of the fat.

(4). It is non-volatile while hot and so produces no injurious gases.

(5). It is comparatively cheap and easy to obtain.

Precautions in Using Sulphuric Acid. Certain precautions should be observed in using sulphuric acid. In measuring or pouring the acid, the operator should keep it away from his face, and when shaking the test bottles, he should point them away from him and at the same time not toward anyone else. If acid is spilled on the flesh or clothes, it should be washed off immediately with cold water and then an alkali solution applied, to completely neutralize any acid remaining. A good washing solution will serve this purpose. Red spots on clothing indicate the presence of acid, which, in time, result in the fabric being destroyed. If taken in time, these spots may be removed by the application of dilute ammonium hydroxide, which in turn must be rinsed off with water. A rubber apron, or some other protection for the clothing, should be used when testing. Where considerable testing is done, the shelves should be covered with lead, as this is the only common metal that is acid resistant.

Foaming. Sometimes foam collects on top of the fat column. This is due to the use of hard water, as it contains carbonates. The sulphuric acid reacts with the carbonates and releases carbon dioxide, which causes the foam. If it is necessary to use hard water, the foaming may be overcome by adding a few drops of sulphuric acid to the water before using.

Speed of the Machine. Some preliminary work by the inventor indicated that in order to obtain as complete a separation of the fat as possible, it was necessary to have a pressure of at least 30 pounds per square inch. To obtain this pressure the machine must revolve at a certain speed in accordance with the diameter of the wheel, the greater the diameter, the less the speed and conversely the smaller the diameter the greater the speed. The diameter of the wheel is the distance from the bottom of a cup lying horizontally to the bottom of the opposite cup lying in the same position. According to Farrington and Woll,[5] the centrifugal force is expressed by the formula,

$$F = \frac{W \times V^2}{32.2r}$$

in which F = centrifugal force, W = weight in pounds of the bottle and contents, V = velocity in feet per second and r = the radius of the wheel in feet. Knowing the force, weight and radius, the velocity can be computed for any sized machine. Based on a centrifugal force of 30.65 pounds per square inch, they produced the figures in the second and third columns of Table 26 which cover practically all sized machines. For example, for a centrifuge with a wheel diameter of 12 inches —

$$30.65 = \frac{.18745 \, V^2}{32.2 \, (.5)}$$

$$30.65 \times 16.10 = .18745 \, V^2$$
$$493.4 \div 0.18745 = V^2 = 2632$$

W = 85 grams = .18745 lbs.

$$V = \sqrt{2632} = 51.31$$

Table 26. Speed of Centrifuge

Diameter of wheel, D.	Velocity in feet per second, V.	Number of revolutions of wheel per minute	*
Inches			
10	46.84	1074	1171
12	51.31	980	1033
14	55.43	909	934
16	59.26	848	859
18	62.84	800	800
20	66.24	759	751
22	69.47	724	711
24	72.56	693	677

* Calculated by B. L. Herrington.

Herrington[6] has brought out the point that the speeds, as given by Farrington and Woll, have been calculated to give equal forces at the bottom of the test bottles, but have not been applied to the more important location, the fat-acid interface. The first centrifuging is considered the most important of the three whirlings and the junction of the fat and acid at this time should be the location for calculating the force and speed. This point is around 1.5 inches from the bottom of the bottle. It is at this surface that the small fat globules in the mixture of milk and acid are collected into an oily layer. Therefore, the force at this point, rather than that at the bottom, should be considered inasmuch as the lower portion is always subjected to the greater force. Thus, the calculated speeds of the different sized machines should be based on the diameter of the wheel as starting about 1.5 inches from the bottom of the cup rather than at the extremity. His calculated speeds, using the original 800 r.p.m. for an 18 inch machine as standard, are shown in the last column of Table 26. He further recommends that the use of small diameter machines, especially those of 10 and 12 inch diameters, be discontinued because the force exerted in the neck of the bottle is much less than in case of larger centrifuges. It may even be negative.

To find the correct speed of a hand machine, first determine the diameter of the wheel and then count the number of revolutions the wheel makes for each turn of the handle. Divide this figure into the number of revolutions required per minute for this sized machine (see Table 26) and the quotient will be the number of turns that the handle must make per minute. For example, if the diameter of the wheel is 12 inches and it makes 14 revolutions to one of the handle, the latter must go 980/14, or 70 turns per minute. For power machines use a speedometer.

Expansion of Fat. The coefficient of expansion of milk fat is 0.0007558 per ml. of fat per degree centigrade. From this figure calculations will show how much the variation in reading a 4 percent milk tempered at different temperatures would be:

Tempered at °F.	Reading %
140	4.000
135	3.991
130	3.982
120	3.967

Thus, a tempering range of 130 to 140° F. would give the same Babcock test reading of 4 percent, but at 120° F. it might be read a tenth lower.

Example: 140° F. vs. 120° F.
60° C. vs. 49° C.
Expansion .0007558 per ml. fat per degree C.
.0007558 × 11 × .8 ml. fat (in 4% column) = .00665 ml. fat

```
  60           .80000 ml.
 —49           .00665
 ─────         ──────
 11° C.        .79335 ml. fat
            ×    .9  sp. gr. fat
               ──────
               0.7140 g fat
```

0.7140 g ÷ 18 g milk = 3.967%

Cleaning the Glassware. The glassware should be washed thoroughly after each test, as any fat left in the test bottles would affect the succeeding test. When through testing, empty the bottles into an earthenware crock, or into a sink lined with lead. These materials are not affected by the acid. The ordinary sinks and drains would soon be destroyed by the acid solution. When emptying the bottles, shake them until the white precipitate, calcium sulphate, at the bottom of the bottles, is removed. This precipitate is not readily removed with water. Then rinse the bottles with hot water, after which fill them about one-half full with hot alkali solution made from a good washing powder.

Use a brush to clean the necks, and to clean out the body of the bottles, shake vigorously when emptying. Give a final rinsing of clear, hot water. Special racks may be used whereby ten or twelve bottles can be shaken at a time. The pipette should be cleaned in the same way, except that the milk should be rinsed out with cold or lukewarm water soon after using, as rinsing with hot water would make the casein stick to the pipette, thus making it difficult to clean.

Earthen Jar for Waste Acid with a Convenient Cover for Draining the Bottles
Courtesy Mo. Exp. Sta.

Test Bottle Bath and Shaker
Courtesy Oakes and Burger Co.

Special test bottle washers will do a quicker and more efficient job of cleaning than by hand. If the cleaning of the glassware is neglected or not properly done, the pieces may become so darkened that the ordinary methods of cleaning will not leave them clear. In such cases the stain can be successfully removed by immersing the glassware for several hours in a special cleaning mixture. The ingredients of this mixture are:

Potassium bichromate	30 grams
Tap water	150 ml.
Concentrated sulphuric acid	230 ml.

First dissolve the potassium bichromate in the 150 ml. of water, placing flask and contents in boiling water to hasten the solu-

Babcock Bottle Washer
Courtesy Will Corp.

tion. When dissolved, add the acid very slowly so that the temperature will not rise rapidly enough to break the flask. In this preparation it is essential to use a container that will withstand high heat as the addition of the acid causes the temperature of the mixture to go higher than that of boiling water. The final step is to cool the mixture and it is then ready for use. A convenient receptacle for the cooled solution is an earthenware crock of sufficient size to contain enough of the mixture to cover the longest piece of glassware.

Calibration. In buying milk or cream by test, the test bottles and pipettes must be checked for accuracy. This testing for accuracy is known as calibration. It is usually done by the experiment station of the particular state. Mercury is used for testing the bottles and alcohol or water for the pipettes. One milliliter of mercury at 20° C. weighs 13.5471 grams, and since the neck of the 8 percent milk test bottle holds 1.6 ml. between the zero and 8 percent graduations, it will take 21.675 grams of mercury to exactly occupy that space. The mercury is weighed out, placed in the bottle, and a cork or rubber stopper inserted into the neck until it reaches the 8 percent graduation. The bottle is then inverted and if the column of mercury just reaches the zero mark, or if it does not extend above or below it by more than the volume of the smallest division, it is certified as correct. The name of the experiment station is then sand blasted on the neck of the bottle. In the same manner cream and skim-milk test bottles are calibrated by using the proper amounts of mercury.

In calibrating the pipette, an accurate burette is used to measure the amount of water or alcohol that the pipette should hold up to the graduation mark on the stem at a temperature of 20° C. Alcohol is preferable to water as it flows more readily and does not stick to the glassware.

Testing Cream

The method of testing cream differs a little from that of whole milk. The sample must be weighed instead of measured, because: (1) cream varies widely in specific gravity, depending on the fat content, which would make a corresponding difference in the weight of a definite amount; (2) it is viscous and so some cream would stick to the pipette; (3) well mixed cream is apt to contain considerable air and carbon dioxide.

Apparatus. The apparatus for testing cream is the same as that for whole milk, except cream test bottles are used in place of the whole milk bottles, and there is the additional requirement of a set of scales and glymol or meniscus remover.

a. Cream scales. The cream scales should be sensitive to 30 milligrams, that is, if this weight is put on one of the pans when at full load, it will deflect the pointer one division. The scales must be set level and on a firm base, and be protected from air currents when in use. Properly marked 9 g. and 18 g. weights should be provided. These should be made of a material which resists corrosion or injury. Weights of the low, squat shape with rounded edges are preferred. These should be checked frequently against standardized weights.[2]

b. Cream test bottle. Both the 9 gram and 18 gram 50 percent bottles are used. The neck of the 9 gram bottle has a capacity of 5 ml. between the zero and 50 percent graduations, while the neck of the 18 gram bottle holds twice that amount. The detailed specifications follow:

Fifty percent, 9 gram, short neck, 6 inch cream test bottle. The total height of the bottle shall be 150-165 mm. (5.9-6.5 inches). The bottom of the bottle shall be flat, and the axis of the neck shall be vertical when the bottle stands on a level surface. The charge of cream for the bottle shall be 9 grams.

Bulb. The capacity of the bulb to the junction with the neck shall be not less than 45 ml. The shape of the bulb shall be either cylindrical or conical. If cylindrical, the outside diameter shall be between 34 and 36 mm.; if conical, the outside diameter of the base shall be between 31 and 33 mm., and the maximum diameter between 35 and 37 mm.

Neck. The neck shall be cylindrical and of uniform diameter from at least 5 mm. below the lowest graduation mark to at least 5 mm. above the highest. The top of the neck shall be flared to a diameter of not less than 15 mm. The graduated portion of the neck shall have a length of not less than 63.5 mm. The total percent graduation shall be 50. The graduation shall represent five percent, one percent, and one-half percent, respectively, from 0.0 to 50 percent. The five percent graduations shall extend at least half-way around the neck to the right; the one-half percent graduations shall not be less than 3 mm. in length; and the one percent graduations shall be intermediate in length between the five percent and the one-half percent graduations and shall project 2 mm. to the left of the one-half percent graduations. Each five percent graduation shall be numbered (thus: 0, 5, 10, —45, 50), the number being placed to the left of the scale. The capacity of the neck for each whole percent on the scale shall be 0.1 ml. The maximum error in the total graduation or any part thereof shall not exceed the volume of the smallest unit of the graduation.

Fifty percent, 9 gram, long neck, 9 inch cream test bottle. The same specifications shall apply to this bottle as to the 50 percent, 9 gram, 6 inch cream test bottle, with the exceptions, however, that the total height of this bottle shall be 210-229 mm. (8.25 to 9.0 inches), that the graduated portion of the neck shall have a length of not less than 120 mm., and that the maximum error in the total graduation or any part thereof shall not exceed the volume of the smallest unit of the graduation.

Fifty percent, 18 gram, long neck, 9 inch cream test bottle. The same specifications shall apply to this bottle as to the 50 percent, 9 gram, 9 inch cream test bottle, with the exception, however, that the charge of cream for this bottle shall be 18 grams.

Each bottle shall bear on the top of the neck above the graduations, in plain legible characters, a mark denoting the weight of the charge to be used, viz., "9 grams" or "18 grams," as the case may be.

Each bottle shall bear a permanent identification number,

Babcock Cream Test Bottle
Courtesy Kimball Glass Co.

placed thereon either by the manufacturer or by the purchaser.

Each bottle shall be constructed of necessary strength to withstand the strain to which it will be subjected in the centrifuge.

c. Water bath. A water bath for cream samples should be provided and have a thermometer and device for maintaining temperature of 38° C.

d. Glymol. Glymol is a high grade white mineral oil, or Diamond paraffin oil, colored red, and must have a specific gravity of not over 0.85. Formerly the oil was colored by means of alkanet root. The cut up root was placed in a cheese cloth and suspended in the oil for one or two days by which time the oil was a deep red color. It is now colored with an aniline dye, National oil red O obtained from the National Aniline and Chemical Company. One ounce of the dye is sufficient to color five gallons of oil. The dye is first dissolved in a pint or quart of oil and then the remainder of the five gallons added to it.

Procedure in Testing Cream. The following are the steps in testing cream for fat:

(1). As in case of whole milk, the sample must be brought

Bottle Cream Test Balance
Courtesy Torsion Balance Co.

Cream Test Balance
Courtesy Torsion Balance Co.

to the right temperature, 60° F. to 70° F., and then thoroughly mixed by pouring from one container to another. In the commercial testing of cream for butterfat, it may be advisable to warm all samples to approximately 95-100° F. before mixing.[2]

(2). Balance the empty test bottles on the scales and then weigh out 9 grams of the well mixed cream. This is done by placing a 9 gram weight on one of the pans after the empty bottles are balanced, and then adding cream to the bottle on the opposite pan until it comes to a balance. The weight is then removed and cream is added to the second bottle until it balances the first. The same procedure is used for scales holding a larger number of bottles. Care should be taken when weighing to see that none of the cream drops on the outside of the bottle or on the pans of the scales.

(3). Add to the charge approximately 9 milliliters of soft water, temperature around 70° F. Cooler water should be added if the cream had a higher temperature when weighed out, so that the mixture will be around 70 degrees F. when the acid is added. The water may be added with sufficient accuracy with the 17.6 ml. pipette, dividing the contents between two bottles. Unless water is added to the cream before the addition of the acid, there is danger of the latter charring the fat because of the large amount of fat in cream.

(4). Add 17.5 ml. of sulphuric acid and mix thoroughly. The mixture should not turn black, as in the case of whole milk, but should remain a brown color.

(5). Centrifuge, add hot water and temper as for whole milk.

(6). When ready to read the tests, take out one bottle at a

time and add glymol to remove the meniscus. The glymol will flatten out the curve in the fat column so that the entire surface will be in a straight line. It may be added with a pipette or with a dropper. In doing so let it flow down the inside of the neck of the bottle until a layer about one-eighth inch in thickness covers the fat column. Its specific gravity is slightly less than that of butterfat and consequently it will remain on top. The glymol must not be dropped on the fat column because the force

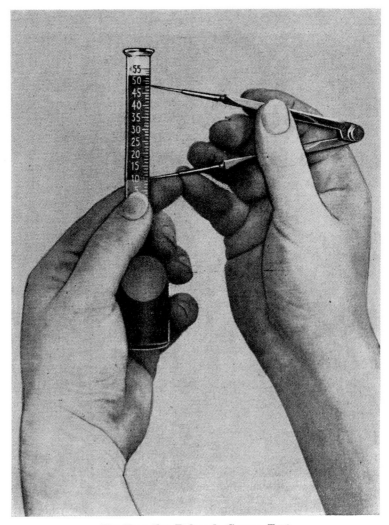

Reading the Babcock Cream Test

of the fall would cause it to mix with the fat and no clear line of demarcation would result. The glymol must not be added while the tests are in the water bath, as the heat might cause the oil and fat to mix. Read the test from the bottom of the fat column to the line between the fat and the glymol.

The meniscus is not included in reading a cream test, because the diameter of the neck of the cream bottle is much greater than that of the whole milk bottle, thus making a meniscus occupying a large volume of space, which is too large a quantity to allow for the small amount of fat left in the bottle. The impurities that are usually present in the fat column of a cream test will make up for the unrecovered fat.

Testing Skimmilk and Whey

Skimmilk is tested in the same manner as whole milk with the following exceptions:

(1). A special double necked bottle is used, one tube of which is for the fat and the other for the introduction of the

A Skimmilk Test Bottle
Courtesy Louis F. Nafis, Inc.

A Skimmilk Test Bottle Showing Completed Test; Reading .06 Percent
Courtesy Mo. Exp. Sta.

milk and acid. The fat tube has a very small bore so that the fat can be measured accurately, as skimmilk contains such a small amount of fat that it could not be read in the whole milk bottle. Some skimmilk bottles are graduated to read to 0.25 of one percent, and others to 0.50 of one percent, the smallest division in both cases being 0.01 percent.

(2). More acid is used, because there are more solids not fat in skimmilk to dissolve. Around 20 ml., or approximately one-fourth inch above the mark on the measure, is the proper amount of acid.

(3). The bottles are whirled longer, ten minutes the first time, instead of five minutes as for whole milk, then two minutes and one minute. Extra whirling is used because the fat globules in skimmilk are small and consequently harder to remove. Care should be taken when mixing the milk and acid to avoid throwing any undissolved curd into the fat tube.

The same directions for testing skimmilk will apply for whey except no extra acid is needed, as the amount of solids not fat is low.

The absence of fat in the graduated neck following centrifuging indicates that insufficient acid was used, the centrifuge did not maintain proper speed, or the temperature was too low during testing.[7]

Testing Cheese

Cheese may be tested satisfactorily by the Babcock method:

(1). Weigh out 4½ grams of the chopped cheese into a 9 gram cream test bottle.

(2). Add 12 to 14 ml. of hot water. This will soften the cheese and thus aid the action of the sulphuric acid in dissolving it.

(3). Add a measure of sulphuric acid and shake thoroughly until all the cheese particles are dissolved. More acid may be used if necessary.

(4). Complete the test as for cream, centrifuging 5 minutes, 2 minutes and one minute. Use glymol and double the reading since only one-half of nine grams is used.

Soft cheeses such as cream and cottage may be tested in the same manner. If it is not possible to get the cheese into the test bottle, first weigh the sample into a beaker, add water and then under constant stirring with a glass rod run the acid into the beaker. The resulting liquid mixture can then be poured into the test bottle. Rinse the beaker with hot water and add this also to the test bottle. Complete the test in the usual manner.

Testing Partially Churned Milk

Often when samples of milk or cream are sent through the mail they arrive with small butter particles floating around. This same condition is encountered at the factories where patrons deliver cans of milk or cream partially churned. This can be prevented by completely filling the receptacle so as not to leave any room for the milk or cream to splash. Again, samples are sometimes partially churned by reckless shaking in their preparation for testing. When churning happens, special means must be used in preparing the sample for testing, as a representative sample cannot be taken until the butter granules are reincorporated.

One method is to warm the sample to 100° F. to 105° F. and hold at this temperature until the butter is melted. Higher temperatures should not be used, as there would then be danger of oiling off. As soon as the butter granules are melted, mix thoroughly and immediately transfer a pipetteful of the milk to the test bottle. Cool the charge in the test bottle to around 70° F. and proceed as for normal milk.

Another method is to use ether to dissolve the butter granules. Add enough ether to the sample to equal 5 percent of its volume, shake until the granules disappear and proceed as for normal milk. Care should be taken in mixing the milk and acid, so as not to lose any of the contents due to the rapid boiling of the ether caused by the heat evolved from the union of the milk and acid. The fat test must be increased by the percent of ether added.

Testing Frozen Milk

Milk containing ice must not be sampled until thawed out, since the ice crystals and the liquid part are not of the same composition. In order to get a proper mixture, it is first necessary to warm the milk until the ice disappears, using a temperature not over 105° F., to avoid oiling off. Then mix well and test as for normal milk.

Accuracy of the Babcock Test. Many studies have been made comparing the results obtained by the Babcock test with those secured by ether extraction methods, such as the Roese-Gottlieb or Mojonnier tests. (See Part III of this chapter.) Operators of milk plants have not been able to account for some of the discrepancies between the amount of fat purchased and the amount sold. Hence the first step was to check the Babcock test with the more sensitive extraction methods.

Some of the Babcock test results obtained agreed closely, a few were below, but most of them were above the gravimetric method. Many of the earlier comparisons were made when the Babcock test bottles were calibrated in 0.2 percent divisions, thus allowing greater latitude in reading the tests. Also the earlier centrifuges were hand operated and unheated, which would have a tendency to give lower results. Another factor that caused at least a slight difference was the question of what constituted a pipetteful of milk. Babcock considered that the pipette was full when the milk touched the mark on the draw tube, while most operators consider a pipette full when the bottom of the meniscus is on the mark. At the present time most tests are run in steam or electrically powered machines, both being heated. A few references dealing with the more recent work will be given here to show what degree of accuracy may be expected of the Babcock test.

Fisher and Walts[8] tested 16 samples of milk in quadruplicates by the Roese-Gottlieb and by the Babcock procedures. The average percent secured by the former was 4.062 and by the latter 4.118 or 0.056 percent higher by the Babcock test.

Fahl, Lucas and Baten[9] compared the Babcock and Mojonnier tests on 513 samples of milk. The Babcock tests were made with the centrifuges at low, 60° F. to 68° F., medium 85° F. to 100° F. and high temperatures 135° F. to 150° F. Mojonnier tests averaged 3.675 percent, and the Babcock at low temperature 3.716, at medium 3.745 and at high 3.757 percent. Thus, the Babcock test ranged from 0.041 to 0.082 percent higher than the Mojonnier.

Hileman, Rush and Moss[10] have compiled data on 15 different sources, which includes their work comparing the Babcock and Mojonnier tests on fluid milk. The average test of 2029 samples by the Babcock test was 4.200 and by the Mojonnier method 4.124 percent. Thus the Babcock procedure was 0.076 percent higher. Compilations on cream testing were also made. The average test of 328 samples by the Babcock test was 35.991 percent and by the Mojonnier method 35.615 percent, a difference of 0.376 percent in the same direction. They conclude that the Babcock test gives a result that is too high in the case of both milk and cream.

Jenness et al.[11] studied the density of milk fat in its relation to the accuracy of the Babcock test. In graduating test bottles a specific gravity of milk fat of 0.9 has been generally used for the computation of the capacity of the neck of the test bottle. Therefore the 8 percent bottle has a volume between the 0 and 8 percent marks of 1.6 ml. (18 grams milk $\times$ 0.08 = 1.44 grs. fat, 1.44 grs. $\div$ 0.9 sp. gr. = 1.6 ml.). The authors found that the density of the fatty materials in the Babcock test was less than 0.9 at 140° F., the average being 0.8918 for milk and 0.8926 for cream. The purified fat was slightly lower than these figures. Calculations showed that this difference in density would increase the fat tests from less than 0.01 to slightly more than 0.05 percent. Hence the use of a 0.9 specific gravity of milk fat in calibrating test bottles introduces an error in the Babcock test which, while not large, can nevertheless be measured.

These citations would indicate that the Babcock test should be adjusted slightly so as to agree more closely with the ether extraction methods. Finer calibrated test bottles could be used, graduated at least to 0.05 in place of 0.1 percent, since with proper reading devices closer readings could be made than are read at the present time. Adjusting the charge in the milk test so that glymol could be used to remove the meniscus would tend to more uniform readings.

II. The Gerber Test

The Gerber test originated in Switzerland and is named for the man who developed it. It was introduced at about the same time as the Babcock test and is now as widely used throughout Europe and other parts of the world as the Babcock test is in this country. The Gerber test is included in this section along with the Babcock and Mojonnier methods since, although it is not officially recognized by the Association of Official Agricultural Chemists, it is now recognized as official[12] by the APHA for milk, cream, and chocolate milk (and as a screening test for homogenized milk and frozen desserts) and is a legal test for fat content in milk in New York, New Jersey, and California and for cream in New York and New Jersey.[13] It is an easy test to operate and has approximately the same accuracy in testing for fat in milk as the Babcock method.

Testing Whole Milk, Chocolate Milk, and Chocolate Drink. These products are tested for fat as follows: Measure 10 ml. of sulphuric acid, specific gravity 1.82 to 1.83, into the empty

Gerber Centrifuge —
24 Bottle, Open
Courtesy New Jersey Dairy Laboratories

Gerber Centrifuge —
24 Bottle, Closed
Courtesy New Jersey Dairy Laboratories

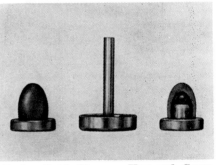

Gerber Lock Stopper, Key and Cross Section of Stopper.
Courtesy New Jersey Dairy Laboratories

Speed Indicator for Gerber Centrifuge. Meniscus drops as speed increases.
Courtesy New Jersey Dairy Laboratories

Gerber Centrifuge — 8 Bottle
Courtesy New Jersey Dairy Laboratories

test bottle, commonly called a "butyrometer." Add 11 ml. of properly mixed milk, letting it run down the side of the bottle,

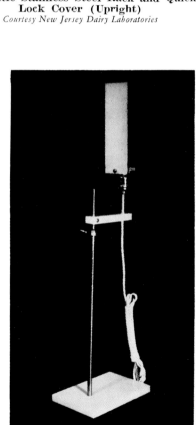

24 Bottle Stainless Steel Rack and Quick Lock Cover (Upright)
Courtesy New Jersey Dairy Laboratories

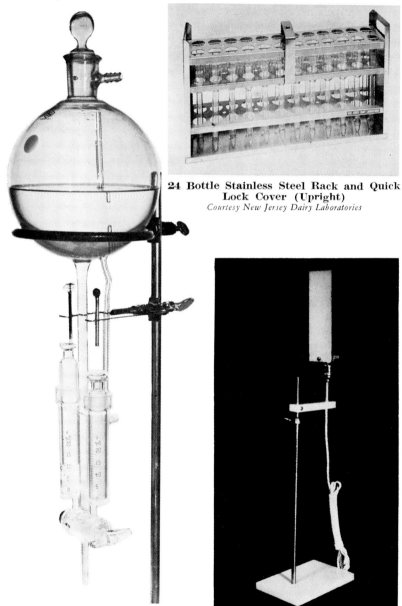

Automatic Feeder for Acid
Courtesy New Jersey Dairy Laboratories

Gerber Test Reader
Courtesy New Jersey Dairy Laboratories

so that it forms a layer on the top of the sulphuric acid without mixing with it. Then add 1 ml. of isoamyl alcohol, specific gravity 0.814 to 0.816. This alcohol is poisonous and so it is advisable to fill the amyl alcohol pipette by immersing it in the reagent rather than by suction. Next insert the rubber stopper and twist it firmly into place, so that it will not be forced out by the expansion during the shaking. Shake thoroughly until all the curd is dissolved, and then centrifuge at a speed of 1,100 revolutions per minute for four minutes. Stop the machine at the end of that time and put the bottles in a tempering bath at 140° F. for five minutes. Read the tests by the use of dividers or by adjusting the stopper until the lower end of the fat column is on the zero point and then note the percentage reading at the

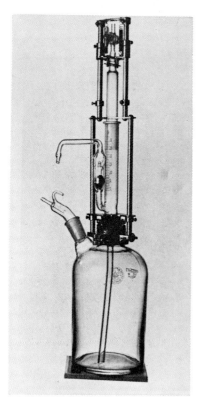

Automatic Syringe for Acid (Pump Type)
Courtesy New Jersey Dairy Laboratories

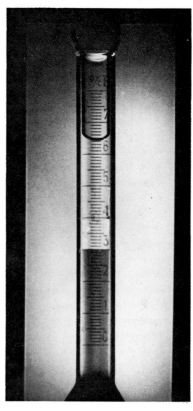

Fat Column in Milk Test Bottle
Courtesy New Jersey Dairy Laboratories

upper end of the column. Read at the very lowest point of the meniscus and not to the top as in the Babcock test. Care should be taken when adjusting the stopper, preparatory to

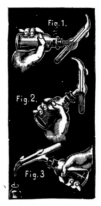

Automatic Acid Measure

reading the test, to guard against its sudden slipping, as it is being twisted into the tube. This results in splashing the fat far up into the neck and in clouding the lower fat line. When this cloudiness happens, centrifuge again for a minute or two and retemper before reading.

Where trunnion-type centrifuges are used, it is necessary to provide heaters which will keep the contents of the butyrometer at 140-145° F. This is unnecessary in the case of disk-type insulated centrifuges since the samples will hold at this temperature during the period of operation.[13]

When large amounts of chocolate flavoring are used in preparing chocolate milk or chocolate drink, it will be advisable to dilute the acid before use in the Gerber test. If this is necessary, slowly add 94 ml. of sulphuric acid having a specific gravity of 1.820-1.825 to 6 ml. of water. The testing procedure is the same as for milk except that a second centrifuging is recommended following the five-minute tempering period in the water bath.

Clean the test bottles in the usual way. These bottles may be easily cleaned, because all parts are readily accessible to a brush.

Testing Cream. The Gerber test for fat in cream is essentially the same as that for milk. Following the addition of 10 ml. of sulphuric acid, a 5-gram sample of cream is weighed directly into the special cream butyrometer which will give direct readings for samples containing as high as 60 or 70 percent fat. Then 5 ml. of distilled water are added, followed by 1 ml. of amyl alcohol. Proceed as in the test for fat in milk.[12]

When cream needs to be warmed as an aid in the proper mixing of the sample, it should be weighed into the butyrometer immediately after thorough mixing and then cooled to room temperature.[13] Water and acid are then added. The acid should be chilled if automatic acid measures result in a rapid addition of acid to the cream-water mixture. In place of adding water to the cream, one may use 17.5 ml. of sulphuric acid having a specific gravity of 1.520.[14] Others recommend combining 10 ml. of sulphuric acid with 10 ml. of water, then adding 15-20 ml. of the mixture to the cream in the butyrometer.[15]

Testing Skimmilk, Buttermilk and Whey. These products are tested in exactly the same manner as milk. Early methods recommended the use of special butyrometers with very small bore. Now, butyrometers with a wider graduated tube and having the same scale length as the milk tubes but with a range of only 0-4 percent are used. Results on these products are generally low so a correction factor should be applied.[14] Where the reading is less than 0.10 percent, 0.05 should be added to it; where the reading is 0.10-0.25, one should add 0.02; where readings exceed 0.25 percent, no correction is necessary. McDowall[14] shows that corrected Gerber tests give results only slightly below the ether extraction method. Greater accuracy is obtained when using Siegfeld tubes, wherein double quantities of product and reagents are used.

The directions for testing skimmilk and buttermilk are as follows: To the skimmilk test bottles add 20 ml. of sulphuric acid, 22 ml. of the milk and 2 ml. of amyl alcohol, insert the rubber stopper, shake thoroughly and then centrifuge at full speed for 3 minutes. Remove the bottles from the tester and warm the samples to 140° F., again centrifuge for 3 minutes and remove for tempering as before. This will be sufficient whirling for the skimmilk, but the buttermilk must be run a third time for 3 minutes. Temper and read in the same manner as given in the directions for testing whole milk.

Testing Frozen Desserts. The non-fat ingredients in frozen desserts make vigorous shaking necessary to liberate fat after the reagents are added. The H_2SO_4 may react with iso-amyl alcohol to increase its solubility in fat, yielding results slightly higher than those obtained with the Mojonnier. This test has considerable merit as a rapid screening test but may not be reliable if critical data are needed.

The test uses the dilute acid described for chocolate milk. Following the addition of 10 ml. of diluted H_2SO_4 to the Gerber ice cream test bottle, five grams of prepared sample are added to the bottle. Then, five ml. of water and 1 ml. of iso-amyl alcohol are added. The test is completed as for milk except that the bottles are centrifuged for five minutes. The test is read to the nearest 0.2%. Then, the bottles may be recentrifuged, retempered, and reread. If the first shaking were insufficient so that the second reading differed by more than 0.2%, the test should be repeated.

Table. 27A. Comparison of Mojonnier, Gerber and Babcock Fat Tests of Milk

Sample No.	Mojonnier	Gerber	Babcock	Variation from Mojonnier	
				Gerber	Babcock
	%	%	%	%	%
1	5.472	5.600	5.625	+0.128	+0.153
2	3.120	3.000	3.200	−0.120	+0.080
3	4.357	4.375	4.500	0.000	+0.143
4	4.010	4.087	4.087	+0.077	+0.077
5	3.152	3.187	3.200	+0.035	+0.048
6	4.270	4.375	4.475	+0.105	+0.205
7	4.650	4.562	4.575	−0.088	−0.075
8	4.742	4.900	4.712	+0.158	−0.030
9	3.280	3.325	3.350	+0.045	+0.070
10	3.957	3.875	3.862	−0.082	−0.095
11	3.097	3.300	3.325	+0.203	+0.228
12	4.660	4.637	4.662	−0.023	+0.002
13	3.440	3.500	3.487	+0.060	+0.047
14	4.602	4.587	4.587	+0.015	−0.015
15	4.152	4.162	4.137	+0.010	−0.015
16	4.022	4.087	4.112	+0.065	+0.090
Average	4.062	4.097	4.118	+0.035	+0.056

Comparing the Babcock and Gerber Tests. Fisher and Walts[8] made an extensive comparison of the results obtained in testing milk and cream by the Babcock and Gerber methods with the Mojonnier test (Roese-Gottlieb) (see Part III of this chapter). Table 27A gives the results obtained in testing 16 samples of milk in quadruplicate by the three methods. A number of authors have

Table 27B. Relative Efficiency of Babcock, Gerber Weight and Gerber Measure Methods When Compared to Mojonnier Test

Method	Percentage of samples checking with Mojonnier within							
	0.25%		0.50%		0.75%		1.00%	
	sweet	sour	sweet	sour	sweet	sour	sweet	sour
Babcock	63	71	94	88	96	100	100	100
Gerber weight	66	58	90	66	100	85	100	88
Gerber measure	8	39	54	52	66	83	88	66

shown that the pipette or butyrometer could be changed to lessen the discrepancy between the results of these tests and those of the Mojonnier. Aas[16] has pointed out that lowering the temperature of the water bath from 148° F. to 122° F. would accomplish the same thing and make for greater ease in handling the test bottles.

Four batches of cream representing low testing, medium testing, high testing and frozen cream were tested for butter fat in both sweet and sour condition by the Mojonnier, Babcock, Gerber weight and Gerber measure methods. The number of determinations was 48 by each test or a combined total of 288. Table 27B shows the relative efficiencies of the Babcock and Gerber methods as compared to the Mojonnier.

The Babcock and Gerber tests for milk and cream (gravimetric method) rank about the same from the standpoint of accuracy. The Gerber test gives highly satifactory results with homogenized milk or with buttermilk resulting from cream which has been standardized with homogenized milk.[14] The regular Babcock test does not give clear readings or accurate results on homogenized products, and even modified Babcock tests (see Chapter V) do not give complete satisfaction. The Gerber test can be used for either normal or homogenized cream and come within the smallest division on the butyrometer (0.5 percent) of results obtained with solvent extraction methods.[13]

McDowall[14] has listed several advantages of the Babcock test over the Gerber method. These include:

(1). The Babcock bottle can be handled without stands or stoppers. In draining, the Babcock bottles can be drained from a rack while the Gerber bottles must be held in the hand.

(2). Milk can be pipetted into the Babcock bottles and then permitted to stand until a convenient time for running. The Gerber bottles receive the acid first and thus must be run immediately after the addition of the milk.

(3). It is easier to weigh cream into the Babcock bottles than into the Gerber butyrometer.

(4). The Babcock test is adapted to reading results with dividers. Since the reader's hands are busy holding the bottle and adjusting the rubber stopper, this is not easy with the Gerber test.

(5). The Gerber test may be influenced by the quality of the amyl alcohol or by permitting the alcohol and acid to be in contact with each other too long before the milk is added.

On the other hand, the Gerber test requires only one period in the centrifuge while the Babcock bottles must be centrifuged three times. The reading of the top meniscus is difficult in the Babcock test while it is so clearly defined in the Gerber bottle that readings to the second decimal place can be made if desired. The Babcock test must be modified to determine the fat content of homogenized products while the Gerber test requires little change. McDowall[14] suggests the Gerber test is better for smaller numbers of samples while the Babcock test, because of the greater ease of handling Babcock bottles, is better adapted to routine laboratory analyses involving large numbers of milk samples.

III. Mojonnier Test for Fat

The Mojonnier test is based on the Roese-Gottlieb method, but it is much more rapid in operation because of the use of special apparatus. This equipment (Mojonnier tester) serves a dual purpose for making both fat and moisture (total solids) tests, which provision is not found in any other apparatus on the market.

Reagents and Their Functions. a. *Water.* Distilled water as nearly chemically pure as possible should be used. It is added

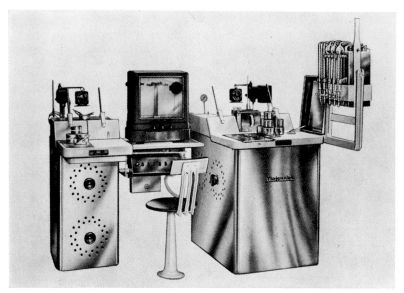
Model D Mojonnier Milk Tester with Balance and Weights
Courtesy Mojonnier Bros. Co.

to concentrated milk products in the flask in order to bring the substances to the fluid condition of whole milk, and to provide a liquid to carry the solids not fat in solution when they are dissolved by the other reagents. Sometimes it is also necessary to add a little water after centrifuging the second extraction in order to bring the dividing line between the ether fat solution and the solids not fat solution up to the desired point, which permits all of the ether fat solution to be poured from the flask without removing any of the other substances.

b. *Ammonia.* The ammonia should be chemically pure, have a specific gravity of 0.8974 at 60° F. and contain about 29.40 percent ammonia gas. Its use is to dissolve the casein which is not in true solution in milk, but is present in the form of minute gelatinous particles evenly distributed throughout the mass. It also neutralizes the acidity of the product. This reduces the viscosity of the mixture, and permits the solvent which is added later to more readily dissolve the fat. The ammonia would also probably tend to destroy colloidal phosphorous compounds, if any are present, and still further reduce the viscosity.

c. *Alcohol.* Ethyl alcohol of 95% strength and a specific gravity of 0.8164 is used. Its purpose is to assist in preventing

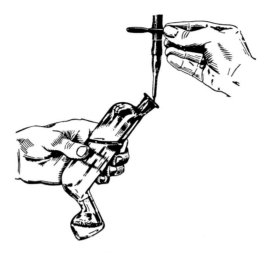

Adding Reagents
Courtesy Mojonnier Bros. Co.

the formation of the characteristic gelatinous mixture which occurs when ether is vigorously shaken with milk. It thus enables the solvent to come in contact with the fat globules during the

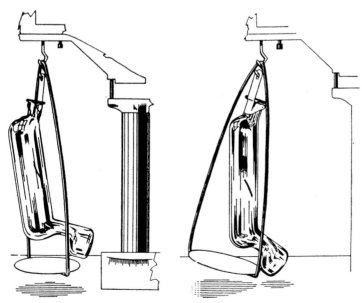

Right Way   Wrong Way
Flask Hanger with Flask Suspended to Balance Arm
Courtesy Mojonnier Bros. Co.

shaking operation, and also allows the ether fat solution to collect in a layer when all of the fat has been dissolved.

d. *Ethyl ether.* The ethyl ether used should be of best commercial quality, contain not more than four percent of water, have a specific gravity of 0.713 to 0.716 at 25° C. and a boiling point of about 35° C. Since it is both inflammable and explosive, it should be stored in a cool and safe place in glass or tinned steel containers. The purpose of ethyl ether is to dissolve the fat and hold it in its own solution.

e. *Petroleum ether.* The petroleum ether used should be of best commercial grade, specific gravity 0.638 to 0.660 at 25° C. and boiling point not over 120° F. to 140° F. It should be handled with the same care and in the same manner as ethyl ether. It serves the purpose of not only extracting fat but of assisting especially in throwing out from the ethyl ether-fat solution the last traces of water. This water holds milk solids not fat in solution and if any of the water is carried over with the ether-fat solution the other solids would be present with the fat when it is finally dried and weighed, thus causing results that would be too high. It throws out of the ethyl ether solution any solids not fat that may have been dissolved therein.

Testing for Fat

Milk, Skimmilk, Buttermilk and Whey. a. *First extraction.*

(1). Mix the sample thoroughly by pouring several times from one vessel to another.

(2). Pipette with a 10 gram pipette a 10 gram sample into an extraction flask, let pipette drain 15 seconds and then blow out the last drop.

(3). Add no water to these samples.

(4). Add 1.5 ml. of ammonia and mix thoroughly.

(5). Add 10 ml. of 95% alcohol, insert cork and shake for one-half minute.

(6). Add 25 ml. of ethyl ether, insert cork and shake vigorously for 20 seconds.

(7). Add 25 ml. of petroleum ether, insert cork and shake vigorously for 20 seconds.

(8). Centrifuge the flasks 30 turns, taking one-half minute.

(9). Pour off the ether mixture containing the extracted fat into previously weighed fat dishes. The empty fat dishes must receive the same treatment before being weighed as they receive before the final weighing, that is, heating in the vacuum

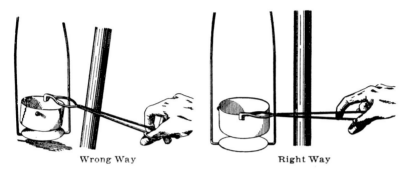

Method of Placing Dish Upon the Balance Pan
Courtesy Mojonnier Bros. Co.

oven at 135° C. for 5 minutes and then cooling in the desiccator for 7 minutes with the water circulating.

b. *Second extraction.*

(10). Add neither water nor ammonia.

(11). Add 5 ml. of alcohol to the residue in the flask and shake 20 seconds.

(12). Add 15 ml. of ethyl ether, insert cork and mix 20 seconds.

(13). Add 15 ml. of petroleum ether, insert cork and mix 20 seconds.

(14). Centrifuge 30 turns.

(15). Pour off the ether-fat solution into the same fat dish used in the first extraction. If necessary to raise the dividing line, add the proper amount of distilled water just before pour-

Correct Procedure When Pouring Ether-Fat Solution into Dish
Courtesy Mojonnier Bros. Co.

ing. Use care to not pour off any of the residue below the ether solutions.

(16). Evaporate off the ether from the fat dishes on the electric hot plate at 135° C. Keep cover over the dishes to draw off the ether fumes. Complete the evaporation in the vacuum

Fat Extraction Flask
Courtesy Mojonnier Bros. Co.

oven at 135° C. for 5 minutes with not less than 20 inches of vacuum.

(17). Cool for 7 minutes in desiccator with the water circulating during this period.

(18). Weigh rapidly, record results and calculate the percentage of fat.

Cream. The procedure for cream is as follows:

(1). Warm the sample to 70° F. and mix by pouring from one vessel to another.

(2). Weigh about 2 grams (to fourth decimal place) of the cream into an extraction flask. If the cream tests over 25 percent fat, about one gram is used.

(3). Add 5 ml. of distilled water to the sample in the extraction flask, 6 ml. if over 25 percent butterfat. Mix thoroughly.

(4). Complete as for milk by using 1.5 ml. of ammonia, 10 ml. of alcohol and 25 ml. each of the two ethers in the first extraction, and in the second, no water or ammonia, but adding 10 ml. of alcohol, and 15 ml. each of the two ethers. For cream testing over 25 percent fat use 25 ml. of each of the ethers in both extractions.

Ice Cream. Weigh about 5 grams of the well mixed sample into an extraction flask, add 5 ml. of distilled water, and complete the test as for cream testing over 25 percent fat. Ice cream which contains fruits or nuts should be mixed in a high-speed blender until the mix is homogeneous.

Evaporated Milk. Warm sample to 70° F. or up to 100° F. if necessary to obtain a homogeneous mixture. Mix thoroughly and weigh about 5 grams of the sample into an extraction flask. Add 4 ml. of distilled water, mix thoroughly and complete as for cream testing over 25 percent fat.

Sweetened Condensed Milk. Warm sample to 70° F. or up to 100° F. if necessary to bring the milk to a homogeneous mixture. Mix thoroughly and weigh about 5 grams into an extraction flask. Add 8 ml. of hot distilled water, mix and complete test as for cream testing over 25 percent fat, except mix alcohol and sample for one minute, and centrifuge 60 turns.

Whole Milk Powder and Skimmilk Powder. Mix sample and weigh about one gram into extraction flask. Add 8.5 ml. of hot water and mix thoroughly. Complete test as for fresh milk, except use 25 ml. of each of the ethers in the second extraction of the whole milk powder.

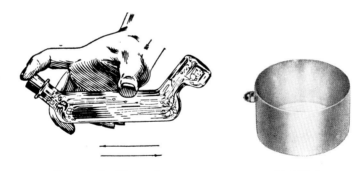

Correct Position of Flask When Shaking.   **Fat Dish**

Courtesy Mojonnier Bros. Co.

Butter and Cheese. Prepare the samples as given in Chapter III. Weigh about one gram into extraction flask by means of the butter boat, add 8 ml. of hot distilled water, 1.5 ml. of ammonia in case of the butter and 3.0 ml. in case of the cheese. Complete both tests as for cream testing over 25 percent fat.

Cottage Cheese and Creamed Cottage Cheese. Dump the sample in a high-speed blender and mix until homogeneous and liquid enough to pour. Then complete the test as directed for milk.

Table 28 — Summary of Methods for Making Fat and Total Solids Tests by Mojonnier Method.

	METHODS FOR MAKING FAT TESTS Keep sample in oven 5 min. at 135°C., and 7 min. in cooling desiccator												METHODS FOR MAKING TOTAL SOLIDS OR MOISTURE TESTS Summary of operations—Weigh directly into dish upon balance			
Product to be tested	How to prepare representative samples	How to weigh fat samples	Size of sample to take for fat test in grams	Reagents to add, and how to shake First extraction					How long to centrifuge	Reagents to add, and how to shake—Second Extraction Add neither water nor ammonia				Approximate size of sample to take for solids or moisture test in grams	Amount of water to add to sample in dish in cc	How long to keep sample in oven and cooling desiccator
				Water	Ammonia	Alcohol	Ethyl Ether	Petroleum Ether		Alcohol	Ethyl Ether	Pet'l'm Ether	How long to centrifuge			
Fresh Milk	Mix thoroughly	Measure with 10 gram pipette. Drain 15 seconds	10	No water	1.5 cc Mix thoroughly	10 cc Shake half minute	25 cc Shake 20 seconds	25 cc Shake 20 seconds	30 turns	5 cc Shake 20 seconds	15 cc Shake 20 seconds	15 cc Shake 20 seconds	30 turns	2	None	10 min. in oven at 100°C. 5 min. in cooling desiccator at room temp.
Skim Milk	"	"	10	"	"	"	"	"	"	"	"	"	"	2	"	"
Whey	"	"	10	"	1.5 cc if whey is acid mix thoroughly	"	"	"	"	"	"	"	"	2	"	"
Buttermilk	"	"	10	"	"	"	"	"	"	"	"	"	"	2	"	"
Cream testing less than 25% B. F.	Mix thoroughly Heat slightly if churned to melt fat	Weigh directly into flask. If necessary use butter boat	About 2	5 cc Mix thoroughly	"	"	"	"	"	"	"	"	"	1	1	"
Cream testing more than 25% B. F.	"	"	About 1	6 cc Mix thoroughly	"	"	"	"	"	"	25 cc Shake 20 seconds	25 cc Shake 20 seconds	"	1	1	"
Ice Cream Mix	"	Use weighing pipette or weigh directly into flask	About 5	5 cc Mix thoroughly	1.5 cc Mix thoroughly	"	"	"	"	"	"	"	"	1	1	"

Courtesy of Mojonnier Bros. Co.

Table 28 — Continued

Product	Preparation		About (g)													Notes
Evaporated Milk	Shake in can or mix in bulk very thoroughly	"	About 5	4 cc Mix thoroughly	"	"	*	*	"	*	"	"	"	1	1	*
Bulk Unsweetened Condensed Milk	Mix very thoroughly	"	About 5	6 cc Mix thoroughly	"	"	*	*	"	*	"	"	"	1	1	*
Bulk extra heavy Unsweetened Condensed Milk	"	"	About 3	6 cc Mix thoroughly	"	10 cc Shake one minute	*	*	"	*	"	"	"	0.5	2	*
Sweetened Condensed Milk	Proceed without diluting. Mix very thoroughly	"	About 5	8 cc hot water. Mix until thoroughly dissolved	"	"	*	*	60 turns	*	"	"	60 turns	0.25	2	90 min in oven at 100°C or 20 min and deduct 0.30% from total. 5 min in cooling desiccator at room temp.
Condensed Milk with Sugar and Chocolate	"	"	About 5	"	"	"	"	"	"	"	"	"	"	0.5	2	*
Condensed Butter Milk (semi-solid)	"	If fluid use weighing pipette otherwise use butter boat	About 3	6 cc hot water. Mix thoroughly	3 cc Mix thoroughly	10 cc Shake half minute	*	*	30 turns	*	15 cc Shake 20 seconds	15 cc Shake 20 seconds	30 turns	0.5	2	10 min in oven at 100°C 5 min in cooling desiccator at room temp.
Skimmed Milk Powder	Pulverize in close grained mortar. Transfer to sealed jar	Use butter boat	About 1	8 cc hot water. Mix thoroughly	1.5 cc Mix thoroughly	"	*	"	"	*	"	"	"	0.3	2	*
Whole Milk Powder	"	"	About 1	"	"	"	"	"	"	"	25 cc Shake 20 seconds	25 cc Shake 20 seconds	"	0.3	2	*
Butter Milk Powder	"	"	About 1	"	3 cc Mix thoroughly	"	*	"	"	*	15 cc Shake 20 seconds	15 cc Shake 20 seconds	"	0.3	2	*

Table 28 — Continued

Product to be tested	How to prepare representative samples	METHODS FOR MAKING FAT TESTS Keep sample in oven 5 min. at 135° C., and 7 min. in cooling desiccator		Reagents to add, and how to shake First extraction					How long to centrifuge	Reagents to add, and how to shake—Second Extraction. Add neither water nor ammonia			How long to centrifuge	METHODS FOR MAKING TOTAL SOLIDS OR MOISTURE TESTS Summary of Operations—Weigh directly into dish upon balance		
		How to weigh fat samples	Size of sample to take for fat test in grams	Water	Ammonia	Alcohol	Ethyl Ether	Petroleum Ether		Alcohol	Ethyl Ether	Pet'l'm Ether		Approximate size of sample to take for solids or moisture test in grams	Amount of water to add to sample in dish in cc	How long to keep sample in oven and cooling desiccator.
Malted Milk	"	Use butter boat	About 0.5	8 cc	1.5 cc Mix thoroughly	10 cc	25 cc	25 cc	"	5 cc	25 cc Shake 20 seconds	25 cc Shake 20 seconds	30 turns	0.3	2	20 min. in oven at 100° C. 5 min. in cooling desiccator at room temp.
Cocoa	"	"	About 0.5	"	"	"	"	"	"	"	"	"	"	0.3	2	"
Milk Chocolate	"	"	About 0.5	"	"	"	"	"	"	"	"	"	"	0.3	2	"
Cheese	"	"	About 1	"	3 cc Mix thoroughly	10 cc Shake one minute	"	"	"	"	"	"	"	0.5	2	"
Butter	"	"	About 1	"	1.5 cc Mix thoroughly	10 cc Shake half minute	"	"	"	"	"	"	"	1	None	10 min. in oven at 100° C. 5 min. in cooling desiccator at room temp.

BIBLIOGRAPHY

1. Herreid, E. O. The Babcock Test; A Review of the Literature. J. D. Sci. 25, pp. 335-370 (1942).
2. Methods of Analysis, A.O.A.C. Published by The Assn. of Off. Agr. Chemists, Washington, D. C. Eighth Ed. (1955).
3. Heineman, B., et al. Procedures for Sampling and Testing Milk by the Babcock Method. Subcommittee Report ADSA. J. Milk and Food Tech. 17:120-121, 125 (1954).
4. Herreid, E. O., et al. Standardizing the Babcock Test for Milk by Increasing the Volume of the Sample and Eliminating the Meniscus on the Fat Column. Subcommittee Report ADSA. J. D. Sci. 33,10:685-691 (1950).
5. Farrington and Woll. Testing Milk and Its Products. p. 57, 23rd Revised Ed. Mendota Book Co. Wis. (1916).
6. Herrington, B. L. A Note Regarding the Speeds of Babcock Centrifuges. J. D. Sci. 27, p. 857 (1944).
7. Milk Industry Foundation. Laboratory Manual. Published by the Milk Ind. Found., Washington (1959).
8. Fisher, R. C. and Walts, C. C. A Comparative Study of Methods for Determining the Percent of Fat in Dairy Products. Storrs Bul. 131 (1925).
9. Fahl, J. R., Lucas, P. S. and Baten, W. D. Factors Involved in Accuracy of Testing Milk Samples. Mich. Agr. Exp. Sta. Tech. Bul. 158 (1938).
10. Hileman, J. L., Rush, K. K. and Moss, C. The Relationship of Errors in the Babcock Test to Losses in Cream Plants. J. D. Sci. 25, pp. 373-379 (1942).
11. Jenness, R., Herreid, E. O., Caulfield, W. J., Burgwald, L. H., Jack, E. L. and Tuckey, S. L. The Density of Milk Fat: Its Relation to the Accuracy of the Babcock Test. J. D. Sci. 25, pp. 949-960 (1942).
12. American Public Health Assn. Standard Methods for the Examination of Dairy Products. Eleventh Ed. (1960).
13. Levowitz, David. The Gerber Test for Milk and Cream. Talk presented at Lab. Section, 85th Ann. Meeting, APHA, Cleveland, Ohio (1957).
14. McDowall, F. H. The Butter Maker's Manual. Vol. 2. The New Zealand University Press, Wellington, N. Z. (1953).
15. Davis, J. G. and MacDonald, F. J. Richmonds Dairy Chemistry. Charles Griffin and Company, Ltd., London (1953).
16. Aas, G. Fat Determination in Milk by Gerber's Method. D. Sci. Abst. 16,11:936 (1954).

REVIEW QUESTIONS

1. When was the Babcock test given to the dairy industry?
2. How many reagents are used in the Babcock test? In the Gerber test?
3. Show how to calculate the volume capacity of an 8 gram milk test bottle between the zero and 8 percent graduations.
4. What advantages has sulphuric acid over other acids for use in the Babcock test?
5. What is meant by a meniscus?
6. What is the cause of curdy tests? Charred tests?
7. What happens if a bottle of sulphuric acid is left unstoppered?
8. What is the cause of foam in a fat column, and how can it be prevented?
9. How does the diameter of the wheel of a centrifuge affect the speed at which it must be revolved?
10. What is the white precipitate in the bottom of the bottles of the completed tests?
11. Why is cream weighed instead of measured when testing it for fat?
12. If cream were measured with an 8.8 ml. pipette instead of weighing out 9 grams, which would give the nearer to the correct reading, a 20% or a 40% cream? Why?
13. Does the Gerber method of testing milk products for fat have any advantages over the Babcock test?
14. What five reagents are used in the Mojonnier test?
15. Why are two kinds of ethers used?

CHAPTER V

Some Other Tests for Fat

SOME OTHER TESTS FOR FAT

The presence of added solids, sugar and chocolate or the process of homogenization causes excessive errors when the regular Babcock method is used to determine the fat content of products having these factors present. Through the years, modifications of the basic Babcock procedure have been suggested in order to obtain readings closely approximating results obtained by the Roese-Gottlieb (Mojonnier) method. Though some of these tests are commonly used in the determination of fat in homogenized products or products which have a tendency to promote charring of the fat column, none have been widely used in the testing for fat in milk or cream.

I. Modified Babcock Tests (Acid Reagents)

A. American Association Test

This test was developed by the American Association of Creamery Butter Manufacturers about 40 years ago to permit a more accurate determination of fat in buttermilk. The Association test employs n-butyl alcohol in addition to the sulfuric acid. In this country, the procedure calls for the addition of 2 ml. of alcohol but New Zealand and British studies[1] indicate greater uniformity of results are obtained if 4 ml. are added. Since part of the alcohol is absorbed in the fat column, this test gives higher readings than the regular Babcock test and is more in line with the ether extraction method, as in the Mojonnier test. However, the ether not only extracts the fat but also includes the phospholipids and consequently both the American Association and the Mojonnier methods give results slightly higher than the true fat content. This is particularly true of buttermilk, since a great share of the phospholipids in the cream, when churned to butter, goes into the buttermilk. A subcommittee report of the American Dairy Science Association[2] stated that the American Association test gave the most satisfactory results for determining the fat content of buttermilk and skimmilk and recommended the adoption of this test for these products.

Testing Buttermilk, Whey and Skimmilk.

In the American Association test two chemicals are used, namely, n-butyl alcohol and sulfuric acid (sp. gr. 1.82-1.83). Directions for testing these products are as follows[2]:

Equipment

Regular Babcock testing equipment and glassware, including 0.5% and 0.25% skimmilk test bottles

Reagents

N-butyl alcohol and sulfuric acid, sp. gr. 1.82 to 1.83

Procedure:

(1). Transfer to the test bottle 2 ml. n-butyl alcohol from a burette.

(2). Add 9 ml. buttermilk, skimmilk or whey and mix well.

(3). Add 7 to 9 ml. sulfuric acid and mix well.

(4). Centrifuge six minutes and add water (140° F.) to the base of the bottle neck.

(5). Centrifuge two minutes and add water to bring the fat into the neck of the bottle.

(6). Centrifuge for two minutes and place the bottles in a water bath at 135° to 140° F. for 5 minutes.

(7). Multiply the test by 2.

B. The Pennsylvania Test

The Pennsylvania method is particularly valuable as a rapid test for fat in dairy products containing sugar or chocolate. The fat content of other dairy products can also be tested by this method although some results[2,3] indicate this test should not be used for homogenized milk, skimmilk, or buttermilk. The Pennsylvania test uses the same equipment as the Babcock procedures for milk or cream but the reagents are changed to provide a clear fat column when sugar or chocolate is present. The test is particularly useful in testing ice cream, chocolate milk, or sweetened condensed milk.

1. Testing Ice Cream

The procedure for testing ice cream by this method is as follows[3a]:

(1). Melt ice cream at room temperature, and if necessary heat to eliminate foam. Warm to about 70° F.

(2). Reduce any large particles in fruit and nut ice cream to a finely divided state with a high speed blender or force through a fine screen.

(3). Weigh 9 grams of the representative sample into a 9-gram, 20 percent ice cream test bottle or an 8 percent milk test bottle. Precautions should be taken to keep the fruit or nut ice cream thoroughly mixed in order to get a representative sample.

(4). Add 2 ml. of 28-29 percent NH_4OH and mix.

(5). Add 3 ml. of n-butyl alcohol and mix thoroughly.

(6). Add 17.5 ml. of diluted H_2SO_4 (sp. gr. 1.72-1.74) and mix. This acid is prepared by slowly adding 3½ parts of Babcock H_2SO_4 by volume to 1 part H_2O. Always add the acid to the water. Since this reaction will generate considerable heat, caution must be used to prevent burning or the breaking of the container.

(7). Centrifuge 5 minutes, add water at 130°-140° F. to the base of the neck, centrifuge for 2 minutes, add water as before, then centrifuge 1 minute. **Do not mix the contents at this stage since this will lower the fat reading.**[4]

(8). Remove from centrifuge and temper for 3 minutes in a water bath controlled at 130° F., add glymol and read.

(9). If the whole milk test bottle is used, multiply reading by 2.

2. Testing Other Dairy Products

The procedures are as follows[4]:

(a) Whole milk and whey. Measure 17.5 ml. of properly prepared sample into a milk test bottle and proceed as above.

(b) Cream. Weigh 9 grams of properly prepared sample into a 9-gram, 50 percent cream test bottle and proceed as above.

(c) Chocolate milk. Weigh 18 grams of prepared product into a milk test bottle and proceed as above.

(d) Sweetened condensed milk. Dilute with an equal amount

of water, thoroughly mix, then weigh 9 grams of the diluted product into a milk test bottle (multiply reading by 4) or into a 9-gram, 20 percent ice cream test bottle (multiply the reading by 2) and proceed as above.

(e) Evaporated milk. Weigh either 4½ or 6 grams of thoroughly mixed evaporated milk into a milk test bottle and proceed as above. Adjust the reading according to the amount of sample used.

(f) Dried whole milk. Add 140 grams of water (125° F.) to 20 grams of sample, shake until thoroughly mixed, cool to 70° F., then weigh 18 grams into a milk test bottle and complete the test as above. Multiply the reading by 8.

(g) Butter. Prepare sample as in Chapter III. Weigh 4.5 grams of thoroughly prepared sample into a 9 gram, 50 percent test bottle, heating the neck of the bottle to bring the butter down into the bulb of the bottle. Complete test as above and multiply result by 2.

(h) Cheese. Prepare sample as in Chapter III. Weigh 4.5 grams of prepared sample into a 9 gram, 50 percent cream test bottle. Add 8-10 ml. of water and heat in a water bath at 150°-180° F. to facilitate dispersion of sample. Proceed as above and multiply reading by 2.

C. Acetic Acid and Sulfuric Acid Method for Ice Cream Mix

This test is used as a factory procedure for ice cream. It is recommended for use on ice cream mix only since melted ice cream samples yield low results.[4] It should not be used for chocolate ice cream mixes. The test is run as follows[4]:

(1). Weigh 9 grams of ice cream mix (70° F.) into a milk test bottle.

(2). Add 13 ml. of glacial acetic acid and mix thoroughly.

(3). Add 9 ml. H_2SO_4 and mix thoroughly.

(4). Place bottle in a water bath at 170° F. for 5 minutes.

(5). Remove from water bath, shake thoroughly, and place in heated centrifuge. Spin for 5 minutes in the centrifuge. (Be sure centrifuge is properly balanced.)

(6). Add water (180° F.) to bring level to lower end of neck.

(7). Spin for 2 minutes, then add water to bring fat into calibrated portion of the neck.

(8). Spin for 1 minute, place in hot water bath (135°-140° F.) for 3 minutes.

(9). Add a few drops of glymol and read test.

(10). Multiply results by 2.

Samples should be retested if they do not agree within 0.1 percent in duplicate samples, if the fat column is indistinct or milky, or if it contains specks or charred particles.[4]

II. Modified Babcock Tests — Alkaline Reagents

A. Minnesota Test for Ice Cream

Since the fat column formed in the American Association test includes slight amounts of the n-butyl alcohol used in the determination, the Minnesota test was devised to recover only the true fat. Consequently its results are lower than the official methods. McDowall[1] gives the results of a comparison of 12 buttermilks, as follows:

Roese-Gottlieb (Mojonnier)	0.745 percent
Minnesota (Original Solution)	0.512 percent
Minnesota (Later Commercial Reagent)	0.198 percent
American Association Test	0.739 percent
Regular Babcock	0.233 percent

The subcommittee report of the American Dairy Science Association[2] recommends this procedure for whey but not for buttermilk or skimmilk. Others[4] have suggested its use for ice cream or mix other than chocolate.

The reagents in the Minnesota test consist of 750 grams of sodium salicylate, 390 grams potassium carbonate, 190 grams of sodium hydroxide. These are completely dissolved in 3 quarts of water and then made up to 3 liters. The mix is available in moisture-free sealed containers ready to mix with water. When the mixture is completely dissolved, 1 liter of isopropyl alcohol is added.[4] The mixture is then ready for use and may be kept for some time in a properly stoppered glass bottle. It is known as the "Minnesota-Babcock Reagent."

The Minnesota procedure for ice cream is as follows:

(1). Weigh 9 grams of melted ice cream into a cream test bottle or a 20 percent ice cream bottle.

(2). Add 15 ml. of Minnesota reagent and mix thoroughly.

(3). Digest in gently boiling water for 12 to 15 minutes. Shake the bottles during this period, once when at least one-half of the mixture has turned dark brown and again 1 minute later. Use care when shaking the bottles to prevent any spurting of the mixture.

(4). When the solution becomes clear and the fat is seen floating on the surface, centrifuge ½ minute.

(5). Add hot water and bring the fat within the graduations, then centrifuge ½ minute.

(6). Temper at 130°-140° F. for 3 minutes, add glymol and read.

B. Nebraska Test for Ice Cream

Two reagents designated for convenience as A and B are used in this test. Reagent A is a mixture of n-butyl alcohol and concentrated ammonium hydroxide made up of 9 parts of n-butyl alcohol and 1 part of C.P. ammonium hydroxide by volume. This reagent is stable when kept in a tightly stoppered bottle. Reagent B is a mixture of equal parts by volume of sulfuric acid (sp. gr. 1.82 to 1.83) and ethyl alcohol (95 percent). This mixture is prepared by pouring the acid slowly into the alcohol in a glass beaker or other glass container **that will withstand a high temperature.** The mixture will become extremely hot and will boil, but no trouble need be experienced in preparing it, if the acid is poured carefully down the side of the beaker where it will run under the alcohol. After all the acid has been added, or during its addition, the mixture should be stirred with a glass rod and finally cooled to room temperature before using. Keep in tightly stoppered bottle.

The procedure in testing for fat is as follows:

(1). Weigh 9 grams of the melted ice cream into an 18 gram milk test bottle.

(2). Add 5 ml. of Reagent A and mix contents of the bottle thoroughly.

(3). Add 30 ml. of Reagent B, temperature around 70° F. When this quantity of Reagent B causes the contents in the bulb of the bottle to rise to the neck, thus making mixing difficult, lesser amounts (28-29 ml.) do not affect the accuracy of the test.[4]

(4). Mix thoroughly until all the curd is dissolved. The contents of the bottle should become clear and free from curd but will not become dark until heated in the water bath.

(5). Place the tests in a water bath at 175°-180° F. for 15 minutes. Shake contents of the bottles at least three times during this heating period.

(6). Centrifuge at regular speed for 5 minutes. Add hot water at 180° F. up to base of neck. Centrifuge 3 minutes and if any curd is apparent, shake the contents of the bottle thoroughly. Add hot water at 180° F. until fat column is within graduations. Centrifuge 1 minute.

(7). Temper at 135° F. to 140° F. for 5 minutes, add glymol and read as in cream test. Multiply readings by 2.

C. Garrett-Overman Ice Cream Test

In this test alkaline reagents are used in place of acid. Two reagents are employed designated as A and B.

Reagent A consists of:
- 75 ml. of C.P. ammonium hydroxide
- 35 ml. of n-butyl alcohol
- 15 ml. of 95 percent ethyl alcohol or denatured grain alcohol

Reagent B consists of:
- 200 grams of trisodium phosphate
- 150 grams of sodium acetate
- 1 liter of water

Trisodium phosphate and sodium acetate of commercial grade are satisfactory. Clean tap water may be used instead of distilled water. The reagent remains in solution at ordinary temperatures but will partially crystallize on becoming cold. If crystallization takes place, the crystals should be dissolved by warming before the reagent is used. The reagent does not seem to deteriorate on long standing.

The procedure in testing the ice cream is as follows:

(1). Weigh 9 grams of the well mixed sample of ice cream into a Babcock milk test bottle (8 or 10 percent) or a 20 percent ice cream test bottle.

(2). Add exactly 2.5 ml. of Reagent A from a burette or pipette. Mix thoroughly.

(3). Add 9 or 10 ml. of Reagent B and again mix thoroughly. The reagent may be measured in an ordinary 9 ml. acid measure or in a 10 ml. pipette.

(4). Place the test bottle in a shallow water bath and heat

the bath to boiling, continuing the heating for several minutes. Shake the contents of the bottle two or three times while heating.

(5). Usually at the end of 15 to 30 minutes the fat will separate and form a clear yellow layer on top of the liquid. The heating must not cease until the fat layer has definitely separated from the dark portion of the liquid and has become clear.

(6). After all the fat has separated, place the test bottle in the centrifuge and whirl 5-2-1 minutes, adding hot water as in the regular Babcock milk test except that the water must not be softened with acid.

(7). Place the bottle in a hot water bath at 130° F. to 140° F. for five minutes.

(8). Read the test, measuring from the bottom of the fat column to the top of the upper meniscus.

(9). If an 8 or 10 percent milk test bottle is used multiply the reading by 2.

III. The Detergent Tests

In 1949, Dr. Philip Schain[5] introduced a new principle in testing for fat in milk. The new test required neither sulfuric acid nor a centrifuge. Instead, separation of the fat from the other constituents was accomplished through the use of two solutions of surface active agents and a water bath. The first solution was used to separate the fat from the other milk constituents and the other to clarify the non-fat milk solids which enabled the separated fat to rise into the neck of the bottle for measurement. In 1950, the originator proposed a simplified method[6] which used no acid, no water bath, no centrifuge, and only one test reagent. Gershenfeld and his coworkers[7,8] modified the test for use with other dairy products. Several early studies, including work by Schain, pointed out that results in this test were too high for milk with high fat content and too low in milk of low fat content. Schain proposed[6] the use of a nomograph, based on the relationship

$$\frac{\text{Measured reading} \times 5 + 3.7}{6} = \text{Percent butterfat.}$$

Sager et al., in 1951, presented a critique[9] of the Schain test and observed that the Schain readings showed no definite pattern of variation from the true fat content. Thus, the use of a nomograph, as proposed by the originator of the test, was questioned. They also showed that the method had "some factor of variability leading to non-reproducibility of results." These workers followed this report with another paper the following year[10] in which a new detergent test, the BDI (Bureau of Dairy Industry) test, was described. Additional reagents were employed and a return to the use of a centrifuge was recommended for greater accuracy in testing. This was followed, in 1955, by a slightly modified version[11] known as the DPS (Dairy Products Section) detergent test which was claimed to improve the accuracy and appearance of the test, thus increasing its usefulness. This report indicated the results of fat analyses in milk and ice cream, using the new method, were within the limits of experimental error when compared with the Babcock and Roese-Gottlieb tests. The DPS test for cream gave readings higher than either Babcock or Roese-Gottlieb results but when the test bottles were silicone-treated, the results were in close agreement with Babcock readings and only slightly higher than Roese-Gottlieb values. It was still found desirable to use a centrifuge in this new method.

A later study[11a] showed that the Babcock test had a closer correlation with the Mojonnier test than either the Schain test or the DPS test. This work also showed the greatest difference between tests occurred with the DPS test.

A report on the development and evaluation of a new detergent procedure[12] was presented at the 1958 meeting of the Laboratory Section of the Milk Industry Foundation. This method is known as the TeSa Test. The originators of the test note that it is a chemical rather than a detergent method but that the TeSa Reagent, essentially a mild alkaline solution, contains certain dispersal agents for its mechanical action. The method has since been approved by the Association of Agricultural Chemists and the Dairy Herd Improvement Association for use in determining the fat content of fresh raw milk.[13] However, the originators believe the test is accurate for most dairy products. The TeSa Reagent is composed of a protein solubilizing agent, two dispersing agents, plus supplemental alkaline, buffering and agitant agents. The reagent mix is supplied as a powder which is mixed with water prior to use. The test does not require the use of a centrifuge and it is reported that the test is sufficiently simple to enable an operator to handle as many as 72 tests per hour even though the test cycle of individual samples is 17 minutes.

A. The DPS Detergent Method for Fat Determination[11]

1. Reagents

(a) Detergent reagent. Dissolve 7.0 g sodium tetraphosphate, 2.0 g $NaHCO_3$, and 3.0 g of either Triton X-100* or Tergitol Dispersant NPX* by stirring, and dilute to 100 ml. with distilled water. (Turbidity caused by heat of solution disappears on cooling to room temperature.) If the solution has been stored for more than two months at room temperature or one year at 10° C., prepare fresh solution.

(b) Methyl alcohol. A.C.S., 50% by volume.

*Triton X-100 (Rohm & Haas Co.) and Tergitol Dispersant NPX (Carbide & Carbon Chemicals Co.) were found suitable and were used in these experiments. Many other detergents were investigated and were found unsuitable. There may be other suitable detergents that were not tried.

2. Procedure for Milk

The test bottle and milk pipette are the same as specified for the official Babcock test, Official Methods of Analysis, 8th Ed., pp. 248-249. Likewise, the centrifuge, calipers, and water bath for "tempering" the test are the same. The sample is prepared as directed on p. 242 of Official Methods of Analysis, 8th Ed., which specifies adjusting the temperature of the milk to 20° C. before sampling. The test is conducted as follows:

Part A. With 17.6 ml. pipette transfer 18.0 g prepared sample to milk test bottle. Blow out milk in pipette tip after free outflow has ceased. Add 5 ml. reagent (a) portion-wise so as to wash all traces of milk into bulb. Shake to mix. Transfer bottle to boiling water bath, with level of water covering level of milk in bottle. (For comfort, a low bath is suggested, as a bath with high sides may direct steam from boiling water onto hands.) After bottle has been in bath about 2 min., shake to remix raised cream and replace in bath for an additional 5 min. Shake to remix. Replace in bath for 5 more min. Shake to remix. Allow to remain in bath for a final 5 min.; then remove from bath without remixing contents. (Exceeding the specified heating time and exact timing of the shaking are not critical except that the bottles should not be heated longer than one-half hour, and they should remain in boiling water for at least 5 min. after the last shaking.)

Part B. While bottle is still hot add 50% methyl alcohol to

bring level of contents to top of graduated scale, allowing alcohol to flow down side of neck. If several tests are being conducted, remove one bottle at a time from boiling water, completing addition of alcohol solution before removing next bottle. Transfer bottle, while still hot, to unheated centrifuge and whirl 2 min. Transfer bottle to water bath at 55-60° C., immerse it to level of top of fat column, and leave until column attains temperature of bath (about 5 min.). Remove bottle from bath, wipe it, and with dividers or calipers measure fat column, in terms of % by weight, from lower surface to highest point of upper meniscus.

3. Procedure for Cream

(a) Procedure with standard Babcock bottles. Reagents (a) and (b) are the same as for milk. The test bottle and other apparatus, preparation of sample, and weighing of sample are the same as in the official Babcock cream test, Official Methods of Analysis, 8th Ed., 15.52 and 15.62-15.63.

Weigh a 9.0 g sample into cream test bottle. Add 10 ml. reagent (a) portion-wise so as to wash all traces of cream into bulb, shake to mix, and proceed as in the detergent method for milk, Part A, beginning "Transfer bottle to boiling water bath." In testing samples containing 40-50% fat, add 15 ml. reagent (a). In testing samples whose approximate compositions are unknown, add 10 ml. reagent (a) at beginning as usual, and then if sharp separation of a layer of some butter oil has not occurred after heating 7-10 min., add another 5 ml. and mix. Continue in the usual manner, completing heating in boiling water for another 10 min. and remixing sample after the first 5 min. Measure fat column bottom to bottom.* Subtract 0.75% from the observed butterfat reading to obtain results, in % by weight, to conform more closely with Babcock test results.

(b) Procedure with silicone-treated bottles. Proceed as above, using Babcock bottles treated with silicone (General Electric Dri-Film, 9987, or equivalent) by saturating cloth with compound and rubbing interior of the neck of the clean, dry bottle with the saturated cloth until oily film is formed. Then rub with dry cloth until surface is clear. Avoid contaminating bulb of bottle with silicone, since longer centrifuging (5 min.) would be required to obtain all the fat. Avoid contact with compound. The silicone film is stable toward neutral and acid cleaners but is removed by strong alkalies.

No correction factor is required when silicone-treated bottles are used.

4. Procedure for Ice Cream

Reagents (a) and (b) are the same as for milk. Melt sample at room temperature or in a water bath at 40-50° C. and mix thoroughly with a rod or spoon. If sample contains insoluble particles, they may be dispersed before weighing by treatment with a high-speed stirrer with cutting blades, e.g., a Waring blender; alternatively, melted sample may be strained through a fine-mesh wire gauze and a correction calculated for weight of material removed. Weigh 9.0 g into ice cream test bottle and proceed as in the detergent method for milk, beginning under Part A, "Add 5 ml. of reagent (a)." Measure fat column, in terms of % of weight, bottom to bottom.*

*Glymol or red reader may be used to flatten the meniscus. Results with reader are usually slightly higher than readings bottom to bottom.

B. TeSa Reagent and Usage for Fat Determination*

1. Preparation of TeSa Reagent

To prepare TeSa Reagent (Pat. pending) Solution, dissolve packaged contents in water and make a total volume of 500 ml. for 78 gm size or 2000 ml. for 312 gm size. The powder should be added slowly to about two-thirds the total water required to avoid loss due to foaming. After reaction has subsided, add remainder of water. Mix and let stand at least six hours to insure complete reaction and solution.

2. Preparation of Samples

Immediately before withdrawing samples for determinations, mix by shaking, pouring, or stirring until uniformly mixed. If sample is very thick, warm to 86°-95° F. and mix. In case lumps of butter have separated, heat sample to 122° F. by placing in warm water bath. Thoroughly mix portions before testing. Avoid overheating sample, thereby causing fat to "oil off." Cool to 68° F. before pipetting.

3. Procedure for Milk
(Raw, Pasteurized, Homogenized, or Composite)

(1). To the 500 ml. of TeSa Reagent solution, prepared, add 100 ml. methanol (commercial grade full strength wood alcohol), if using 2000 ml. add 400 ml. methanol.

*Instructions for Butterfat Testing with TeSa Reagent by Technical Industries, Inc., Fort Lauderdale, Fla.

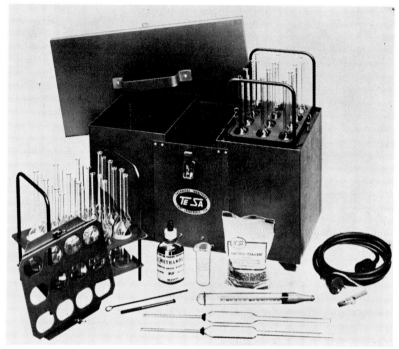

TeSa Apparatus
Courtesy Technical Industries, Inc.

(2). Pipette 17.6 ml. of milk sample into side arm of TeSa bottle.

(3). Add 15 ml. of TeSa Reagent-methanol solution through side arm of TeSa bottle and mix at once with swirling motion.

(4). Place in boiling water bath deep enough to cover the liquid level in the bottles. Allow bottles to remain 10 to 12 minutes after boiling resumes.

(5). Remove and fill side arm of TeSa bottle to within ¼" of top with water from boiling bath. Let stand for 5 to 7 minutes at room temperature.

(6). Transfer to tempering bath 135° to 140° deep enough to cover the liquid level in the neck of the bottle.

(7). After 2 to 3 minutes tempering, remove from bath and read fat percentage indicated in center of column from bottom to upper meniscus to bottom of lower meniscus. This may be done by dividers or by adding water slowly to bring fat column to zero, or by lowering rod in side arm to raise fat column to zero for direct reading.

4. Procedure for Chocolate Milk and Low Fat Content Frozen Desserts

Proceed exactly as directed for milk except using 20 ml. of TeSa Reagent without methanol and boil for 15 minutes with frequent swirling (8 to 10 times) during the first 12 minutes. Either 9 or 18 grams of the sample may be used, if 9 grams used, add 9 ml. of water before adding Reagent, then multiply reading by 2. Centrifuge 30 seconds before tempering and reading.

5. Procedure for Cream, Ice Cream or High Fat Frozen Desserts

Proceed exactly as directed for milk except using 9 grams of sample into special TeSa Cream bottle, add 9 ml. of water washing all sample into bottle. Add 20 ml. TeSa Reagent without methanol and boil for 15 minutes with frequent swirling (8 to 10 times) during the first 12 minutes. Add hot water through side arm of TeSa Cream bottle to bring top of fat just into bottom of neck and allow to remain in boiling water bath for remaining three minutes without further swirling or agitation. Centrifuge 30 seconds before tempering and reading. Add reading oil as required.

General Precautions

To insure best results under all conditions, make TeSa Reagent solution fresh each two weeks and keep solution from extreme heat, cold or direct sunlight.

Do not add methanol to liquid reagent until the day of usage. While the liquid reagent-water mix is stable for at least two weeks, the reagent-methanol mixture should not be used after the third day.

It is important that the 10 minute time in boiling bath be at full temperature. Do not include time while water is not at vigorous boil.

In all cases using a TeSa Milk or Cream Test bottle, the fat column should be read in terms of percent by weight, from bottom of upper meniscus to bottom of lower meniscus.

If lower fat surface is not clear enough to read well (sometimes observed in frozen desserts containing gums) add three drops of 40% methanol through fat column, allow to settle, retemper and read as directed.

Sepascope
Courtesy Whitman Laboratories

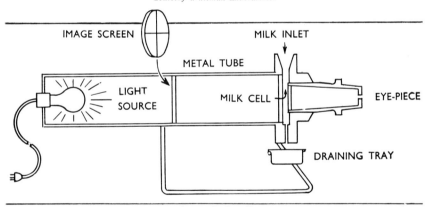

Construction of the Sepascope
Courtesy Whitman Laboratories

IV. Optical Methods

Ashworth[14] observed that one can estimate the fat content in milk if its turbidity is known and that this ability is of practical significance in determining the fat content of skimmilk and other low fat products where the use of the ordinary Babcock method is limited. The turbidity values could also be used to determine homogenization efficiency in homogenized milk.

Since that time several optical methods have been proposed for control of fat in milk. Deackoff and Rees[15] applied the basic

Ashworth technique to evaluate the performance of homogenizing valves, especially the newer high efficiency valves. They observed that wave lengths near 1000 millimicrons gave a high degree of resolution and duplicability. Commercial skimmed milk gives 100 percent transmission when diluted with distilled water and referred to this distilled water. They noted that, with the spectrophotometer method, they were able to reach an apparent maximum of 87 percent transmission. Since other methods were ineffective above 70-75 percent transmission, this gave them considerably greater precision in evaluating valve design.

Interfering turbidity caused by casein particles was eliminated chemically through the use of a calcium chelating agent by Haugaard and Pettinati.[16] Dilution of the samples and use of a favorable wave length of light ($0.6\ \mu$) left only fat content and average globule size to influence scattering. Photometric readings on known samples were used to construct a nomograph and then commercial samples were tested by the method and compared with Babcock readings. Eighty-six percent of the optical results were within 0.10 percent of the Babcock test and the maximum error noted was 0.17 percent. The authors believe the test is applicable to any dairy product having less than 40 percent fat content, providing the actual fat concentration of the diluted sample (on which the optical determination is made) does not exceed 1.0 percent fat.

Recently an optical device known as a Sepascope* has been introduced which may be attached to a separator, enabling the operator to tell the approximate fat content of the skimmilk at a glance. The instrument is based on the principle that the amount of light passing through a layer of skimmilk is dependent on the thickness of the milk layer and the fat content of the product. As the fat content increases, less light will pass through a given layer of milk. An image placed between the light source and the milk layer will be less distinct as the fat content increases. By adjusting the thickness of the milk layer so that the image becomes visible, a relationship is established to the fat content of the product which can be read on a simple dial indicator. Although useful for daily control of fat losses in the plant, this instrument is not intended to take the place of routine testing for fat in the skimmilk.

*Distributed in USA by Whitman Laboratories, Inc., Norwich, N. Y.

BIBLIOGRAPHY

1. McDowall, F. H. The Buttermakers Manual, Vol. 2, p. 1500. N. Z. Univ. Press, Wellington, N. Z. (1953).
2. Hansen, H. C., et al. Final Report of the Subcommittee on the Determination of the Percentage of Fat in Buttermilk, Skimmilk and Whey. J. D. Sci. 28:325-327 (1945).
3. Trout, G. M., and Lucas, P. S. A Comparison of the Babcock, Gerber, Minnesota, Pennsylvania and Mojonnier Methods for Determining the Percentage of Fat in Homogenized Milk. J. D. Sci. 30:145-159 (1947).
3a. Swope, W. D. The Pennsylvania Method for Determining the Percentage of Fat in Dairy Products. Penn. Bull. 412 (1941).
4. Milk Ind. Found. Laboratory Manual, Published by the Milk Industry Foundation (1959).
5. Schain, P. The Use of Detergents for Quantitative Fat Determinations. I. Determination of Fat in Milk. Science 110:121-122 (1949).
6. Schain, P. A Single Solution Detergent Method for Determining the Butterfat Content of Milk. Milk Dealer 40,3:49-50, 54,56,58 (1950).
7. Gershenfeld, L., and Ucko, B. Fat Determinations in Milk. J. Milk and Food Tech. 13,3:175-76 (1950).
8. Gershenfeld, L., and Rosenthal, M. H. Fat Determinations in Milk and Milk Products. J. Milk and Food Tech. 14,1:17-18 (1951).
9. Sager, O. S., et al. A Study of the Schain Butterfat Test. Proc. MIF Convention, Lab. Section (1951).
10. Sager, O. S., and Sanders, G. P. A BDI Detergent Test for Butterfat in Milk and Other Dairy Products. Proc. MIF Convention, Lab. Section (1952).
11. Sager, O. S., et al. A Detergent Test for the Milk Fat Content of Dairy Products. I. Milk, Cream, and Ice Cream. J. Assoc. Off. Agr. Chem. 38:931-940 (Nov. 1955).
11a. Hoover, S. R., et al. A Comparison of Detergent Tests for Butterfat in Milk with Official Methods. J. D. Sci. 41,3:398-408 (1958).
12. Jaquith, B. The TeSa Butterfat Test Kits and Reagent. Proc. MIF Convention, Lab. Section (1958).
13. Technical Industries, Inc. TeSa Butterfat Test Method (Evaluation Data and Reports) (1959).
14. Ashworth, U. S. Turbidity as a Means for Determining the Efficiency of Homogenization. J. D. Sci. 34:317-320 (1951).
15. Deackoff, L. P., and Rees, L. H. Testing Homogenization Efficiency by Light Transmission. Milk Dealer, 46:61 (July 1957).
16. Haugaard, G., and Pettinati, J. D. Photometric Milk Fat Determination. J. D. Sci. 42,8:1255-1275 (1959).

REVIEW QUESTIONS

1. What reagents are used in the American Association Test for fat?
2. Why does the American Association Test give higher fat readings than the Babcock Test?
3. The Pennsylvania Test is especially valuable for determining fat in what dairy products? Why?
4. Why is the fat test obtained with the Minnesota Method lower than with the Babcock Method?
5. What are the advantages of the Schain Test? Disadvantages?
6. What reagents are used in the Alkaline Test for butterfat in place of the acid used in the Babcock Method?
7. What is a "Sepascope"?

CHAPTER VI

Testing for Solids in Dairy Products

The procedure officially recognized by the Association of Official Agricultural Chemists for determining the total solids content (or moisture content in the case of powdered milk) is much too time consuming for product control in milk plant operations. Through the years, a number of substitute methods have been developed to make testing for milk solids faster, easier, and less expensive. These methods make use of alternate systems for heating samples, heating under vacuum, as well as direct reading of solids (or moisture) content from appropriate scales. Though the basic principle is commonly that of evaporating the moisture from a known quantity of sample, weighing the residue and computing it as percent of the original, certain new techniques have been suggested for the purpose. These include titration procedures, the use of plastic disks of known density, refractive index measurements, etc. Numerous formulas based on lactometer readings and fat content have been proposed for rapid estimations of total solids or solids-not-fat content. Some of these formulas will be discussed in Chapter VII.

The AOAC official procedure and several of the more common laboratory procedures for determining the total solids in milk and other dairy products will be presented in this chapter.

A. AOAC Official Method for Total Solids in Milk and Cream[1]

Weigh 2.5-3.0 grams of prepared sample into a weighed flat bottom dish of not less than 5 cm. diameter. When ash is to be determined on the same portion, use 5 grams of sample in a platinum dish. Heat on a steam bath for 10-15 minutes, exposing maximum surface of the dish bottom to live steam. Then heat for 3 hours in an air oven at 98-100° C. Cool in a desiccator, weigh quickly, and report the percent residue as total solids.

This procedure, slightly modified, is used in the official method of analyses for total solids (or moisture) in other dairy products.

Weighing the Solids Sample
Courtesy Mojonnier Bros. Co.

Dish Contact Maker

The official methods of analysis (AOAC) also recognize the lactometer procedure (see Chapter VII) as an approximate method for estimating total solids.

B. The Mojonnier Test for Total Solids
Milk, Skimmilk, Buttermilk and Whey.

(1). Prepare the solids dish by heating in the vacuum oven at 100° C. for 10 minutes and then cooling in the desiccator for 5 minutes with the water circulating.

(2). Mix the sample by pouring several times from one container to another.

(3). Weigh about 2 grams into a prepared and weighed solids dish plus cover.

(4). Add no water but spread the milk in a thin film over the entire bottom of the dish.

(5). Place the dish on the hot plate at 180° C., press on dish with contact tool to insure uniform evaporation and heat until the first traces of brown appear.

(6). Transfer the dish to the vacuum oven at a temperature of 100° C. for 10 minutes at not less than 20 inches of vacuum.

(7). Cool in desiccator for 5 minutes with water circulating during this time.

(8). Weigh the dish and contents with cover on the dish.

(9). Calculate the percentage of solids.

Cream.

(1). Weigh about 1.0 gram of the sample if testing less than 25 percent butterfat and about 0.50 gram if testing over 25 percent.

(2). Add 1 ml. of distilled water to sample in the dish if fat test is less than 25 percent, and 1.5 ml. if fat test is over 25 percent.

(3). Spread the diluted cream in a thin film evenly over the bottom of the dish and complete as for milk.

Ice Cream and Evaporated Milk. Proceed exactly as for cream testing less than 25 percent butterfat.

Sweetened Condensed Milk.

(1). Weigh about 0.25 gram sample into tared and weighed solids dish.

(2). Add 2 ml. of hot distilled water to dissolve the sugar and dilute the sample.

(3). Spread evenly over the bottom of the dish.

(4). Heat on hot plate at 180° C. until first traces of brown appear.

(5). Dry in the oven for 20 minutes under not less than 20 inches of vacuum.

(6). Cool in desiccator for 5 minutes and then weigh.

(7). Calculate the percent of total solids and deduct 0.30 percent from the result.

Complete drying requires 90 minutes in the oven, but 30 minutes with the deduction is just as accurate and saves considerable time.

Whole Milk Powder and Skimmilk Powder. Weigh about 0.3 gram directly into a dish, add 2 ml. of hot water and spread the diluted sample over the bottom of the dish and proceed as for milk.

Butter. Weigh about 1.0 gram of the prepared sample into a dish, add no water and proceed as for milk.

Cheese.

(1). Weigh a solids dish with a blunt pointed glass rod that can be used to break up any cheese lumps.

TESTING FOR SOLIDS IN DAIRY PRODUCTS

(2). Weigh about 0.5 gram of cheese into the dish.

(3). Add 1.5 ml. of hot distilled water and spread the cheese and added water over the bottom of the dish. Use glass rod to break up any lumps.

Leave glass rod in cup during the entire operation.

(4). Complete as for milk except dry the sample in the oven for 20 minutes.

Cottage Cheese.

(1). Dump cheese in a blender and run until cheese is smooth and sufficiently liquid to pour.

(2). Make determination using the same procedure as for milk except permit a 20-minute drying period in the vacuum oven.

C. Dietert Method

The Dietert method of determining total solids is based on the fact that moisture can be removed from milk solids more rapidly if the solids dishes are placed in a system of flowing warm air. It has been shown[2] that this is a satisfactory procedure when the temperature and air volume remain constant for all samples.

Dietert Total Solids Determination in Milk Products[3]

Instructions for Dietert-Detroit Solfat Determinator total solids determination in milk products.

Apparatus:
   No. 295A Speed Oven
   No. 5000-43 Solids Dishes (6)
   No. 5000-51 Counterpoise
   No. 5000-0194 Tongs
   No. 5150 Speed Desiccator
   No. 5000-22 Cover
   No. 5000-44 Pipette (2) (1 gr.)
   Analytical Balance and Weights

Reagents: Distilled Water.

Procedure:

(1). Place dry, clean solids dishes in the Speed Oven, set thermostat at 275° F. and heat for ten minutes.

(2). Remove dishes from oven by means of the tongs and place in closed Desiccator for five minutes.

Dietert-Detroit Solids Determinator
Courtesy Harry W. Dietert Co.

(3). Transfer cooled dish to the left hand balance pan using tongs as before. Place cover, which has been stored in balance case, on the pan and weigh to the fourth decimal place using the counterweight on the right hand pan.

(4). Bring the sample to a temperature of 68-70° F. and pipette the amount indicated in table below into the dish. Replace the cover and quickly reweigh again to the fourth place.

(5). Solid and powdered samples are weighed into a dish by means of a spatula, after rubbing in a mortar to break up lumps.

Product to be tested	Approx. Size of sample in grams	Amt. of Dist. water to add cc.	Length of time in Solfat Oven (min.)
Fresh Milk	2	None	20
Chocolate Milk	2	None	20
Skimmilk	2	None	20
Whey	2	None	20
Buttermilk	2	None	20
Cream	1	1	20
Ice Cream Mix	1	1	20
Evaporated Milk	1	1	20
Plain Cond. Milk	1	1	20
Sw. Cond. Milk	0.25	2	30
Milk Powder (skim, whole, buttermilk)	0.3	2	20
Malted Milk	0.3	2	20
Cocoa	0.3	2	20
Butter	.5	2	20
Cheese	1.0	0	10

(6). Remove the cover and add the required amount of distilled water to the sample. Mix and spread evenly over bottom of pan by rocking the dish gently. Mixing is hastened by using hot distilled water.

(7). Place the sample and dish into the Speed Oven at 275° F. for the time specified in the table. 275° F. is the maximum temperature to be used. It may be as low as 255° F. without apparently affecting the results.

(8). Remove the dish from Speed Oven and place in desiccator for five minutes to cool.

(9). Transfer dish to balance, cover and quickly weigh as before.

(10). The calculation for total solids on such samples is as follows:

$$\frac{\text{Weight of solids} \times 100}{\text{Weight of Sample}} = \text{Percent total solids}$$

(11). A quantity of fat and solids dishes may be prepared and held in the desiccator prior to use. They must, however, be weighed just prior to adding the samples.

(12). After a solids test is completed, the dish should be filled with water and a few drops of ammonia added. The ammonia will break up the residue but will not affect the monel dish. When the residue has softened, it can be easily brushed out in running water.

(13). All dairy products acquire some shade of brown when dried in the total solids test. The definite temperature and time settings on the Speed Oven permit any degree of browning to be reproduced at will. If it is desired to have the solids test in the Speed Oven check some other method which in the hands of a particular analyst produces a greater or less degree of browning, the time and/or temperature of the Speed Oven may be altered to give the desired matching of results.

To obtain a lighter brown (higher average solids) the Speed Oven should be set at 255° F. and the time in the table observed. A deeper brown may be obtained by increasing the time period five minutes with the temperature at 275° F. which is the safe upper limit. One exception is sweetened condensed milk which can be tested at 302° F. for twenty minutes safely.

(14). The times and temperatures shown in the table are the average result of tests against A.O.A.C. and other methods made in several dairy and university laboratories.

D. Cenco Moisture Balance

In this procedure, infra-red radiation is used to accomplish rapid drying. The tester has a sensitive torsion balance which permits rapid weighing. Moisture percentages may be read directly from the balance scale, eliminating the need for computing or calculating to obtain results. Comparative studies[4] have shown that the Cenco procedure is less time consuming and less costly when compared with the Mojonnier method. It was reported that the results closely approximated those obtained by the Mojonnier procedure and variance between sample results were generally less than that of the Mojonnier. These studies indicated that the Cenco instrument should be periodically standardized against a procedure of known accuracy in order to adjust the voltage output for correct determinations. The authors also caution that errors of 0.2-0.3 percent can be made if the operator does not read the scale from the same angle each time a determination is made.

The recommended method is as follows[4]:

Cenco Moisture Balance
Courtesy Central Scientific Co.

Procedure

(1). Place an 11 cm. filter paper on the sample pan and a cotton fiber disc on top of the filter paper.

(2). Put the sample pan on its support in the balance, close the lamp housing, adjust the voltage output from the variable transformer to 100 volts, and move the toggle switch to the "on" position.

(3). Leave the infra-red lamp on until the pointer comes to rest indicating that the sample bed is at constant weight. Rotate the knurled wheel so that the 100% line on the graduated scale is lined up with the index line. Move the zero adjusting knob up or down so that the pointer is in line with the index line, then loosen the zero adjusting knob and let it rest on the bottom of its slot. The instrument is then prepared to receive the sample.

(4). Move the toggle switch to the "off" position, open the lamp housing, and while the sample bed is cooling, rotate the knurled wheel until the 0% line on the graduated scale is lined up with the index line. Allow about one minute for the sample bed to cool, then add approximately 5 g. of properly prepared sample to the absorbent cotton mat, distributing it evenly over the entire surface. The exact amount of sample is that amount required to line-up the pointer with the index line.

(5). When the sample is weighed, close the lamp housing, move the toggle switch to the "on" position, and rotate the knurled wheel until the 80% line on the graduated scale is lined up with the index line. This keeps the sample pan level and allows for even drying.

(6). In 4 to 4½ minutes, the movement of the pointer will indicate that 80% of the initial sample weight has been evaporated as moisture. At that time reduce the voltage from the transformer to 90 and allow drying to continue until complete i.e., until the pointer remains motionless for ½ to 1 minute. During the final stages of drying continuously adjust the position of the pointer to line up with the index line by rotating the knurled wheel.

(7). When the sample is dried read the percent of moisture from the graduated scale opposite the index line.

E. Toluene Distillation Method

Moir[5] gave a brief critical survey of basic analytical methods for controlling the quality of milk powder at the XIV International Dairy Congress in 1956. He pointed out that the estimation of moisture content in powdered milk is a rather difficult matter because:

(1). The powder is extremely hygroscopic.

(2). During heating in the presence of moisture, reactive groups of the protein and lactose combine and lose water which was originally an integral part of the chemical compounds.

(3). Lactose (which is present as a mono-hydrate) loses some of its water by crystallization when heated at temperatures above 100° C.

Moir observed that the moisture determination might not be correct just because replicate analyses agreed very closely.

The Toluene Distillation Method is recommended by the American Dry Milk Institute for determining moisture in powdered milk.[6]

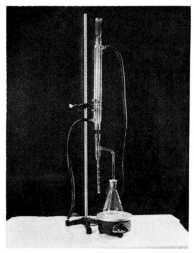

Toluene Distillation Apparatus
Courtesy Vt. Agr. Exp. Sta.

Apparatus and Reagents:

Balance — torsion or similar type, approximately 500 gm. capacity, and 0.1 gm. or better sensitivity.

Condenser — Liebig, with a water jacket at least 300 mm. in length.

Condenser brush — over-all length 24"; length of bristle tuft $3\frac{1}{2}$"; diameter of bristle tuft $\frac{1}{2}$".

TESTING FOR SOLIDS IN DAIRY PRODUCTS

Distillation trap — 4 ml. graduated in 0.1 ml. divisions.

Electric heater — with rheostat control to maintain a distillation rate of approximately 4 drops per second under conditions of the test.

Flask — Erlenmeyer, Pyrex, 300 ml. narrow mouth.

Rubber tubing — flexible, suitable for water connections.

Stands and clamps — suitable for holding equipment.

Stoppers — cork. Not necessary if condenser, trap and flask are fitted with ground glass connections.

***Toluene, technical** — moisture free. Blank determinations should be made to assure suitability.

Procedure:

Transfer a 50 gm. sample to a clean, dry 300 ml. Erlenmeyer flask as quickly as possible and immediately pour approximately 75-100 ml. toluene into the flask to cover the sample. Rinse down any milk particles on the inside of the flask when introducing the toluene.

Insert a dry distillation trap to the flask and fill the trap with the toluene.

Shake the flask with the sample thoroughly before connecting the trap and flask to the condenser. Make sure cold water is circulating through the condenser. Heat the contents to boiling making sure that the sample does not scorch on the bottom of the flask. The amount of heat should be so regulated that the toluene will condense into the trap at a rate of about 4 drops per second.

Forty-five minutes after distillation has begun and without interrupting distillation, dislodge water droplets in the condenser tube by means of a condenser brush. While the brush is in the upper part of the condenser, flush the tube with 10 ml. of toluene.

Read the moisture level in the trap to the nearest half of a scale division (0.05 ml.). Be sure that the meniscus between the toluene and water may be dislodged by a long, stiff wire inserted down through the condenser into the trap.

Continue the distillation for an additional 15-minute period (making a total time of 60 minutes) and again dislodge water droplets from the condenser tube as before and note the water

***Note:** Toluene is readily inflammable and the utmost care must be exercised to keep supplies of it away from flames and heating elements.

level in the trap. If this reading agrees with the previous reading within 0.05 ml. water in the trap, discontinue the distillation and report the determined moisture content. The milliliters of water in the trap multiplied by 2 equals the percentage of moisture in the sample.

For some products a 60-minute distillation period is not sufficient.

If readings fail to agree as required, continue the distillation for additional 15-minute periods until successive results agree within a half scale division.

It is extremely important to clean the condenser and distillation trap thoroughly in order to minimize the adherence of water to the glass surface during distillation. If equipment is dirty or greasy, difficulty will be experienced in dislodging water droplets and results may be unreliable. Wash glass equipment by soaking in and scrubbing with a solution of a suitable detergent; then rinse with clean water. If this fails to remove dirt and grease, soak glassware over night in a cleaning solution and rinse equipment well with clean water and dry thoroughly. Distillation traps require thorough cleaning after each determination and condensers usually once a week.

F. Other Methods of Determining the Total Solids Content of Dairy Products

The urgent need for a rapid, yet accurate, method for the determination of milk solids has resulted in the introduction of a number of techniques to accomplish this evaluation. Some are modifications of long recognized procedures, others are relatively new in design.

Ystgaard and co-workers[7] studied the use of the Brabender Semi-Automatic Moisture Tester for determining total solids in milk. This tester, supplied by the Brabender Corporation of Rochelle Park, New York, hastens the drying of the sample by using forced draft ventilation. Samples may be weighed automatically without removing them from the oven and the moisture content is obtained by a direct reading on an illuminated scale. They found that 10 gram samples held at 110° C. for 3 hours provided optimum conditions for solids determinations in milk. However, results were significantly lower than those obtained by either the AOAC official procedures or the Mojonnier method.

Semi-Automatic Rapid Moisture Tester
Courtesy Brabender Corp.

The Karl Fischer method may be used for determining moisture in milk powders. This is a procedure in which the dried milk sample is first suspended in anhydrous methanol. The moisture is then determined by titration with the Karl Fischer reagent (commercially available or may be made from the ingredients). The sample is over titrated until the color changes from yellow to dark brown, then back-titrated with standard water solution to eliminate the excess reagent. The percent moisture is obtained by use of appropriate formulae.

Work in Australia[8,9] indicates that the refractive index can be used to rapidly determine total solids in skimmilk concentrates but that the refractive index measurement is no more reliable than the specific gravity value as determined by the picnometer in estimating total solids in sweetened condensed milk.

Dr. N. S. Golding of Washington State College has developed a method[10] which removes the milk fat from skimmilk and determines its specific gravity in one process. Specific gravity is then converted to percentage of solids-not-fat by reference to a conversion table. Plastic disks of different color and different but known density are placed in milk under controlled conditions and the number of disks that float are a measure of the specific gravity of the sample. Although the author believes the sphere test holds considerable promise for density determinations in whole milk, it must be made more accurate before it can be recommended for solids-not-fat percentage in skimmilk.

Two methods, which appear to merit consideration, were discussed at the International Dairy Congress held in Rome in 1956. A method long used in the Netherlands to determine the moisture content of dried milk consists of drying samples by means of a current of pre-dried air in an apparatus designed by Meihuizen in 1919. Results of this method agree with those obtained by the Fischer titration.[11]

Australian workers[12] discussed a method based on a principle used with blood for many years. They noted that copper sulfate solutions of known specific gravity ranging from 1.026 to 1.036 in a graded series could be prepared. If a drop of milk is dropped into these graded solutions, the milk droplet will remain relatively intact for about 20 seconds during which time its behavior may be noted. It will sink in those solutions whose specific gravity is lower than that of the milk droplet and rise in those solutions which are heavier than the sample. The milk droplet will remain stationary in the copper sulfate solution having the same specific gravity for at least 10 seconds. Emphasizing that the work was still in preliminary stages, the authors were of the opinion that the method has distinct possibilities as a practical rapid test.

BIBLIOGRAPHY

1. Assoc. Off. Agr. Chemists. Official Methods of Analysis. 8th Ed. (1955).
2. Doan, F. J. Comparative Accuracy of the Dietert Solfat Determinator. Proc. Milk Ind. Foundation, Lab. Section (Oct. 1950).
3. Harry W. Dietert Co. Instructions for Dietert-Detroit Solfat Determinator (May 1, 1955).
4. Mickle, J. B., et al. Methods of Determining the Total Solids in Fluid Milk. Okla. Exp. Sta. Tech. Bull. T-67 (1957).
5. Moir, G. M. The Analysis of Dried Milk. Proc. XIV Int. Dairy Congress. Vol. 3, Part 2, pp. 307-316 (1956).
6. Amer. Dry Milk Inst., Inc. Standards for Grades for the Dry Milk Industry. Bull. 916 (1960).
7. Ystgaard, O. M., et al. The Use of the Brabender Semi-Automatic Moisture Tester for the Determination of Total Solids in Milk. J. D. Sci. 34:695-698 (1951).
8. Lawrence, A. J. Use of the Refractometer to Determine Total Solids in Sweetened Condensed Milk. Aust. J. D. Tech. 8,3:92-93 (1953).
9. Lawrence, A. J. The Determination of Total Solids in Skimmilk Concentrates by the Refractometer. Aust. J. Dairy Tech. pp. 6-7 (Jan.-Mar. 1955).
10. Golding, N. S. Plastic Hydrometers for Measuring the Density of Skim and Whole Milk. Proc. XV Int. Dairy Congress. Vol. 3, Sect. 5, pp. 1566-1571 (London, 1959).
11. Kruisheer, C. I., and Eisses, J. Determination of the Moisture Content of Dried Milk and Other Milk Products by Means of the Apparatus of Meihuizen. Proc. XIV Int. Dairy Congress. Vol. 3, Part 2, pp. 212-220 (Rome, 1956).
12. Macdermott, C. J., et al. Some Preliminary Work on a Rapid Test for Solids Not Fat in Milk. Proc. XIV Int. Dairy Congress. Vol. 3, Part 2, pp. 286-300 (Rome, 1956).

REVIEW QUESTIONS

1. What is the basic principle of the Dietert Total Solids Test? How does it differ from the Mojonnier Solids Test?
2. Why should ammonia and water be added to the solids dishes upon completion of the test?
3. How is rapid drying accomplished in the Cenco Moisture Tester?
4. What is a basic advantage of the Cenco Moisture Balance?
5. Why is it difficult to obtain an accurate moisture test on powdered milk?

CHAPTER VII

The Lactometer and Its Uses

Specific Gravity of Milk. The specific gravity of a substance is the ratio of its weight to the weight of an equal volume of another substance taken as a standard. Water is the standard for solids and liquids, while hydrogen is the standard for gases. The specific gravity of water is considered as 1.0 at 60 degrees F.

The average specific gravity of normal whole milk is 1.032. This means that if a certain amount of water weighs 1,000 pounds, the same quantity of milk would weigh 1,032 pounds (1,000 × 1.032). Whole milk varies in specific gravity from 1.029 for low testing milk to 1.034 for rich milk. The average specific gravity of skimmilk is 1.036, while that of cream varies from around 0.93 to 1.0 depending on its richness in fat. Pure butterfat has a specific gravity of nearly 0.9 at a temperature of 140 degrees F.

Hydrometers. The specific gravity of liquids may be determined in several ways such as by means of a hydrometer, a Westphal balance or a pycnometer. The hydrometer works on the principle "that if a body floats in a liquid it is buoyed up by a force equal to the weight of the liquid it displaces." Thus, a body would not sink as far in a heavy liquid as it would in a light one, because it would take a smaller volume of the former to equal the weight of the body than it would take of the latter. A special form of hydrometer is used for taking the specific gravity of milk. It is called a lactometer. Two kinds of lactometers are generally used, the Quevenne and the New York State Board of Health. They are abbreviated Q. and B. of H. respectively.

The Quevenne Lactometer. The Quevenne lactometer consists of a long, slender stem connected to the body, which is a large air chamber. This chamber causes the instrument to float. Attached to the lower end of the body is the bulb which is filled with shot or mercury to cause the lactometer to sink to the proper level and to float in an upright position in the milk. Inside of the lactometer is a thermometer tube extending from the bulb up into the upper part of the stem where the scale is located. It is important to know the temperature of the milk when the lactometer reading is taken. This thermometer does not record temperatures above 100 degrees F. and so the lactometer must not be put into liquids above this temperature. When cleaning the lactometer, use cool water and then wipe it with a dry cloth. Just below the thermometer is the lactometer scale, with graduations ranging from 15 at the top to 45 at the bottom. Each division is known as a lactometer degree.

The sensitiveness of the lactometer depends on the relative size of the stem to that of the body. A lactometer with a small

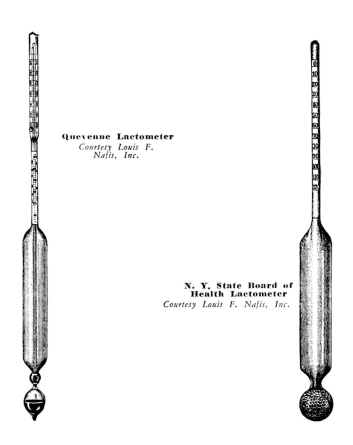

Quevenne Lactometer
Courtesy Louis F. Nafis, Inc.

N. Y. State Board of Health Lactometer
Courtesy Louis F. Nafis, Inc.

stem and a large body will give more accurate results than one with a large stem and small body.

The average Quevenne reading of normal whole milk is 32; of low testing milk 29; of rich milk 33 or 34; and of skimmilk 36. The Quevenne reading may be changed to specific gravity by prefixing 1.0. Thus, milk having a Quevenne reading of 32.5 would have a specific gravity of 1.0325. This ease of changing the lactometer reading to specific gravity is the chief advantage of the Quevenne lactometer.

Board of Health Lactometer. The Board of Health lactometer is similar in form to the Quevenne, except that the ordinary sized instruments have no thermometers. The scale on the Board of Health lactometer ranges from 0 to 120. It reads zero in water at 60 degrees F. A reading of 100 on the Board of Health lactometer equals 29 on the Quevenne, and so one degree on the former is equal to 0.29 degrees on the latter. To change from the Board of Health reading to the Quevenne, multiply the Board of Health reading by 0.29, and conversely, in changing from the Quevenne reading to the Board of Health divide the Q. reading by 0.29.

The average reading of normal whole milk on the Board of Health lactometer is 110; of low testing milk 100; of rich milk 114 to 117; and of skimmilk 124.

To obtain the specific gravity of a sample of milk from a Board of Health reading, it is first necessary to change the reading to the Quevenne by multiplying it by 0.29 and then prefixing 1.0.

The advantage of the Board of Health lactometer is its greater sensitivity, due to a larger scale. It will detect smaller adulterations of milk than the Quevenne.

Effect of Temperature on the Density of a Liquid. Heat affects the density of a substance. A rise in temperature of a liquid will cause it to become thinner and therefore to have less weight per unit of volume, and conversely when cooled the weight per unit of volume is greater. For this reason lactometer readings must be taken at a standard temperature, which is 60° F. Since it is not always convenient to have the sample of milk exactly at 60° F., the reading may be taken anywhere between 50° F. and 70° F., provided a correction is made. An average cor-

THE LACTOMETER AND ITS USES

rection factor for the Quevenne lactometer is 0.1 lactometer degree for each degree in temperature that the sample of milk varies from the standard. Add the correction to the lactometer reading when the temperature is above 60° F. and subtract when it is below. For example, if the lactometer reading of a sample of milk is 31.5 at a temperature of 68° F., at 60° F. it would be 31.5 plus 0.8 (8 × .1), or 32.3, which is the true reading. The specific gravity for this milk would then be 1.0323. The correction factor does not hold beyond temperatures of 50° F. and 70° F.

Like the Quevenne, the Board of Health reading must be corrected to the standard temperature of 60 degrees F. The correction in this case is 0.3 of a lactometer degree for each degree of temperature that the milk varies from the standard. For example, if the Board of Health reading is 109.5 at a temperature of 55° F., subtract 1.5 (5 × .3), which gives a corrected reading of 108.

Reading the Lactometer. In taking a lactometer reading of milk, first mix the sample thoroughly, then pour the milk into a glass cylinder which is large enough to let the lactometer float freely, insert the lactometer and as soon as the instrument comes to a rest read the division at the top of the meniscus. The older lactometers were graduated to be read through the meniscus film and on a level with the milk, but the newer instruments have been marked to read correctly at the top of the meniscus so as to facilitate the reading. The reading should be taken with as little delay as possible, as the fat globules soon rise and therefore the lower part of the lactometer would be resting in partially skimmed milk, and this fact would give a higher reading. Note the temperature of the milk when reading the lactometer and correct the lactometer degrees to the standard temperature.

The Westphal Balance. The specific gravity of liquids may be determined more accurately by means of the Westphal balance than by a lactometer. The outfit consists of the balance, a plummet displacing 5 grams of distilled water at 60° F., a glass cylinder for holding the liquid, and a set of riders weighing 5, 0.5, 0.05 and 0.005 grams.

In determining the specific gravity of milk, temper the sample to 60° F., mix by pouring several times from one vessel to another, being careful not to incorporate any excess air, fill the cylinder with the milk and suspend the plummet in the sample so that it is completely covered. Then add the various riders to the beam until the buoyant force of the milk is counter-

Westphal Balance
Courtesy Fisher Scientific Co.

balanced. Read the specific gravity direct according to the positions of the riders. The 5 gram rider indicates tenths on the beam, the 0.5 gram rider, hundredths, the 0.05 gram rider, thousandths and the 0.005 gram rider ten thousandths. For example, if the large rider was on notch 10 and the other three, respectively, on notches 3, 2 and 8, the specific gravity would be 1.0328.

The calculations of this specific gravity may be presented as follows:

```
5.00  gram rider on notch 10 = 5.    × 1  = 5.
0.50  gram rider on notch  3 = 0.5   × .3 = .15
0.05  gram rider on notch  2 = 0.05  × .2 = .01
0.005 gram rider on notch  8 = 0.005 × .8 = .004
                                            ─────
                                            5.164 grams dis-
                                                  placed by
                                                  plummet.
```

Since the plummet displaces 5 grams of distilled water at 60° F., the specific gravity of the milk would be 5.164 ÷ 5 or 1.0328.

The Pycnometer. The pycnometer is a small conical shaped flask with a thermometer ground to serve as a stopper, and has a side tube with capillary opening for overflow of expanding liquids. It must be carefully cleaned and dried before each determination. To perform this operation properly, first clean the instrument with a glass cleaning solution such as potassium bichromate in sulphuric acid, next rinse with water, then rinse several times with alcohol and finally with ether. Allow ether fumes to escape and the pycnometer to become perfectly dry before use.

The use of the pycnometer in determining the specific gravity of milk, while requiring special attention to details of procedure, gives a high degree of accuracy.

Jenness and Herreid[1] found that it was necessary to consider the factor of buoyancy when weighing the pycnometer. To reduce this effect of buoyancy they used a counter-poise so constructed that it had nearly the same displacement as the pycnometer, and thus eliminated the use of the regular small weights which would not have a buoyant effect comparable to that of the pycnometer.

In determining the specific gravity of milk, the first step is to ascertain that of water. Obtain the weight of the properly prepared empty pycnometer, then fill it with recently boiled water cooled to around 55° F., place in tempering bath of about the same temperature and allow to warm up to 60° F. Adjust the water to the proper level, insert stopper, remove pycnometer, wipe dry and weigh. Follow the same procedure with the sample of milk, and take care to have cleaned and dried the pycnometer before adding the milk. From the relative weights of the milk and water, the specific gravity of the milk can be calculated.

Jenness and Herreid[1] used the following procedure in obtaining the specific gravity of purified milk fat: The samples of fat were transferred to the pycnometer at about 30° C. with 10-ml. Mohr pipettes. Next the thermometer was inserted and the instrument placed in a water bath at 37.5° C. ± 0.1 for 15 minutes to permit the excess fat to escape through the capillary tube. The tube was capped and the pycnometer carefully dried with a clean cloth or with ether and cooled to room temperature of approximately 21° to 22° C. in a desiccator and weighed. The determinations of the fatty materials from the Babcock test were made in the same manner except they were evacuated at 23 to 25 inches for three minutes to remove bubbles of air. All determinations were made at 37.5°/37.5° C. with water as the base.

Calculating the Percent of Solids in Milk

The federal government and each of the fifty states have laws regulating the minimum amount of fat and solids-not-fat (or total solids) that milk must contain. This ruling is to protect the consumer from purchasing milk that has been adulterated by skimming or watering or milk that is naturally low in solids content. The purpose of the federal law is to regulate interstate shipments of milk, while the state law regulates the milk sold within its boundaries. The federal standard[2] is 3.25 percent fat and 8.25 percent solids-not-fat. The standards vary in the different states, some are below and some are above the federal standard. When milk is shipped out of the state, it must satisfy both the federal standard and that of the state to which it is shipped.

Formulae for Solids in Milk. Formulae relating to the solids-not-fat content of milk to its fat content and specific gravity have been used for nearly a century. Prior to the introduction of the Babcock and Gerber tests for fat, it was quicker to determine total solids and specific gravity of a sample of milk and then determine fat by formula. The recent emphasis on the nutritional value of the non-fat solids has caused a new interest in these relationships.

Herrmann[3] discussed the basis for and the history of the published methods for making indirect estimates of the solids-not-fat content of milk. He lists 70 different formulae which were introduced in the period 1882-1951 and shows the calculated value of the solids-not-fat content for many of these under three different fat and lactometer situations. It is obvious from this study that results vary over a rather wide range. Ystgaard and co-workers[4] likewise obtained differing results when several of these formulae were compared for accuracy with the total solids determination by the Mojonnier.

The work of Ystgaard indicated that previous heat treatment of the sample was a factor in the accuracy of solids estimations based on fat and lactometer readings. Work in Scotland[5] demonstrated that the ratio of fat, casein, and lactose to each other gave an additional source of error and one much more difficult to anticipate. They considered casein content to be particularly important in this respect. Later work[6] showed that the ratio of casein to lactose bore a linear relationship to the error in the solids content as measured by specific gravity. Lawrence[7] has shown that lactometric procedures for determining solids content can be applied to the milk from individual cows with considerable accuracy. He found 95 percent of the results of over 2000 cow samples within 0.25 percent of the gravimetric figure.

Several formulae now in use include:

1. $\dfrac{L + F}{4} = \%$ S.N.F. Troy's Formula.

2. $\frac{1}{4} L + .2F = \%$ S.N.F. Babcock's Formula.

In these formulae, L stands for the corrected Quevenne lactometer reading and F for the percent of fat. If the Board of Health lactometer is used, the reading must be changed to the Quevenne before using the formula. The percent of fat must not be expressed decimally, nor must specific gravity readings be used in these formulae. The total solids equal the sum of the fat and the solids-not-fat.

These formulae give slightly different results. An example will show how they work. Suppose a sample of milk has a fat test of 3.6 percent and a lactometer reading of 31.5 at 65 degrees F. First, correct the reading to 60° F., which will make it 32, and then substitute in the formula:

1. $\dfrac{32 + 3.6}{4} = 8.90\%$ S.N.F., or 12.50% total solids.

2. $8 + .72 = 8.72\%$ S.N.F., or 12.32% total solids.

Sharp and Hart[8] state that a sample of milk must have special treatment before taking a lactometer reading in order to get the fat in a uniform physical condition each time a determination is made. The sample should be warmed to 45° C. for 30 seconds, then cooled to 30° C. and the lactometer reading made at this temperature. Under these conditions they advocate the following formula for total solids:

154   CHEMISTRY AND TESTING OF DAIRY PRODUCTS

3. $\text{Fat \%} \times 1.2537 + \left(0.268 \times \dfrac{\text{Q. Lact.}}{\text{sp. gr.}}\right) = \text{Total Solids}$

Some laboratories have found this formula of Sharp and Hart to give results that are too low, the average deviation from the Mojonnier test ranging from —0.53 to —0.76 percent. Herrington[9] has given a reason for these lower results. He points out that the originators of the formula used the Westphal balance for determining the specific gravities at temperatures of 30° C., and since the Quevenne lactometer gives specific gravities based upon water at 60° F. an allowance would have to be made for the expansion of glass if used at 30° C. Therefore, Quevenne readings cannot be substituted directly into the formula wherein higher temperatures are used without some correction. He has calculated that to overcome the effect of the expansion of the lactometer, the formula should be modified to be as follows:

$\text{Fat \%} \times 1.2537 + \dfrac{268 \, (\text{Q. Lact.} + 3)}{\text{Q.} + 1000} = \text{Total Solids}$

This formula will give results for total solids about three-fourths percent higher.

More recently, Watson[10] has proposed a lactometer method for determining milk solids based on the fact that milk fat is completely in the liquid state at 102° F. A special lactometer is used which can be read to 0.2 degree (0.0002 when expressed as specific gravity). In this test the milk sample is heated rapidly to a temperature of 102° F. in water held at 115° F. The flask is loosely stoppered to prevent excess evaporation. The sample is transferred to the lactometer cylinder held in a water bath at 102° F. The lactometer is preheated to 102° F. for not less than 3 minutes, then wiped dry and slowly immersed in the milk. The reading is taken at the top of the meniscus and total solids determined, as follows:

4. $\text{Total Solids (percent)} = 1.33 \, F + \dfrac{273 \, L}{L + 1000} - 0.40$

F=Fat percentage according to the Babcock Test.
L=Lactometer reading in degrees.

The author notes that the ordinary 60° F. lactometer could be used to make determinations at 102° F. if 6.6 lactometer degrees are added to the reading before applying the formula. Results of this method have shown that individual cow samples could be expected (95 percent level) to fall within -0.21 to $+0.31$ percent of the gravimetric values for total solids. Mixed herd samples will show even less variation from gravimetric results.

In a critical analysis of the methods now being used in determinations of the physical and chemical characteristics of milk, Gould and Armstrong[11] have presented in summary form the reported accuracy of several methods for determining the total solids content of milk and skimmilk (see Table 29). They observe that lactometric analyses, if properly done, display an accuracy on most milks at least comparable with that of the Babcock Test for fat. Since readings with the Quevenne lactometer are generally low for skimmilk while results with high fat milks show certain inconsistencies, they note a need for further study in these areas.

Table 29. Reported Accuracy of Various Methods for Determining the Total Solids of Milk and Skimmilk*

Method	No. of samples	Difference from standard percent	Source
Brabender (Drying)	12	-0.19	Ystgaard, et al.
Brabender (Drying)	46	± 0.147	Davey and Patton
Infra-Red Lamp (Drying)	1448	-0.19	Mickel, et al.
Infra-Red Lamp (Drying)	686	-0.11	Lowinstein
Karl Fischer (Moisture titration)	40	$+0.03$	Mickel, et al.
Oxidimetry	36	-0.014	Leggatt
Oxidimetry	30	± 0.150	Davey and Patton
Watson (Lactometer)	60	± 0.12	Robinson, et al.
Watson (Lactometer)	99	± 0.05	Watson
Watson (Lactometer)	101	± 0.06	Watson
Watson (Lactometer)	1454	± 0.20	Madden, et al.

*From data summarized by Gould and Armstrong.[11]

Relationship of Fat to the Solids Not Fat. A certain though not direct relationship exists between the amount of fat and the amount of solids not fat in normal milk. Milk that is low in fat is also low in solids not fat and vice-versa, milk that is rich in fat is also high in solids not fat. Table 30 shows this relationship. It is based on an average of over 100,000 analyses of milk,[12] which should give a very close approximation of the proportion of the fat to the solids not fat.

Table 30. Relation of Fat to Other Solids in Milk and to the Specific Gravity

Fat	Solids Not Fat	Total Solids	Specific Gravity*
%	%	%	
2.8	8.19	10.99	1.03052
2.9	8.23	11.13	1.03060
3.0	8.27	11.27	1.03068
3.1	8.31	11.41	1.03076
3.2	8.35	11.55	1.03084
3.3	8.39	11.69	1.03092
3.4	8.43	11.83	1.03100
3.5	8.47	11.97	1.03108
3.6	8.51	12.11	1.03116
3.7	8.55	12.25	1.03124
3.8	8.59	12.39	1.03132
3.9	8.63	12.53	1.03140
4.0	8.67	12.67	1.03148
4.1	8.71	12.81	1.03156
4.2	8.75	12.95	1.03164
4.3	8.79	13.09	1.03172
4.4	8.83	13.23	1.03180
4.5	8.87	13.37	1.03188
4.6	8.91	13.51	1.03196
4.7	8.95	13.65	1.03204
4.8	8.99	13.79	1.03212
4.9	9.03	13.93	1.03220
5.0	9.07	14.07	1.03228
5.1	9.11	14.21	1.03236
5.2	9.15	14.35	1.03244
5.3	9.19	14.49	1.03252
5.4	9.23	14.63	1.03260
5.5	9.27	14.77	1.03268
5.6	9.31	14.91	1.03276
5.7	9.35	15.05	1.03284
5.8	9.39	15.19	1.03292
5.9	9.43	15.33	1.03300
6.0	9.47	15.47	1.03308

*Calculated by the author using formula $\frac{1}{4}L + .2F = \%$ S.N.F. to obtain lactometer reading and then prefixing 1.0 to obtain specific gravity.

It will be noted in Table 30 that, with an increase in the fat there is an increase in the solids not fat, but they do not increase in the same proportion, that is, milk containing 4.5 percent fat does not have one and a half times as much solids not fat as 3.0 percent milk. However, there is about the same increase between 3 and 4 percent milk as between 4 and 5 percent milk, etc.

Example of Why Rich Milk Has a Higher Lactometer Reading Than Low Testing Milk. It is sometimes supposed that Holstein milk is heavier than Jersey milk, but this is not correct. Jersey milk is slightly heavier. Milk that is high in fat has a higher lactometer reading than milk that is low in fat, for milk that has a high percent of fat has a high percent of solids not fat, and the increase in these solids not fat raises the lactometer reading

THE LACTOMETER AND ITS USES

more than the increase in fat lowers it. The difference in weight between high and low testing milk would only amount to about half a pound per forty quart can in favor of high testing milk. It must be remembered that the above discussion refers to normal whole milk, and not to standardized milk.

The following example will show why rich milk is heavier than low testing milk: Assume that a sample of milk tests 4 percent fat and has a Board of Health lactometer reading of 110. Now, if the 4 percent of fat were removed, the resulting skimmilk would have a lactometer reading of 124. Thus, the 4 percent of fat lowered the reading by 14 lactometer degrees (124 to 110). One percent fat would then lower the reading 3.5 degrees (14.0 ÷ 4). This shows what effect the fat has in lowering the lactometer reading. Next consider what effect the solids not fat have in raising it.

If the solids were removed from the above skimmilk, water would be left and water has a Board of Health reading of zero. The lactometer reading of the skimmilk was 124. Therefore, the solids in the skimmilk increased the reading 124 lactometer degrees (124 — 0). The skimmilk contained 9.0 percent solids not fat $\dfrac{(L + F)}{4}$ which raised the lactometer reading 124 lactometer degrees. One percent S.N.F. would then raise it 13.8 degrees (124 ÷ 9.0). With these figures, 3.5°, that one percent of fat lowers the lactometer reading, and 13.8° that one percent of solids not fat raises the reading, consider what the difference in lactometer readings between a 5.0 percent milk and a 3.5 percent milk would be. Referring to Table 30, the percent of solids not fat which goes with the corresponding percent of fat can be obtained, and then the problem may be set down and worked as follows:

```
    5.0% Fat                    9.07% S.N.F.
    3.5% Fat                    8.47% S.N.F.
    1.5% Fat difference          .60% S.N.F. difference
    1.5  ×   3.5 = 5.25 lact. deg. lowered.
    0.60 ×  13.8 = 8.28 lact. deg. raised.
```

The above solution shows that there is a net gain in the lactometer reading of the 5.0 percent milk over that of the 3.5 percent milk by 3.03 degrees (8.28 — 5.25).

Detecting Skimming and Watering

Methods of Adulterating Milk. The common methods of adulterating milk are skimming, watering and a combination of both skimming and watering. Adulterating milk was practiced quite widely before the advent of the Babcock test, but since that time it has not been so commonly done. Since milk is commonly sold by test, there is less advantage in adulterating than when it was customary to sell it by weight or volume alone. However, a look at the following example will show why the temptation to practice adulteration may still be strong to some people.

Let us assume a farmer delivers 300 pounds of milk which tests 4.4% fat, and he is paid $4.50 per hundred for 3.7% milk with a fat differential of 9 cents. Then he would receive $5.13/cwt. for his milk or $15.39. However, if he adds enough water to lower the test to 3.4%, he would then receive $16.41 for the adulterated milk, an increase of $1.02.

a. 300 lbs. of 4.4 milk at a price of $4.50 + .09
 4.50 + (4.4 − 3.7 × 9) = 4.50 + .63 = $5.13/cwt.
 3 × $5.13 = $15.39

b. 300 lbs. of 4.4 milk + 88 lbs. of water = 388 lbs. of product testing 3.4%
 388 lbs. at a price of $4.50 + (3.7 − 3.4 × 9) =
 (4.50 − .27) = $4.23/cwt.
 3.88 × 4.23 = $16.41
 $16.41 − $15.39 = $1.02 per day increase

Detection of adulteration is commonly based on the fact that in normal milk there is a certain relationship between the amount of fat and the other solids. Any skimming or watering of milk would change this relationship.

In connection with the Babcock test the lactometer may be used to detect adulterated milk. This use of the lactometer may best be illustrated by giving examples of each method of adulteration.

Example of Skimming. Since fat is the lightest portion of milk, the removal of any part of it will cause an increase in the lactometer reading. Consequently, if a sample of milk has a low fat test with a high lactometer reading or high percent of solids not fat for that class of milk, it indicates skimming. The following example illustrates this method of adulteration:

	% Fat	Q. Reading	% S.N.F.
Normal	4.8	31.5	9.08
Same sample after skimming	3.0	33.4	9.10

$1.8 \div 4.8 \times 100 = 37\frac{1}{2}\%$ fat removed by skimming

In the above problem it will be seen that the lactometer reading of the skimmed milk is two degrees higher than the normal and that the fat has been lowered from 4.8 to 3 percent which is a percentage reduction of $37\frac{1}{2}$ percent. The percent of solids not fat remains practically the same, which is to be expected since none of it has been removed. As long as the percent of fat alone has been lowered and not the solids not fat, the method of adulteration is by skimming only.

Suppose that in the above problem the normal or control sample was not on hand, how could the method and amount of adulteration be determined judging by the skimmed sample alone? In the first place the low fat test with the high lactometer reading indicates skimming; and secondly, by comparing the percent of solids not fat with the figures in Table 30, the percent of fat which corresponds with this amount of solids not fat can be obtained, and then the amount of adulteration computed. In this case the percent of solids not fat was 9.10 percent which according to the table corresponds to milk containing approximately 5 percent of fat, which means the original milk must have contained around this amount of fat. Thus the fat was reduced by skimming from approximately 5 percent to 3 percent, which is a reduction of about 40 percent. This corresponds closely to the results secured in the first case with the control sample on hand.

Example of Watering. The solids not fat cannot be removed from milk through any ordinary means. Their content can be reduced, however, by diluting the milk with a liquid of lower specific gravity such as water. While watering does not actually remove the solids, yet the result is the same as if they were removed because of the greater volume through which the same amount of solids must be distributed.

When water is added to milk all the solids are reduced in the same proportion. For example, if the water lowers the percent of solids not fat by one-fourth, it will also lower the fat by one-fourth. Thus, the percent of fat that is removed by watering milk can be ascertained by the change in percent of solids not fat.

Since water is lighter than milk the addition of it would lower the specific gravity of milk, and so a low lactometer reading and a low percent of solids not fat indicate watering, especially if a high percent of fat is present. The following is an example of watering:

	% Fat	Q. Reading	% S.N.F.
Normal	4.5	30.9	8.85
Watered	3.6	24.7	7.08
	0.9		1.77

$0.9 \div 4.5 \times 100 = 20\%$ $1.77 \div 8.85 \times 100 = 20\%$
Fat removed by added water   S.N.F. removed by added water

In studying the above example, it will be noted that the watered milk has a fair percent of fat, but has a very low lactometer reading and a low percent of solids not fat. This shows that water has been added to the milk. If the percent of fat had been lowered by skimming, the lactometer reading of the adulterated milk would have been higher than the normal and the percent of solids not fat would have remained unchanged.

It will also be seen that the percent of solids not fat and the percent of fat are both reduced by 20 percent. This is in accordance with the rule that if water is added to milk all the solids are reduced in the same proportion. It should be borne in mind that when both the solids not fat and the fat are reduced the same amount, the adulteration was done by watering alone. If the percent of fat has been lowered more than the solids not fat, the additional fat lost would have been due to skimming. This would be a case of both skimming and watering, which will be taken up later. It should also be remembered that the amount of fat removed by watering is calculated from the reduction in solids not fat. This removal of fat cannot be figured solely by the reduction of fat, because, judging by the fat content alone, the method of adulteration cannot be told, since part of the fat may have been removed by skimming.

Skimming and Watering. In the preceding examples of adulterated milk, it was seen that skimming raises the lactometer reading and watering lowers it. By using a combination of both skimming and watering, it is possible to get a lactometer reading and percent of solids not fat that correspond with the fat test. The following problem will illustrate the results of adulterating milk by both skimming and watering:

	% Fat	Q. Reading	% S.N.F.
Normal	5.6	32	9.4
Watered and sk.	3.0	29	8.0
	2.6		1.4

2.6 ÷ 5.6 × 100 = 46.4%   1.4 ÷ 9.4 × 100 = 14.9%

46.4% Total percent of fat removed
14.9% Fat removed by watering
──────
31.5% Fat removed by skimming

In studying the above example we know that the milk has been watered because the percent of solids not fat has been lowered. It was lowered from 9.4 percent in the normal milk to 8 percent in the adulterated, which is a reduction of 14.9 percent. Since 14.9 percent of the solids not fat was removed by watering, 14.9 percent of the fat was removed by the same means, because when water is added to milk all the solids are reduced in the same proportion. However, it is seen that more than this amount of fat was removed, the total amount being 46.4 percent. This means that the additional amount of fat removed by skimming would be the difference between the total percent of fat removed and the percent removed by watering, which in this case would be 46.4 minus 14.9, or 31.5 percent. Whenever the percent of fat removed and the percent of solids not fat removed are the same, the adulteration is watering only, but when the percent of fat removed is greater than the percent of solids not fat removed, we know that the milk was both skimmed and watered.

Making a further study of the example, suppose that only the adulterated sample was on hand. We see that it contains 3 percent fat and 8 percent solids not fat. Judging by this composition alone, the sample would pass as normal milk of low solids content, because the fat and solids not fat are approximately in the right proportion, but if it was known that this milk came from a Jersey herd or some other high testing breed, it would immediately be seen that it was adulterated, since both the fat and solids not fat are too low for that grade of milk. The procedure to follow in that case would be to get a sample of normal milk from the herd and check its composition with the adulterated sample. The approximate amount of adulteration could then be determined in the same way as illustrated in the preceding problem.

BIBLIOGRAPHY

1. Jenness, R., et al. The Density of Milk Fat. J. D. Sci. 25:949-960 (1942).
2. U. S. Dept. of Health, Education and Welfare. Public Health Service-Milk Ordinance and Code-1953 p. 3.
3. Herrmann, Louis F. Indirect Estimates of the Solids-Not-Fat Content of Milk. USDA Agric. Marketing Service (1954).
4. Ystgaard, O. M., et al. Determination of Total Solids in Normal and Watered Milks by Lactometric Methods. J. D. Sci. 34:689-694 (1951).
5. Waite, R., and Abbot, J. The Measurement of Milk Solids-Not-Fat by Density Methods. Proc. XIV Int. Dairy Congress 3,2:699-710 (Rome, 1956).
6. Waite, R. Further Measurements of Milk Solids-Not-Fat by a Density Method. Proc. XV Int. Dairy Congress 3:1559-1565 (London, 1959).
7. Lawrence, A. J. Lactometric Determinations of Solids-Not-Fat in Milks of Individual Cows. Aust. J. D. Tech. 13,3:144-145 (1958).
8. Sharp, P. F., and Hart, R. G. The Influence of the Physical State of the Fat on the Calculation of Solids from Specific Gravity of Milk. J. D. Sci. 19:683-697 (1936).
9. Herrington, B. L. The Calculation of Total Solids by Means of the Sharp and Hart Equation. J. D. Sci. 29:87 (1946).
10. Watson, Paul D. A Lactometer Method for Determining the Solids in Milk. USDA Agr. Res. Ser. Bull. ARS-73-10 (1956).
11. Gould, I. A., and Armstrong, T. V. Let's Get Up-To-Date on Our Physical and Chemical Determinations. Proc. Lab. Sect. Milk Ind. Foundation (1958).
12. Jacobson, M. S. Butterfat and Total Solids in New England Farmers' Milk as Delivered to Processing Plants. J. D. Sci. 19:174 (1936).

REVIEW QUESTIONS

1. If a quart of water weighs 2.08 pounds, what would a quart of Babcock sulphuric acid weigh?
2. What advantage has the Quevenne lactometer over the Board of Health?
3. How far will the B. of H. lactometer sink in water? How far will the Quevenne sink?
4. What is a pycnometer?
5. Name the solids-not-fat in milk.
6. In normal mixed herd milk does the specific gravity increase or decrease with an increase in the fat test? Explain.
7. How does adulterating milk by skimming affect the lactometer reading?
8. If milk is watered will the amount of fat be lessened as much as the solids-not-fat? Explain.
9. Why is it possible to adulterate milk by both skimming and watering, and still leave the composition apparently normal?
10. Which has the greater percentage of solids-not-fat, whole milk or skimmilk? Holstein or Jersey milk?

CHAPTER VIII

Tests for Milk Quality

Modern milk handling methods require milk and other dairy products to remain fresh for ever lengthening periods of time between production of milk at the farm and consumption of these dairy products in the home. The trend to every-other-day pickup of milk from farms, to five or six-day operation of milk plants, and to home deliveries on a two or three times weekly basis means that milk must be of increasingly better quality if it is to maintain its desirable properties until it is consumed.

Present day processing methods and a more critical consuming public have caused the food industries to re-evaluate their quality standards. Quality means different things to different people and standards are ever changing, slowly but surely. Freeman[1] has pointed out that our concepts of milk quality depend on many factors, among them the following: (a) Educational background of the individual, (b) economics and social status of the individual, (c) scientific information available, and (d) the ordinances, laws, regulations, etc., under which the dairy industry operates. Until very recently, quality evaluations for raw milk were based primarily on bacteria counts. Regulations were established to keep milk clean and properly refrigerated in order that the bacteria in milk might be kept to a minimum. If milk was free from adulteration and had a low bacteria count, it was generally considered to be of good quality.

Today, dairy products are subjected to much more rigid quality standards. The milk industry has been a leader in establishing programs to give the consumer a product that is pure, of good flavor, of attractive appearance, and of desirable keeping quality. These programs emphasize rigorous laboratory examination of milk and dairy products to ensure that this quality is maintained.

The following tests have been selected from the many which are known and regularly used in control laboratories. They were chosen for their simplicity and economy as well as to present a study of the many aspects of milk quality control. Tests for bacteria and adulteration are covered in other chapters and will not be presented here.

Phosphatase Test for Pasteurization

The introduction of the phosphatase test for checking on the proper pasteurization of milk has been a valuable contribution to the milk control officials. Before its development, no satisfactory method was available for determining, whether or not, milk had been heated to the proper temperature and held for the proper length of time, or whether it had been contaminated in later operations.

The principle of the test is based on the inactivation of the enzyme, phosphatase, which is present in all milks. Its inactivation temperature is slightly above that required to destroy the most resistant disease organism likely to be found in milk, as for example, *Mycobacterium tuberculosis*. Since pasteurization destroys all disease organisms, and if the enzyme is inactivated, the disease bacteria must of necessity have been subjected to a temperature and time sufficient to destroy them.

Kay and Graham[2] of England were the first to present a phosphatase test (1935). The principle of the test is to determine the degree of inactivation of the enzyme by measuring the amount of phenol liberated from a phosphoric ester to which a small sample of milk has been added. The principal steps in conducting the test are: (1) The addition of small portions of milk under test to a large excess of phosphoric ester; (2) incubating the mixture for a definite period during which time the enzyme if active will hydrolyze the ester and liberate phenol, and (3) stopping the action and then measuring colorimetrically the amount of phenol produced. They used a Lovibond tintometer for measuring the color. The limit set was 2.3 Lovibond blue units as a standard, as no sample of milk pasteurized at 145° F. for 30 minutes exceeded this color.

Gilcreas and Davis[3] of this country used the Kay and Graham test, but modified it in the method of reading by preparing color standards containing known amounts of phenol instead of using a Lovibond tintometer. The limit set was less than 0.04 mg. phenol per 0.5 ml. of milk, as indicative of proper pasteurization, 143° F. for 30 minutes.

Both these tests require an incubation period of 24 hours, which is a disadvantage. However, they are sensitive to small errors. Burgwald[3] found that both tests detected milk which had been heated to only 142° F. for 30 minutes, milk which had been held for only 25 minutes or less at 143° F., and pasteurized milk to which had been added 0.1 percent of raw milk.

In 1951, Kosikowsky[4] published information on a phosphatase test which contained only one buffer substrate concentration and one precipitant concentration for all dairy products. This method had several inherent advantages, including an ability to give accurate determinations on samples of unknown age or origin in addition to the obvious advantage of being readily adaptable to use with the many different dairy products. The test was presented as three different procedures to permit a long but extremely sensitive determination, a short (one hour) method, and a rapid field test.

The Official Methods of Analysis (AOAC)[5] list two procedures which are accepted for official determinations. These are the method of Gilcreas, using the Folin-Ciocalteau phenol reagent, and the Sanders and Sager Method, which uses Gibbs reagent in its determination. However, these procedures are rather complex and thus seldom run in quality control laboratories.

Scharer Field Test for Pasteurization of Milk. Scharer[6] modified the phosphatase test by shortening the incubation period to one hour in place of 24 hours in the laboratory method, and used, in place of Folin's reagent, 2,6 dibromoquinonechloroimide solution (Gibb's reagent) for measuring the free phenol. This laboratory method was found to be as accurate as the Kay-Graham and Gilcreas-Davis tests. He also developed a field test which can be completed in 10 to 15 minutes. This test has lately been improved so that it is nearly as sensitive as the longer laboratory methods. The necessary reagents and glassware are packed in a compact kit, which can be carried direct to any plant for checking the pasteurization efficiency. No technical training on the part of the operator is necessary to obtain accurate results. It is a popular test and is quite generally used. Detailed directions for using the improved rapid field test follow:

Scharer Field Kit
Courtesy Will Corp.

(a) Reagents

1. Neutral n-butyl alcohol (Neutral Butanol).

2. B. Q. C. tablets (yellow). These consist of 2,6 dibromoquinonechloroimide, and when used are dissolved in 5 ml. of ethyl or methyl alcohol.

3. Buffer substrate tablets (white). These contain the phosphoric ester, disodium phenyl phosphate. To prepare the tablets for use, crush one tablet into a test tube and dissolve in 5 ml. of distilled water. Then add 2 drops of the B. Q. C. solution and allow 5 minutes for color development. Next extract the indophenol with 2 to 2.5 ml. of n-butyl alcohol. Allow to stand until alcohol layer has separated at top of tube. Remove alcohol layer with medicine dropper and discard. Dilute remainder to 50 ml. The solution is then phenol free.

(b) Procedure

(1). Add 0.5 ml. milk to 5 ml. of the buffered substrate in a test tube.

(2). Stopper the tube with gum rubber stoppers and shake briefly.

(3). Incubate for 10 minutes in a water bath at 36° C. to 44° C. (97° F. to 111° F.) or if no water bath be available, incubate in vest pocket for a 15 to 20 minute period.

(4). Next add 6 drops of the B. Q. C. solution and mix well. Allow to stand 5 minutes and then compare with color standards. Appearance of the blue color indicates inadequate pasteurization.

The Cornell Phosphatase Test[4]
(a) Reagents* and Materials

1. Carbonate buffer substrate: Dissolve 11.50 g. anhydrous Na_2CO_3, 10.15 g. anhydrous $NaHCO_3$ and 1.09 g. pure disodium phenylphosphate in water and make up to 1 liter (pH=9.80).

2. Trichloracetic-hydrochloric acid precipitant: Dissolve 25 g. trichloracetic acid crystals in water, make up to 50 ml. with water; add 50 ml. conc. HCl (approximately 36 percent) and mix thoroughly.

3. Sodium carbonate solution (8 percent): Dissolve 80 g. anhydrous Na_2CO_3 in water and make up to 1 liter.

4. Copper sulfate-Calgon solution (for milk and all dairy products except ripened cheese): Dissolve 500 mg. $CuSO_4.5H_2O$ and 20 g. sodium hexametaphosphate crystals (tech.) in water and make up to 1 liter.

5. Calgon solution (10 percent) (for ripened cheese only): Dissolve 100 g. sodium hexametaphosphate crystals (tech.) and make up to 1 liter (pH of solution approximately 6.3).

6. 2,6-Dibromoquinonechloroimide solution (BQC): Dissolve 50 mg. BQC in 10 ml. absolute ethyl or methyl alcohol and store in dark bottles.

7. Color standards.

Make the following solutions as preliminary to making standards.

(a) Stock phenol solution: Dissovle 1 g. phenol crystals in water and make up to 1 liter.

(b) Buffer solution: Make 1 liter of carbonate buffer containing 11.50 g. Na_2CO_3 and 10.15 g. $NaHCO_3$.

(c) Diluted phenol solution: Using 4 ml. of stock phenol solution (7a) make up to 500 ml. with buffer solution (7b). This solution contains 8γ phenol per millimeter.

Preparation of color standards. Place in clean 16x150 mm. test tubes 0.5 to 5 ml. portions of diluted phenol solution (7c). Add enough buffer (7b) so that total volume of liquid in each tube is 10 ml. Then add 1 ml. of copper sulfate-Calgon solution (reagent No. 4) to each tube. Finally, add two drops of BQC and four drops USP chloroform and mix. Let stand for 15 min. at 37° C. Seal tubes with paraffin wax and store in refrigerator. Tubes

*Unless otherwise directed, use C.P. chemicals and distilled water.

containing 0.5, 1.0, 1.5, 2.5 and 5.0 ml. portions of diluted phenol solution will produce standards of 4.0, 8.0, 12.0, 20.0 and 40.0γ, respectively, after final color development.

Color standards of 2.0 and 5.0γ can be easily obtained by making 4.0 and 20.0γ solutions without color development and diluting these with sufficient buffer solution (7b). Color is then developed in 10-ml. portions by adding copper sulfate-Calgon solution and BQC.

To obtain alcohol color standards, 5 ml. of butyl alcohol are added to a duplicate series of final aqueous color standards. The tubes are inverted 10 times to extract color and the cork stoppers are sealed with wax.

(b) Procedure
I. Long Method. 18 to 24 hr. at 32 to 37° C.

1. Sampling and incubating. For milk and other fluid dairy products[a] 1 ml. of milk or product is transferred to a 25x150 mm. test tube. This is followed by the addition of 10 ml. of warm (40° C.) carbonate buffer substrate and four drops of USP chloroform.

For cheese and other solid dairy products[b] a representative 0.5-g. portion is placed in a 25x150 mm. test tube. The cheese is macerated thoroughly with a glass rod, 1 ml. of warm (40° C.) carbonate buffer substrate is added and the cheese or other solid dairy product is stirred into a paste. Then 9 ml. more of the buffer substrate and four drops of USP chloroform are added and mixed thoroughly.

a. Includes all fluid dairy products such as milk, cream, chocolate milk, buttermilk, ice cream mix, evaporated milk, whey, etc.
b. Includes all solid dairy products such as cheese of all types, butter, milk powder, etc.

A piece of parchment paper is fitted over the tube using a rubber band and the milk or cheese solution is incubated at 32 to 37° C. for 18 to 24 hrs.

2. Precipitation. After incubation 1 ml. of trichloracetic-HCl precipitant is slowly added to the tube. The resulting protein precipitate is filtered off through Whatman No. 42 paper (11 cm.).

3. Color development. For milk and all dairy products except ripened-type cheese, 5 ml. of the clear filtrate is pipetted into a 16 x 150 mm. test tube. One ml. of $CuSO_4$-Calgon solution and 5 ml. of 8 percent Na_2CO_3 are added. Then two drops of

BQC solution are placed in this solution. The tubes after mixing are inserted in a water bath at 37° C. for 15 min. Color development is measured after this interval against suitable color standards or in a colorimeter.

For ripened cheese only, including fresh curd and green cheese, the same color development procedure as for milk is used except that 1 ml. of plain 10 percent Calgon solution is substituted for the 1 ml. of $CuSO_4$-Calgon solution.

4. Interpretation of results. All final color readings are, after consideration of control values, multiplied by a factor of 1.2 to convert γ phenol per 0.5 ml. or 0.25 g. Any value over 5 γ per 0.5 ml. milk or other fluid dairy product or over 5 γ per 0.25 g. cheese or other solid dairy product is tentatively considered to indicate either underpasteurization or the presence of raw milk products, or a combination thereof, using the long method.

II. Short Method. 1 hr. at 37° C.

The short method, even to the extent of using the same size samples (1 ml. milk, 0.5 g. cheese), is conducted in the same way as the long method except that (a) incubation is carried out for 1 hr. at 37 to 38° C., and (b) chloroform is omitted.

In interpreting the results all final color readings, after consideration of control values, are multiplied by a factor of 1.2 to convert γ phenol per 0.5 ml. or 0.25 g. Any value over 2.0 γ per 0.5 ml. milk and other fluid dairy products or per 0.25 g. cheese and other solid dairy products is tentatively considered to indicate underpasteurization or the presence of raw milk products, or a combination thereof, using the short method. Where values of 1.0 to 2.0 γ per unit of sample are attained, it is advisable to confirm using the long method.

Rapid field test. 10 min. at 37° C.

Samples are prepared and incubated with buffer substrate exactly as for the short (1-hr.) test. After 10 min. at 37° C., 1 ml. of incubated solution is removed and placed in a small cup of a white spot plate. One drop of BQC is added and color development noted after 3 min. against that of properly pasteurized samples. Dark grey or blue indicates underpasteurization or the presence of raw products. (Note: Remaining 10-ml. portion in test tubes can continue to be incubated for the 1-hr. or 18-hr. test without loss of accuracy if confirmation of the rapid field test is necessary.)

Controls for long and short methods. One-half ml. of milk or 0.25 g. of cheese, preferably from pasteurized stock, are placed in test tubes, heated to 170° F. for 15 sec. in a water bath and then cooled immediately. These heated controls are tested by the same method as employed for samples of unknown history. All portions of the samples being tested should be heated.

Alcohol extraction. If necessary, butyl alcohol extraction may be used in either the long or short method, especially on critical values. Five ml. of N-butyl alcohol is added to the test tube containing the aqueous colored solutions, and the latter is inverted ten times. The clear layer appears without centrifuging and is compared against alcohol standards. Alcohol extraction is preferred where a colorimeter is not used, as aqueous phenol standards for all methods deteriorate as a rule relatively rapidly.

Factors Affecting the Phosphatase Test on Milk. Certain factors[3] have been studied as to their effect on the phosphatase test of milk. Age of sample was found to have no apparent effect, provided it was stored at a temperature that would prevent any appreciable growth of bacteria. Samples were held as long as 50 days with no effect on the test. When samples soured, however, they reacted as underpasteurized when using Folin's reagent, but not with Gibb's reagent. Certain bacteria[7] decompose protein into such decomposition products as tyrosine and tryptophane which react with the Folin's reagent but not with Gibb's. However, the number of organisms must be very large to have any effect, usually millions per milliliter.

Several preservatives in milk have been reported as affecting the test, but not mercuric chloride, which is the one ordinarily used. Carbolic soaps, phenol or similar disinfectants must be avoided when making a phosphatase test as free phenol will be present.

There is considerable recent evidence that certain bacteria may produce phosphatase or free phenol in stored milk and cream samples which have been pasteurized by HTST methods. Consequently, positive phosphatase tests on aged samples must be regarded as somewhat inconclusive. Eddleman and Babel[8] observed that ultra-high temperature pasteurization resulted in varying degrees of phosphatase reactivation, depending on such factors as temperature of heating, holding time, and fat content. Fram[9] showed that phosphatase reactivation occurs with milk samples heated by HTST systems but not when pasteurized by holder methods (145° F. for 30 minutes).

Sediment Test of Milk

One of the quality tests for milk is the sediment test. It shows the care and sanitation that have been exercised in the production of the milk. Under the best conditions some particles of foreign material will get into the milk pail, but the presence of any appreciable amount of dirt is due to careless methods. The sediment test will bring to the attention of the producer this adverse condition.

In making the test, a one pint sample should be taken from the well mixed contents of a can of milk or from the recently dumped milk in a weigh vat. Warm the sample to 70°-100° F. and put through a sediment tester such as the vacuum sediment tester. The milk is drawn through a cotton lintine disk of $1\frac{1}{4}$ inch diameter, fuzzy side toward the milk and the foreign particles are there collected. When installed in the sediment tester, a circle of $1\frac{1}{8}$ inch diameter is exposed. The disks are then usually scored according to photoprints which may be obtained from the American Public Health Association, 1790 Broadway, New York 19, New York.

Normal fresh milk will pass through the sediment disks easily but milk that is partially coagulated due to high acid, freezing or excessive heat will not pass through readily and sometimes cannot be tested for sediment. Some prefer to sample for sediment directly off the bottom of unagitated cans. This can be readily accomplished by means of a plunger type tester but it must be recognized that any agitation of the cans prior to sampling will give variable results.

The advent of the farm bulk tank has created new problems in sediment control. It has been shown[10] that a gallon of mixed milk from a bulk tank will give equivalent results to a pint sample taken from a 10-gallon can using the off-bottom technique. These workers also reported that the sediment tester could be modified to permit the use of a smaller sample if desired. By using stainless steel disks, they reduced the size of the opening from $1\frac{1}{3}$ inches to 0.64 inches for quart samples and 0.44 inches for pint samples. In this manner, the disks could be referred to the previously mentioned standards for comparison.

Several new methods for obtaining and testing bulk milk samples for sediment have been proposed[11] but as yet there has been no general acceptance of any single method for determining sediment scores on bulk cooled milk. Several problems have been

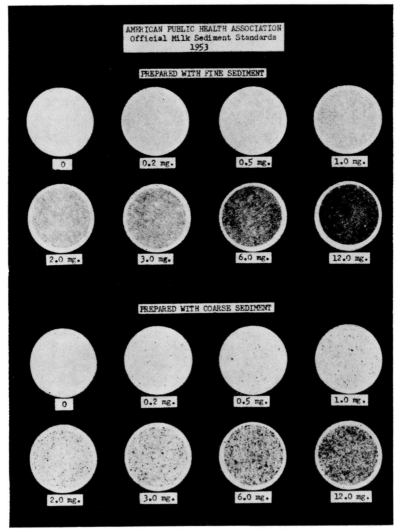

Range of sediment as prepared for official grading. This reproduction is not suitable for control work in the field. Original photo prints are available from the American Public Health Assn., 1790 Broadway, New York 19, N. Y.

encountered in bulk milk which were not evident in can milk and thus efforts to date must be considered as exploratory in nature.

Milk Sediment Tester
Courtesy Sediment Testing Supply Co.

Viscosity

As pointed out in Chapter II, the viscosity of milk has little practical importance for the quality control laboratory. However, many laboratories like to test the viscosity of their cream because many housewives still mistakenly believe that cream viscosity or thickness is an indication of the richness of the cream. Though there have been a number of different methods introduced to determine the viscosity of dairy products, two very simple tests remain as favorites. Both of these tests measure relative thickness, and results are usually recorded on a simple comparison basis.

A. Pipette Method

A measuring pipette of convenient size, such as 50-100 ml., and marked at some arbitrary spot below the bulb, is used to make the comparison. The sample to be tested is tempered to a standard temperature in the range of 60°-70° F. The pipette is then tempered by drawing water of the same temperature into the pipette several times. After the water is discharged, the sample is drawn to the upper mark of the pipette. The time required for the sample to discharge to the lower mark is recorded in seconds and comparisons made on a day-to-day basis.

B. The Borden Flow Meter[11]

This method was first published by Nair and Mook[12] in 1933 and has been popular as a simple viscosity test since that time.

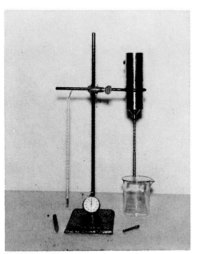

Borden Flow Meter
Courtesy The Borden Company

It has been improved by substituting a stainless steel reservoir for the original glass half pint bottle but the dimensions of the outlet tube and tips are the same as in the original model. Tips are available in different bore sizes to handle products of widely varying viscosities.

a. Preparation of Sample

1. Do not handle the cream sample to be tested more than is necessary. Before transferring to the flowmeter gently invert the sample a few times to insure uniformity in the cream. A half pint of sample is sufficient to make one test.

2. Test each sample in duplicate and average the results for the most accurate readings. The instrument may be chilled in water to the approximate testing temperature. The sample should be brought to a temperature a few degrees below the testing temperature by gentle stirring with a chemical thermometer and immersion, if necessary, in a dish of water at 65-70° F. for brief intervals.

3. It is customary to test light cream at 50° F., heavy cream and chocolate drinks at 60° F. and buttermilk at 70° F.

b. Procedure

1. Place the flowmeter in an upright position in a ring stand over a convenient receptacle and insert the pointed end of a skewer into the opening of the efflux tube.

2. Grasp the instrument tightly in the left hand and with the index finger over the overflow tube, pour the sample into the reservoir.

3. If desired, the cream may be strained by pouring through a 20 mesh screen as it leaves the sample bottle.

4. With the index finger still over the overflow tube, stir the sample gently with a thermometer until the desired testing temperature is obtained. The heat from the hand is usually sufficient to raise the temperature slightly.

5. Remove the thermometer. Allow the excess of cream to flow out of the overflow tube into a small beaker.

6. Remove the skewer and record the exact time in seconds required for a continuous stream of cream to flow out of the flowmeter. This value is usually referred to as "body-flow time."

7. Cream that has passed through the flowmeter once will not recover its original viscosity unless it is held in the cooler for several hours.

NOTE: Borden Flowmeter available from Borden Food Products Division, The Borden Co., 350 Madison Ave., New York 17, N. Y.

Feathering and Oiling-Off of Cream in Coffee

Dairy control laboratories need to continually check their coffee cream to be sure that it will not feather or oil-off in coffee. These defects indicate the cream was improperly processed or has excessive age. Feathering is evidenced by the formation of flocculent particles of coagulated milk proteins. The defect is quite objectionable for though the defect may occur with fresh cream, the housewife associates the problem with cream which has soured. Many consumers also object to the presence of visible oil droplets on the surface of coffee, caused by a partial churning of the fat globules during processing and handling. The following tests may be used to determine if a given lot of cream is likely to feather or oil-off when added to coffee.

Test for Feathering of Coffee Cream[11]
Apparatus
1. Torsion balance and weights.
2. Test tubes with corks or aluminum test tube covers.
3. Pipette, 9 ml., cream.
4. Pipettes, 1 ml. graduated.
5. Erlenmeyer flask, 150 ml.
6. Boiling hot water bath.
7. Bunsen burner or hot plate.
8. Test tube rack.
9. Fahrenheit thermometer fitted with stopper for test tube.
10. Test tube tongs.
11. Graduate, 100 ml.

Reagents
1. Nescafe 100 percent pure "Instant" soluble coffee, pH 5.0.
2. Citric acid stock solution (3.2 gm $C_6H_8O_7 \cdot H_2O$ per liter).
3. Sodium citrate stock solution (11.2 gm $Na_3C_6H_5O_7 \cdot 2H_2O$ per liter).
4. Distilled water.

Procedure*
1. Weigh 3.0 gm of Nescafe soluble coffee into Erlenmeyer flask and dissolve in exactly 197 ml. of distilled water. Mix thoroughly to insure complete solution. Adjust pH of solution to 5.0 ± 0.1 with dilute NaOH if necessary. Most samples of Nescafe have a pH of 5.0.

2. Pipette 9 ml. portions of coffee solution into 6 test tubes.

3. Pipette increasing amounts of citric acid from 0.1 to 0.5 ml. into 5 of the test tubes; add sufficient distilled water to each tube to total a 1 ml. addition to each tube.

4. Add 1 ml. distilled water to the 6th tube and place the Fahrenheit thermometer in this tube.

5. Shake the test tubes sufficiently to mix the contents; stopper loosely and place in the boiling water bath.

6. Heat the test tubes until the contents indicate a temperature of 205-212° F.

*From Laboratory Manual, Sealtest Division, National Dairy Products Corp., New York, N. Y.

7. Remove each tube individually from the water bath and add 1 ml. of cream to be tested. Twist or rotate each tube to mix the contents and place in rack at room temperature.

8. After the contents of all tubes have had a chance to settle (generally less than 1 minute) observe for feathering or curdling. The highest concentration of citric acid which fails to cause feathering is recorded as the feathering value in accordance with the table below.

9. If feathering is observed in the test tube containing 0.1 ml. citric acid, repeat the test using sodium citrate solution. The lowest concentration of sodium citrate at which feathering fails to occur is recorded as the feathering value.

10. Record feathering in accordance with the following tables.

Table 31. Feathering Values with Citric Acid.

Tube Containing		
ml. Citric acid solution	ml. Distilled water	Feathering value
0.5	0.5	+5
0.4	0.6	+4
0.3	0.7	+3
0.2	0.8	+2
0.1	0.9	+1
0	1.0	0

Table 32. Feathering Values with Sodium Citrate.

Tube Containing		
ml. Sodium citrate	ml. Distilled water	Feathering value
0.1	0.9	−1
0.2	0.8	−2
0.3	0.7	−3
0.4	0.6	−4
0.5	0.5	−5

Table 33. Interpretation of Feathering Value.

Value	Condition of Cream
+3 to +5	Highly stable*
+1 to +2	Stable
0	Moderately stable
−1 to −2	Slightly unstable
−3 to −5	Unstable, definitely unmarketable

*A value of +5 is mandatory for cream served to patrons who prepare coffee by adding boiling hot coffee (210° F.) to cream.

Oiling-Off Test for Cream[11,13]
(Massachusetts State College Method)
Apparatus

1. Babcock skimmilk test bottle, of type with stem extending down into bottle proper to within ½ inch of bottom and stem should have no side hole in base of neck as in some styles.

2. Babcock centrifuge.

3. Pipette, 1 ml.

Procedure

1. Pipette 1 ml. of cream into a skimmilk test bottle.

2. Add water at least 200° F. (93° C.) to within ½ inch of the base of the neck. Mix thoroughly while adding.

3. Centrifuge for not longer than 10 seconds at the standard Babcock centrifuge speed.

4. Remove the bottle from the centrifuge and tap the side of the bottle with the finger to break the very thin film of cream which has collected on the top. Do not agitate.

NOTE: If the machine is whirled longer than 10 seconds a thick film will form on top which will be difficult to break and the pieces will clog the graduated neck.

5. Add water at 200° F. (93° C.) or above and bring the column up to the top of the neck and centrifuge for 5 minutes.

6. Read the oil layer as soon as the bottle is removed from the centrifuge. Each small division of the graduated scale should be read as 1.

Interpretation: Creams yielding a reading of 3 or more with this test will show a noticeable separation when used in coffee.

Protein Stability

It is sometimes desirable to know whether milk will stand heat treatment without causing coagulation. This is of utmost importance in evaporated milk plants because of the high temperatures employed in the sterilizing process. Testing for protein stability is useful in milk plants as a means of detecting abnormal milk (colostrum, late lactation, etc.) or pasteurized milk which has undergone excessive bacterial deterioration during storage and thus might be approaching the end of its useful life. One test which is simple and yet effective in determining protein stability is the Storrs Test which follows.

The Storrs Test for Protein Stability[14]

This test for protein stability includes mixing increasing amounts of N/10 HCl with portions of the milk, boiling the mixtures for a specified length of time and then examining the samples for coagulation. While originally devised as a control test for the enzymatic treatment of milk, it need not be limited solely to that application.

Equipment and Reagents

10-ml. volumetric pipette
1-ml. pipette, graduated in 0.1 or 0.05 ml.
Supply of test tubes, 16 x 150 mm. (Pyrex)
Test-tube rack
Water bath
N/10 hydrochloric acid

Procedure

Arrange and number a series of test tubes as follows, adding N/10 HCl to each by means of the 1-ml. pipette in the amounts shown:

Tube no.	N/10 HCl (ml.)
0	0.00
5	0.05
10	0.10
15	0.15
20	0.20
etc., as needed	

(Note: HCl in the amounts 0.05, 0.15, 0.25 ml., etc., can be estimated satisfactorily with a pipette graduated in 0.1 ml.)

The tube numbers correspond to 100 x the ml. of N/10 HCl added to each tube. This eliminates the decimal point as well as any need for further interpolation of results.

Add to each tube by means of a volumetric pipette 10 ml. of the milk to be tested. All tubes are then placed in a water bath maintained at the boiling temperature. After 10 minutes the tubes are removed from the boiling water and examined for coagulation by tipping. The number of the first tube in the series which shows coagulation represents the end-point and is recorded as the stability number of the milk. The average stability number for fresh milk has been found to be about 60-70.

Keeping Quality

Modern processing methods in the dairy industry have created a need for a rapid test to estimate the keeping quality of milk.

Recent studies by Doan and coworkers[15,16] have indicated that the Neotetrazolium Reduction Test has considerable merit in estimating shelf life of bottled milk. These workers have shown that this method can be applied without modification to either bulk or bottled cream and to concentrated milk as well as to milk. It has limited value in checking the keeping quality of cottage cheese or cheese curd since the surface spoilage common to these products is not detected by the test. The test is based on the reduction of the neotetrazolium dye (p-p'—Diphenylenebis -2-[3,5-diphenyltetrazolium chloride]) to a pink color during incubation. A positive result on a sample of milk at least 3 days old would indicate spoilage which could be detected organoleptically within a 3-4 day period.

The Neotetrazolium Test[15]
Reagents and Equipment

1. p-p'—Diphenylenebis -2-(3,5-diphenyltetrazolium chloride) (neotetrazolium). 0.2 percent aqueous solution.

2. Incubation tubes, either Thunberg oxidation tubes or tubes so fitted and set up as to be easily evacuated and to be capable of maintaining the vacuum during the incubation period.

3. A means of evacuating the tubes such as a laboratory water aspirator or vacuum pump.

4. Thermostatically controlled water bath incubator (37° C.).

5. Mercury manometer or a vacuum gauge to test the apparatus and assure satisfactory evacuation of the tubes before incubation.

Procedure

1. Pipette 0.5 ml. of neotetrazolium (0.2% aqueous solution) into a clean dry incubation tube (preferably sterile).

2. Pipette 5.0 ml. of milk to be tested into the tube and mix.

3. Temper the contents of the tube to 0° to 5° C.

4. Evacuate the tube at 15 mm. pressure for 3 minutes and then seal the tube.

5. Temper to 37° C. and incubate at this temperature for 4 hours after which observe for a change in color of the milk from white to pink.

Interpretation of Results

Reduction of the dye to a definitely discernible pink color at the end of 4 hours incubation at 37° C. is the criterion for a positive test. Such a result obtained on bottled milk after a minimum of 3 days refrigerated storage, would predict spoilage within a 3 to 4 day period.

Rancidity in the Milk Supply

The increased use of pipeline milkers, farm bulk tanks, and the longer storage of raw milk have caused an increased incidence of rancidity in the milk supply in some parts of the country. This, together with the greater awareness of the role of milk flavor on the consumption of milk, has given considerable importance to this flavor defect and methods for its detection and control. Herrington[17] discussed the enzyme lipase, the rancidity problem and the methods which have been used to evaluate the extent of fat deterioration in an excellent review article.

Workers at Minnesota have modified several older methods of determining the extent of lipolysis to develop a simplified titration technique[18] which has shown considerable value as a general laboratory procedure. The Thomas Test was evaluated and further modified by Slade[19] at the Vermont Experiment Station. He found the test was not closely correlated to the organoleptic method of testing for rancidity but was valid in detecting lipolytic activity in raw whole milk if the sample were tested within 24 hours after collection. The test was also shown to be of value in detecting increases and decreases of lipase activity in farm bulk-cooled milk.

The Thomas Test for Hydrolytic Rancidity in Milk[18,19]
Apparatus and Reagents

1. Centrifuge: A Babcock centrifuge or any centrifuge that will receive an 18 gm., 8 percent milk test bottle may be used. An unheated centrifuge is satisfactory.
2. Boiling Water Bath: Water in the bath should be at a depth sufficient to cover the base of the test bottles.
3. Tempering Bath: This should be controlled to within a temperature range of 150° F. to 160° F. (54.4° C. to 60° C.).
4. Glassware: The glassware required consists of the following — a standard 18 gm., 8 percent milk test bottle; a 1 ml. tuberculin type syringe with a No. 19 needle; a 50 ml. graduated syringe with a No. 15 needle or a standard 17.6 ml. milk pipette; a 50 ml. Erlenmeyer flask; and a 5 ml. microburette.
5. B.D.I. Reagent: (The Bureau of Dairy Industry Reagent)

Thirty gm. of Triton X-100 (a nonionic surface active agent manufactured by Rohm and Haas Co., Philadelphia, Pa.) and 70 gm. of sodium tetraphosphate are made up to 1 liter with distilled water. This reagent is used as it will not change the physical or chemical properties of the milk fat; also it quickly and effectively de-emulsifies the fat in milk.

6. Alcoholic KOH: To 50 grams of KOH add 50 ml. of distilled water and swirl until completely dissolved. (This gives a 50 percent KOH solution.) Next mix 1.08 ml. of 50 percent KOH in 500 ml. of ethanol and standardize until an 0.02N alcoholic KOH solution is established. This solution should be standardized frequently against standard potassium acid phthalate or other suitable standards. 0.1N potassium acid phthalate is made up by placing a sufficient quantity of pure potassium acid phthalate crystals in a weighing bottle and drying at 100° C. (212° F.) for 2 hours. Cool in desiccator containing concentrated H_2SO_4. Weigh out exactly 10.2068 grams and transfer to a 500 ml. volumetric flask containing about 200 ml. of recently boiled and cooled (70° F. to 75° F.) (21.1° C. to 23.8° C.) distilled water. Dissolve crystals until solution is complete. Dilute to the 500 ml. mark with recently boiled distilled water.

7. Indicator Solution: This is prepared by dissolving 1 gm. of phenolphthalein in 100 ml. of absolute ethanol.

8. Absolute Ethanol.

9. Petroleum Ether: Boiling point range 30° C. to 60° C. (86° F. to 140° F.).

10. Aqueous Methyl Alcohol: This consists of equal volumes of chemically pure methanol and distilled water.

Recovery of the Fat

1. Place 35 ml. of milk in an 18 gm., 8 percent milk test bottle. The milk may be added conveniently by means of the 50 ml. syringe or the 17.6 ml. milk pipette.

2. Next, add 10 ml. of BDI reagent and mix thoroughly. (It is important to use only 10 ml. of BDI reagent in the 35 ml. of whole milk as the acid degree value of the fat will decrease with an increase of the amount of reagent used.)

3. Place the bottle in a gently boiling water bath. After 5 minutes

and again after 10 minutes in the bath, agitate the contents thoroughly.
4. After a total time of 15 minutes in the boiling water bath, centrifuge for 1 minute. Sufficient aqueous methyl alcohol is then added to bring the top of the fat column to the 6 percent graduation. Centrifuge for an additional 1 minute period.
5. Place the bottle in the tempering bath (150° F. to 160° F.) (54.4° C. to 60° C.) for 5 minutes. The water level must be at or above the top of the fat column.

Titration

1. Transfer 1 ml. of the tempered milk fat from the test bottle to a 500 ml. Erlenmeyer flask using the 1 ml. syringe and a No. 19 needle.
2. Dissolve the fat in 10 ml. of petroleum ether and 5 ml. absolute ethanol. Add 10 drops of the indicator solution.
3. Titrate to the first definite color change with the standardized alcoholic KOH solution (0.02N) using the 5 ml. microburette.
4. Express results in terms of acid degree value. (Ml. of 1N base required to titrate 100 gm. of fat.) Acid degree is the number of milliliters of normal alkali required to neutralize the free acid in 100 gm. of fat.

Homogenization Efficiency

Homogenized milk is defined[20] as milk "which has been treated in such a manner as to insure break-up of the fat globules to such an extent that, after 48 hours of quiescent storage, no visible cream separation occurs on the milk, and the fat percentage of the top 100 milliliters of milk in a quart bottle, or of proportionate volumes in containers of other sizes, does not differ by more than 10 percent of itself from the fat percentage of the remaining milk as determined after thorough mixing."

The general trend to homogenization in the market milk industry makes it mandatory that milk plants perform some test to ensure that the milk is properly homogenized. The most common method is to collect bottled products from the filler, hold them in the refrigerator in a quiescent state for 48 hours and then run fat tests on the top and bottom portions of the milk to determine whether excessive fat separation has occurred. This may be done conveniently by carefully pouring off the top 100 milliliters into a beaker, then thoroughly agitating this portion of the milk and the remainder which is in the bottle, finally completing the test by making duplicate Babcock or Gerber analyses for fat. Thus, if the top milk contains 4.0 percent of fat, the remainder must test at least 3.6 percent fat. (The difference, 0.4 percent, is 10 percent of 4.0 percent.)

Recently there has been some interest in optical tests for homogenizer efficiency. Work by Deackoff and Rees[21] in this country was improved by Goulden[22] in England. The former workers used wave lengths near 1000 millimicrons with a Beckman Model B Spectrophotometer and found a high degree of resolution and duplicability. A simple direct reading of the percentage of light transmitted through a highly diluted sample (using distilled water as a reference) determined the degree of homogenization. The higher the percentage of light transmitted, the higher was the degree of homogenization. It is of interest to note that commercial skimmilk had 100 percent transmission when tested as below, thus the only factor influencing transmission was the nature of the fat globules in the sample. The authors proposed the method as an ideal way for research workers to investigate such variables as valve design or multiple passes on homogenization efficiency. It also gives the dairy laboratory an opportunity to immediately check the degree of homogenization without waiting for the 48-hour storage period. Goulden worked with undiluted samples and found this method satisfactory if the milk samples were tested at a thickness of about 0.1 millimeter, and wave lengths near the infra-red region were used. However, he found this method was not as sensitive to changes in homogenization conditions as the test proposed by Deackoff and Rees.

Deackoff and Rees Test for Homogenization Efficiency[21]
Equipment and Reagents

Standard commercial spectrophotometer (using 1020 millimicron light source).

Ammonium hydroxide (5 Normal).

Procedure

1. Pipette 1 ml. of whole milk into container at 70° F.
2. Treat with 5 ml. of 5N NH_4OH.
3. Dilute with 250 ml. of water at 120°-130° F.
4. Allow to stand for 30 minutes.

5. Cool to 77° F.
6. Pour sample into cuvette and read percent transmission using distilled water as a reference. Use 1020 millicron light source.

Interpretation

Samples of commercial homogenized milk gave readings of about 70 percent transmission. Deackoff and Rees report a remarkable improvement in quality (particularly taste) and shelf life when transmissions exceeding 70 percent are obtained. Using this technique, the authors reported a single technician was able to test 30 samples in a two to three-hour period, as compared with a full day's work to obtain results by the top and bottom method.

BIBLIOGRAPHY

1. Freeman, T. R. Milk Quality. J. Milk and Food Tech. 15:162-166 (1952).
2. Kay, H. D., and Graham, W. R. The Phosphatase Test for Pasteurized Milk. J. Dai. Rsh. 6, pp. 190-203 (1935).
3. Burgwald, L. H. A Critical Review of the Phosphatase Test. Bul. Int. Assoc. of Milk Dealers. pp. 366-389 (1940-41).
4. Kosikowsky, Frank V. The Effectiveness of the Cornell Phosphatase Test for Dairy Products. J. D. Sci. 34:1151-1158 (1951).
5. Methods of Analysis A.O.A.C. Published by the Assn. of Off. Agr. Chem., Washington, D. C. Eighth Ed. (1955).
6. Burgwald, L. H., and Giberson, E. M. An Evaluation of the Various Procedures for Making Phosphatase Tests. J. Milk Tech. 1, p. 11 (1938).
7. Scharer, H. A Rapid Phosmonoesterase Test for Control of Dairy Pasteurization. J. D. Sci. 21, pp. 21-33 (1938).
8. Eddleman, T. L., and Babel, F. J. Phosphatase Reactivation in Dairy Products. J. Milk and Food Tech. 21:126-130 (1958).
9. Fram, H. The Reactivation of Phosphatase in H.T.S.T. Pasteurized Dairy Products. J. D. Sci. 40:19-27 (1957).
10. Liska, B. J., and Calbert, H. E. A Study of Mixed Milk Sample Methods of Sediment Testing Using Off-the-Bottom Standards and Equipment for Possible Use with Farm Bulk Milk Holding Tanks. J. Milk and Food Tech. 17:82-85 (1954).
11. Milk Industry Foundation. Laboratory Manual. Pub. by Milk Industry Foundation, Washington, D. C. Third Ed. (1959).
12. Nair, J. H., and Mook, D. E. Viscosity Studies of Fluid Cream. J. D. Sci. 16:1-9 (1933).
13. Jenkins, H., and Mack, M. J. A Study of Oiling-Off of Cream in Coffee. J. D. Sci. 20:723-735 (1937).
14. Storrs, Arnold B. A Test for the Protein Stability of Milk. J. D. Sci. 25:19-24 (1942).
15. Day, E. A., and Doan, F. J. A Test for the Keeping Quality of Pasteurized Milk. J. Milk and Food Tech. 19:63-66 (1956).
16. Csenge, J. L., and Doan, F. J. The Neotetrazolium Test for Deterioration of Stored Cream, Concentrated Milk and Cottage Cheese. J. Milk and Food Tech. 21:223-225 (1958).
17. Herrington, B. L. Lipase: A Review. J. D. Sci. 37:775-789 (1954).
18. Thomas, E. L., et al. Hydrolytic Rancidity in Milk. Am. Milk Rev. pp. 50-52,85 (January, 1955).
19. Slade, J. W., Jr. A Study of the Titration Test of Thomas, et al. as a Simplified Means of Measuring the Degree of Rancidity Present in Normal Raw Milk. M. S. Thesis. University of Vermont (1958).
20. Milk Ordinance and Code. 1953 Recommendations of the Public Health Service.
21. Deackoff, L. P., and Rees, L. H. Testing Homogenization Efficiency by Light Transmission. The Milk Dealer (July, 1957).
22. Goulden, J. D. S. A Rapid Optical Test for Homogenizer Efficiency. J. D. Res. 25:320-323 (1958).

REVIEW QUESTIONS

1. What governs a person's concept of milk quality?
2. What is the purpose of the phosphatase test?
3. What are the limits of accuracy of the phosphatase test?
4. What are the advantages of the Cornell Phosphatase Test?
5. What factors affect the phosphatase test on milk?
6. What causes cream to "feather" when added to coffee?
7. What causes cream to "oil-off" when added to coffee?
8. What factors affect the protein stability of milk products?
9. What is the value of the Neotetrazolium Test?
10. When is milk properly homogenized?

CHAPTER IX

The Adulteration of Milk and Milk Products

In 1856, a Massachusetts legislature passed a law prohibiting the adulteration of milk. It was the first law in this country pertaining directly to milk.[1] Later, in 1864 the Massachusetts courts upheld the validity of the act and ruled that a milk dealer was responsible for an adulterated product regardless of his knowledge or ignorance of the situation. Modern courts still hold to this principle.[1] Laws prohibiting the use of harmful preservatives or of preservatives which promote fraud have been upheld by the courts in a number of rulings since 1901. In 1953, the Secretary of the Department of Health, Education, and Welfare published a statement of policy which made the presence of antibiotics in foods intended for human consumption, or the addition of such drugs, directly or indirectly, to the foods, an adulteration within the meaning of the Federal Food, Drug, and Cosmetic Act. Any food thus adulterated is prohibited from shipment in interstate commerce.[2]

The most common form of milk adulteration has been adding water to milk or removing cream from milk. This form of adulteration has been a problem throughout the history of the market milk industry and continues to be a problem today. The fact that water may have been added accidentally still constitutes adulteration of the product.

While the reduction of fat percentage by the addition of water, the skimming of cream, or by a combination of the two are the common methods of adulterating milk, dairy literature makes it readily apparent that many other types of adulteration have been used as the industry developed. In general, adulteration is practiced to either substitute cheaper ingredients for more expensive ones or to deceive the buyer into thinking the product is more valuable or of better quality than is actually the case. Local, state, and national regulatory bodies have adopted strict measures to prevent the use of any additive to foods which might promote fraud and/or which might be detrimental to the health of the consuming public.

The food industries are now faced with a problem relating to technological progress in these industries rather than to any deliberate attempt to deceive the public. Welch[3] has pointed out that in the processing of food many foreign substances, such as antioxidants, coloring materials, flavors, coatings, stabilizers, emulsifiers, etc., are commonly used. New cleaning materials, sanitizers, and lubricants are employed which may leave traces of residue in the product. Food packages incorporate many new plastics, enamels, coatings, etc., which are further possible sources of adulteration. The Food and Drug Administration is quite concerned that some of these agents, which are used in the preparation and distribution of food, may be inherently toxic or may have an accumulative effect on health. There is a possibility that certain combinations of them may be found to have a synergistic toxic effect. There is no easy way to determine the effect any given additive may have on human health.

Substitution of Cheaper Ingredients

Added Water. As we have noted, adding water to milk is probably the most common form of milk adulteration. In Chapter VII, the use of the lactometer for determining whether or not milk had been skimmed or watered was discussed. While the lactometric method is the most common means for testing milk samples for added water, the lactometer cannot detect low levels of dilution. Results are questionable when samples contain less than 10% of added water.

The Official Methods of Analysis of the AOAC accepts a refractive index determination as an approved method in testing for added water. This test is based on the property of milk serum to deflect rays of light passing through it. If the milk was watered, the deflection would be altered and the change could be observed on a Zeiss immersion refractometer. The serum is obtained by acidification of the milk or by treating milk with a solution of copper sulfate. However, these tests have questionable accuracy and thus have been quite generally discarded.[4]

The method most commonly accepted for official analyses of

milk to determine if water has been added is based on the fact that the freezing point of the sample is a sensitive way of detecting such adulteration. The present official method makes use of the Hortvet Cryoscope, a procedure first approved in 1923. Actually the freezing point of milk as determined by the Hortvet method is not the true freezing point of milk but rather a value based on empirical procedure designed for handling ease and reproducibility of results.[5]

The literature since 1953 indicates considerable question concerning the usefulness of freezing point data as determined by the Hortvet technique. This is partly due to the fact that the instrument is expensive but also that the procedure is time-consuming and lacking in reproducibility of results.[6] A most important problem, however, is a disagreement concerning the proper freezing point base. As mentioned in Chapter II, the weight of evidence indicates that the freezing point should be near $-0.543°$ C. rather than the presently accepted level of $-0.550°$ C. Some recent workers, however, contend that the value of $-0.550°$ C. is satisfactory and the higher values result from careless production techniques.[7]

The introduction of electrical "thermistors" in 1939 led to an adaptation of a basic method for determining the osmotic pressure of blood to establish a new technique for making freezing point analyses on milk. While the instrument is expensive, the results are rapidly obtained and are reproducible within a limit of $0.001°$ C. Using this instrument, known as the Fiske Cryoscope, and precooling samples in an ice bath, Levowitz[4] was able to analyze about 30 samples per hour. The use of this instrument has created new interest in freezing point techniques for determining added water and the detailed procedure is presented in the 11th Edition of Standard Methods for the Examination of Dairy Products.[9] The reader is directed to comprehensive discussions on the subject by Robertson,[8] Dahlberg[5] and Levowitz.[4]

The Immersion Refractometer. The Zeiss immersion refractometer is an instrument for measuring the refractive index of a liquid. The serum of milk has the ability to deflect rays of light passing through it and if the milk has been watered the deflection will be altered and the change can be observed on the refractometer. This procedure is more accurate than the lactometer method as it will detect the addition of small amounts of added water. The instrument, however, is relatively expensive.

The test, using the copper serum method, is made as follows: Make a copper sulphate solution by dissolving 72.5 grams of $CuSO_4 \cdot 5 H_2O$ per liter and adjust this solution if necessary to read 36 at 20° C. on the scale of the refractometer, or to a specific gravity of 1.04433 at 20°/4° C. Add one part by volume of this copper sulphate solution to four parts by volume of milk. Shake well and filter to obtain the clear serum. Determine the refractometer reading on the serum at 20° C. A reading below 36 indicates added water.

Added Water in Milk — Thermistor Cryoscope[9]

Apparatus

a. Assembled cryoscope — Consists of:
   1. Bath to permit rapid adjustment of sample to near freezing temperature,
   2. Stirrer for test portion of sample,
   3. Automatic supercooling control,
   4. Freezing mechanism, and
   5. Thermistor probe, to permit interpretation of temperature readings on scale, range 0.001° to −1° C., with successive 0.001° C. graduations at 1.2 mm intervals (calibrate instrument as per directions).

b. Sample tubes — Without lip, 16x100 mm (older models may require tubes 16x100 mm with lip).

c. Rubber stopper — Or other closures for each tube to prevent evaporation or contamination of contents.

d. Test tube rack — To hold tubes vertical.

e. Bath — 2-3 in. deep to hold ice water and cracked ice.

f. Pipettes — Clean and dry, 2 ml. (disposable pipettes are convenient).

Standardization of Cryoscope

Use either of the following standards:

a. Sucrose solutions 7 and 10% — 7 and 10 g. sucrose, analytical grade, each made up to 100 ml. with water, or

b. 7 and 10% sucrose solution equivalents — 0.6892 and 1.0206 g. NaCl, analytical grade, in 100 g. water.

   Adjust dial to 7% at −0.422° C. and 10% at −0.621° C. as

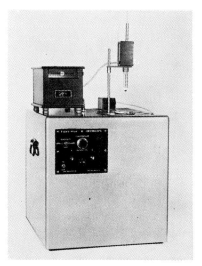

Advanced-Fisk Milk Cryoscope
Courtesy Advanced Instruments, Inc.

follows (detailed steps are in instruction manual for each instrument, and earlier models may vary slightly from directions below):

1. Rotate calibrator A maximum clockwise.
2. Freeze 7% standard, temperature set at $-0.442°$ C. Balance with calibrator B only.
3. Freeze 10% standard. Balance with temperature knob only.
4. Record temperature as X. Compute $Y = \dfrac{621-X}{3.7}$
5. Freeze 7% standard, temperature set at $-0.422°$ C. Then rotate A calibrator Y turns counterclockwise. Balance with calibrator B only.
6. Freeze 10% standard. Balance with temperature knob only.
7. Repeat Steps 3 to 6 until instrument reads within ± 1 unit of standard values.

Determine freezing point of standards, using the same test procedure as for unknowns. Need for standardization and recalibration of instrument will probably be infrequent. A new curve or complete recalibration will usually be needed only after replacing a damaged thermistor.

Procedure

Although slight improvements have been incorporated in successive models of the cryoscope, it is essential to use the same technique with each instrument throughout tests, both on standard solutions and on milk samples. Use thoroughly mixed samples of milk having titratable acidity not over 0.18% without further preparation.

Transfer about 2 ml. sample to test tube. Precool sample by placing tube in ice bath. Place test tube with temperature probe, stirrer and freezing assembly in instrument bath.

Cool sample. Switch to Position 1, or "COOL" or "STIR." Set temperature knob to expected value for sample or to base freezing point. Stop bath circulation at prescribed point by switching to Position 2, or "SUPERCOOL," or by lifting bubbler tube out of bath. Freeze sample as galvanometer spotlight switches to prescribed point by switching to Position 3 or "FREEZE."

Rotate temperature knob to adjust galvanometer spot to zero. To obtain exact point, depress high-sensitivity switch and adjust further to zero the galvanometer spot when it reaches its extreme swing to the right.

Note — On newer models, use alternative procedure as follows: Leave "temperature" at base freezing point. Depress high-sensitivity switch and observe galvanometer spot move to right of zero. The reading corresponds with scale graduations indicating directly the approximate percentage of added water in sample.

With both old and new models read temperature knob as the true freezing point of sample, T.

Precautions

First determinations should be pilot runs to establish freezing point range. Normally only second and third runs are needed to establish most accurate freezing points. Use identical technic on all tests. Closely control exact supercooling step. Wait for maximum right-hand movement of galvanometer before reading scale.

Periodic checks include:
1. Calibration against standard solutions
2. Bath temperature (normal range $-4°$ to $-7°$ C.)
3. Position of thermistor (black speck at tip should be near center of 2 ml. sample)

4. Bath level
5. Motion of stirrer and freezer-vibrator

Calculations

Calculate added water as follows:
One percent of water raises the freezing point of milk about T/100.
By common usage,

$$\% \text{ Added Water} = 100 \frac{(T - T')}{T}$$

Where T = Base Freezing Point of Authenticated Samples
T' = True Freezing Point of Test Sample.
For example, if
Authenticated sample freezes at — 0.542° C,T, and
Unknown milk freezes at — 0.520° C,T' then

$$\% \text{ Added Water} = \left[\frac{-0.542 - (-0.520)}{-0.542} \right] 100$$

= approximately 4.0%.

Substitution of Vegetable Fats for Milk Fat in Dairy Products

In recent years technological advances in the dairy industry combined with marked differences in value of milk fat and other fats have caused some operators to substitute cheaper fats for milk fat in dairy products. Considerable interest[10,11,12] in the problem of detecting foreign fats in milk fat has been shown since 1951.

Keeney[11] has observed that this type of adulteration is not new — that references relating to the subject date back to the turn of the century. However, in the period since 1950, there has been renewed interest in preventing the fraudulent use of vegetable fats and other animal fats in butter and ice cream. More recently, Robertson[13] has reported many cases where vegetable fats were used in a product resembling Mozzarella cheese.

Bhalerao and Kummerow[12] presented a critical discussion of some of the techniques which have been recommended to detect the presence of foreign fats in butterfat. They discussed methods based on: (1) butyric acid content, (2) urea fractionation of butterfat, (3) unsaponifiable matter, (4) unsaturated fatty acids, (5) iso-valeric acid content, (6) addition of indicator substances to hydrogenated fats, (7) glyceride composition of butterfat and (8) the formation of pink colored hydroxamic acid complexes from the esters of the lower fatty acids of milk fat. They prepared deliberately adulterated samples in the laboratory and found that the only method useful for detecting as little as 10% of a substitute fat was the one based on the glyceride structure

Table 34. Limitations of the Various Methods Suggested for Detecting Adulteration of Butterfat at 10% Level.[12]

Method	Butterfat	Variation for butterfat	Butterfat + coconut oil	Butterfat + Crisco	Butterfat + oleo oil	Butterfat + synthetic fat[a]	Butterfat + mixture[b]	Butterfat + re-arranged lard[c]
Butyric acid method (M%)	10.1	9.6–11.3	9.3	9.2	9.0	10.0	9.3	10.1
Tocopherol method (γ per g. fat)	31.6	up to 50γ per g. fat	39.4	166.0	31.5	36.1	71.0	32.0
Phytosterol acetate method	113.5° C.	113–116	115.1° C.	121.6° C.	114.0° C.	114.3° C.	117.6° C.	113.2° C.
Critical temp. of dissolution method	44° C.	42–46° C.	40° C.	47° C.	46° C.	44° C.	41° C.	46° C.
Polenske method	1.8	1–2.6	2.9	1.6	1.6	2.3	2.6	1.6
Glyceride structure:								
A. Trisaturated glycerides of alcohol soluble fraction of fat (%)	28.5	26.3–29.6	34.2	25.8	26.1	29.0	32.6	26.3
B. nD at 40° C. of alcohol-insoluble acetone-soluble fraction of fat after iodination	1.4732	1.4713–1.4732	1.4700	1.4800	1.4785	1.4783	1.4746	1.4780

a. Synthetic fat mixture of 25% coconut oil + 25% beef fat + 50% lard rearranged with 7.5% tributyrin.
b. Mixture of 75% coconut oil + 25% hydrogenated cottonseed oil.
c. Lard rearranged with 7.5% tributyrin.

(see Table 34). Synthetic fat mixtures could be prepared having essentially the fatty acid composition as milk fat and thus would not be detected by other methods. On the other hand, milk fat has a characteristic glyceride structure which no one has been able to duplicate.

Despite the above observations concerning the inadequacy of the tests for butyric acid in detecting adulteration by synthetic mixtures of fats, it is reported[13] that certain refinements of the butyric acid technique are being given consideration by the Association of Official Agricultural Chemists since the use of this method in conjunction with other tests will detect vegetable fat mixtures added to butterfat even if the mixture contained added tributyrin.

Keeney[14] has proposed a rapid screening test for milk fat adulteration which is based on the observation that pure milk fat when crystallized from alcohol, yields fat crystals which entrains and holds small air bubbles when the mixture of crystals and solvent is shaken. Milk fat crystals thus become buoyant and float in the solvent. Milk fat adulterated with other fats does not have this property and thus the crystals sink. The test is a qualitative method, and confirmatory tests of a chromatographic nature will tell the extent of adulteration.

Improving Quality or Appearance

Preservatives in Milk. The addition of preservatives to improve keeping quality of dairy products or even to delay spoilage for a considerable period of time was a problem for regulatory bodies in the early history of dairying. Since most preservatives may be detected by comparatively simple tests, their use was soon discouraged by legal means and their use is uncommon today. However, these tests hold interest for students in dairy testing because of their simplicity in chemical analysis.

Formaldehyde Test. Small amounts of formaldehyde in milk, one in 200,000 parts, may easily be detected. It may be done in connection with the Babcock test. Pipette the regular amount of milk into a test bottle, add a few drops of ferric chloride solution ($FeCl_3$) and then add a measure of sulphuric acid. Shake the bottle gently with a rotary motion but not enough to mix the contents, and if formaldehyde is present a lavender colored ring will appear at the junction of the milk and acid.

Usually it is not necessary to add the ferric chloride, because ferric salts are nearly always present as impurities in the sulphuric acid used for testing. The test will not function if too much formalin has been added to the milk.

"Boiled Milk" Tests. Sometimes it is desirable to know if a sample of milk has been heated to 175° F. or above. The two tests commonly used are as follows:

(1). Place about 15 ml. of the milk in a test tube, add a few drops of hydrogen peroxide (H_2O_2), and then add a few crystals of paraphenylenediamine hydrogenchloride, $C_6H_4(NH_2)_2 \cdot 2$ HCl. Shake thoroughly and if the milk has been boiled or heated to at least 175° F., it will remain white, otherwise it will turn blue.

(2). Use the same amount of milk as in the first test, add two or three drops of a boiled starch solution, five or six drops of a 10 percent solution of potassium iodide and finally a drop of hydrogen peroxide. Mix thoroughly. The results will be the same as in test No. 1.

The above results are due to the action of the enzyme, peroxidase. It has the property of being able to liberate nascent oxygen from the peroxide, and this free atom, in the first test, acts directly on the paraphenylene compound and turns it blue. In the second test the free oxygen liberates iodine from the potassium iodide, and this iodine turns starch blue. The heated milk shows no color because the enzyme has been inactivated by the heat.

Test for Carbonates. An excess of sodium carbonate in milk may be shown as follows: To 10 ml. of the milk in a test tube, add 10 ml. of 95 percent alcohol and 2 drops of a one percent aqueous solution of rosolic acid and mix. A rose color indicates carbonates present, whereas normal milk shows only a brownish coloration. Rosolic acid is an indicator changing at pH 7.0 to 8.0, which shows the milk is made slightly alkaline by the carbonates.

Test for Boric Acid. To 25 ml. of milk add lime water until the milk is alkaline, evaporate to dryness on a steam bath, and burn to an ash in a small porcelain or platinum dish. To the ash add a few drops of dilute hydrochloric acid; then add a few drops of water and place in this water solution a strip of turmeric paper. The paper will turn a cherry red color if either borax or boric acid is present. A further check is made by adding a few drops of ammonium hydroxide to the same paper, in which case it will turn a dark blue-green.

Some Additions to Improve the Physical Characteristics of Milk

Another type of adulteration which appears from time to time is the use of agents which will restore or increase such physical characteristics as consistency or viscosity and the addition of coloring agents to restore lost color or to improve the appearance of a cream layer.[15,16,17] This type of adulteration generally occurs only in isolated cases and holds limited interest today.

Smith and Doan[18] reported an interesting study based on the illegal practice of using reconstituted superheated condensed skimmilk in standardizing high test milk to expand the cream volume. They found maximum cream volume could be obtained if the ratio of fat to superheated protein ranged between 2.85 and 4.0. Milk thus adulterated exhibited normal values for most of the common means of analyses. Detection was made possible by the fact that the cream layer exhibited subnormal fat to casein and fat to solids-not-fat ratios. The adulterated milk gave more sediment on centrifuging. It also had a higher ratio of casein to albumin and globulin.

Gnagy[19] presented a spectrophotometric method for the detection of eleven different stabilizers used in soft curd cheeses. Bundesen and Martinek[20] identified the various gums in concentrations as low as 0.1 gram per 100 grams of product. They used trichloroacetic acid to precipitate the protein and then treated the protein-free solution with alcohol to precipitate the gums. The alginates, which precipitate with the proteins upon the addition of trichloroacetic acid, were separated from the proteins with a saturated solution of magnesium sulfate.

Coloring Matters[21] — **Official Test.** About 150 ml. of milk is warmed in a casserole over a flame. Add about 5 ml. acetic acid (1+3) and heat slowly, with stirring, to nearly boiling. The curd is gathered into one mass with a stirring rod and the whey is poured off. If the curd breaks up into small particles, the material may be strained through a sieve or colander to separate the curd from the whey. Press the curd to free from adhering liquid and transfer to a small flask. Add about 50 ml. ether, macerate while keeping flask tightly corked. Shake at intervals and allow to stand for several hours and preferably over-night. Decant the ether extract into an evaporating dish, evaporate the ether, and test the fatty residue for annatto. This is done by neutralizing the residue with NaOH and pouring on a moistened filter paper. Paper will absorb color and when washed with a gentle stream of water, it will remain dyed a straw color. Dry the filter paper, add one drop of $SnCl_2$ solution and again dry. The presence of annatto is confirmed if the paper turns purple.

The curd of uncolored milk is perfectly white after a complete extraction with ether, as is also that of milk which has been colored with annatto. If the extracted fat-free curd remains distinctly orange or yellowish in color, the presence of a *coal-tar dye* is indicated.

Many times if a lump of fat-free curd in a test tube is treated with a little HCl, the color will change to pink and the presence of a dye similar to aniline yellow or butter yellow or possibly one of the acid azo yellows or oranges is indicated. These may be identified by following the proper official methods.

Sometimes the presence of coal-tar dyes can be detected by treating about 100 ml. of milk with an equal volume of HCl in a porcelain casserole, giving the dish a slight rotary motion. The curd becomes pink if certain coal-tar dyes are present.

Accidental Adulteration

Detecting Foreign Residues in Milk. The increased use of chemicals in the control of insect pests, in the control of mastitis and other diseases of dairy animals, and in the cleaning and sanitation of milking utensils has caused a serious problem for the dairy industry and the regulatory bodies supervising the production of milk. Residues of certain of these chemicals have been found in the milk supply and are considered to be adulterants. Some are relatively easy to detect, others require long and expensive procedures.

The dairy industry has been concerned about residues of sanitation chemicals for thirty years or more. The usual concern is based on the belief that their presence in milk suppresses bacterial growth. Since bacteria counts are used as an indication of the sanitary quality of milk, it is believed that these residues could thus be used to cover up improper milk production practices. There is also some feeling that the residues from quaternary ammonium compounds may hinder desirable bacterial multiplication in cheese making. While there is considerable controversy in the literature concerning the role of sanitation chemicals on milk quality and in cheese making, there is no question that these are not normal constituents of milk and therefore should not be found there.

In recent years the dairy industry has come to use considerable quantities of antibiotics in the treatment of animal diseases, notably mastitis. When such treatments are made, the milk from the treated animal is not supposed to be used for human consumption for at least 72 hours following the last treatment. If such milk is added to milk from untreated cows before the 72-hour period is completed, residual antibiotics may be detected in the entire supply. These antibiotics may enter the milk directly, as in intermammary treatments for mastitis, or indirectly through the blood stream of an animal which has been intravenously injected with antibiotic preparations for some pathological condition. There are also reports that large amounts of antibiotic supplements in grain may result in the detection of residues in the milk.

The use of certain ingredients in barn and livestock sprays has been prohibited because residues of these chemicals may pass into the milk. Likewise, certain sprays for crops which will later be eaten by milk cows have been prohibited or their use permitted only at certain periods of development because they may be recovered in the milk supply. The presence of antibiotics or pesticides in milk or other dairy products has serious health implications and hence they must be used only as specified.

There are many tests for measuring the concentration of chlorine or quaternary ammonium compounds in sanitizing solutions. Simple tests to detect their presence in milk will be given below. Penicillin apparently has greater influence on public health than other antibiotics in common use and thus has held the main interest in developing test procedures for residual antibiotics. Recently, there have been attempts[22,23] to find some dye which may be incorporated into antibiotic preparations which will (1) pass into the milk supply for the same period of time as penicillin and (2) be immediately detected by simple laboratory methods. While results look promising, none of the dyes has been accepted to date and the Food and Drug Administration now question whether or not these dyes may not contribute their own residue problem.[24] Many tests, based on bacteriological techniques, have been suggested for laboratory analysis of residual antibiotics. However, being dependent on the growth of certain bacteria, these tests generally require more time to obtain results than the industry cares to wait. The test given below is the method recommended by the Food and Drug Administration for the detection of penicillin in milk. Present tests for pesticide residues are too detailed for general laboratory analysis.

Quaternary Ammonium Compounds in Water and Milk[25]

(a) Reagents

(a) Tetrachloroethane (technical grade).

(b) Lactic acid (reagent grade), adjusted to 50 percent by weight in distilled water.

(c) Eosin yellowish dye (Biological Color Commission Index 768) solution prepared at rate of 0.5 mg. dye per 1 ml. of distilled water.

(d) 4.0 molar sodium hydroxide, analytical reagent.

(e) Citric acid (monohydrate, analytical reagent) buffer prepared by adjusting a 25 percent solution of citric acid in distilled water to pH 4.5 with 50 percent sodium hydroxide (analytical reagent).

(f) Anionic surface active solution prepared at rate of 0.1 mg. of 100 percent pure Aerosol OT (dioctyl sodium sulfosuccinate) per milliliter of distilled water to provide final concentration of 0.01 percent of active Aerosol OT (Pure Aerosol OT obtained from American Cyanamid Co.).

(b) Procedure

Place the following in a test tube: 5 ml. tetrachloroethane, 2 ml. 50 percent lactic acid, and 5 ml. milk sample. Cork tube and shake vigorously for one minute. Add 2 ml. of 4.0 molar sodium hydroxide to tube and invert six times to mix contents. Rubber stoppers have proven unsatisfactory. Cork stoppers or glass-stoppered tubes suitable for centrifuging are preferable.

Separate tetrachloroethane from milk solids and water by centrifugation. About 5 minutes are required at 3200 r.p.m. in a 10-inch centrifuge. Three distinct layers should result following centrifuging. The top, aqueous layer should be clear; the middle, solid layer should consist chiefly of white, precipitated protein; and the lower layer should be clear tetrachloroethane containing any QAC originally in the milk.

Remove the top layer by decantation or with an aspirator. Separate as much of the tetrachloroethane as possible from the curd. Place a 2-ml. aliquot of the tetrachloroethane in a clean tube containing 0.5 ml. of pH 4.5 buffer and 0.2 ml. of eosin solution. Cork tube and shake vigorously. A pink to red color in the tetrachloroethane indicates presence of QAC. Titrate to a colorless end point with 0.01 percent Aerosol OT solution. Vigorous shaking is required during the titration.

Sensitivity of the above described method for QAC in milk can be increased considerably by increasing the ratio of milk to solvent. However, the percent of recovery of total QAC present is reduced. The ratio of 5 ml. of sample to 5 ml. of tetrachloroethane provided recoveries approaching 100 percent with three out of four QACs tested. Preliminary trials indicated that application of photometric methods using a wave length of 545 mμ. might prove suitable for increasing sensitivity.

Test for Hypochlorites and Chloramins. This test may be used for detection of sodium or potassium hypochlorite and similar sterilizing powders or liquids commonly used in water to rinse out utensils. To 5 ml. of the milk add 1.5 ml. of potassium iodide solution (7 grams of KI in 100 ml. of distilled water). In the presence of a small amount of hypochlorite, the milk becomes yellow or brownish in color.

A Rapid Disc Assay Method for Detecting Penicillin in Milk[26]
(a) Reagents

Culture media. For carrying the test organism and for performing the assay, use a medium of the following composition (medium No. 1): Peptone 6 Gm., pancreatic digest of casein 4 Gm., yeast extract 3 Gm., beef extract 1.5 Gm., dextrose 1 Gm., agar 15 Gm., and distilled water to make 1,000 ml. The final pH after sterilization is 6.5 to 6.6. For carrying the test organism use a medium of the same composition as medium No. 1 except that it contains in addition 300 mg. of $MnSO_4 \cdot H_2O$ per liter (medium No. 2). Medium No. 1 in dehydrated form may be purchased from commercial suppliers.

Working standard. Dilute the penicillin working standard with sterile 1 percent phosphate buffer pH 6.0 (2.0 Gm. dipotassium phosphate and 8.0 Gm. monopotassium phosphate per liter) to make a stock solution containing 100 units per ml. This stock solution should be stored in the refrigerator.

(b) Procedure

Preparation of sample. The sample is undiluted milk.

Preparation of test organism. The test organism is *Bacillus subtilis*, ATCC 6633. Maintain the test organism on nutrient agar medium No. 1 and transfer to a fresh slant every month. Inoculate a fresh slant of agar medium No. 1 with the test organism and incubate at 37° C. for 16-24 hours. Wash the culture from the slant with sterile physiological saline onto the surface of a Roux bottle containing 300 ml. of medium No. 2 Incubate at 37° C. for 5 days. Wash the resulting growth from the agar surface with 50 ml. of sterile physiological saline, centrifuge, and decant the supernatant liquid. Reconstitute the sediment with sterile physiological saline and heat-shock the suspension by heating for 30 minutes at 70 C. Maintain the spore suspension in the refrigerator. This spore suspension will keep for several months. Add about 2 ml. of the spore suspension to each 100 ml. of agar medium No. 1 for the test.

The amount of inoculum to be used should be determined in the following manner at the time a new spore suspension is prepared.

Prepare plates with varying amounts of inoculum. For example, 1 ml., 2 ml., and 3 ml. of spore suspension per 100 ml. of agar. To each plate add control discs prepared from milk to which penicillin has been added to give a concentration of 0.05 units of penicillin per ml. Incubate the plates as described for the test. At the end of the incubation period observe the inoculum that showed the best response (considering both sensitivity and discernibility of the zone of inhibition) to the control discs. Use this inoculum for preparing plates to be used for the examination of milk samples.

Preparation of plates. Add 10 ml. of inoculated agar (prepared as above) to each 20 x 100 mm. Petri dish. Plastic dishes may be used. Distribute the agar evenly in the dishes, cover with porcelain covers glazed only on the outside and allow to harden. Store prepared plates in the refrigerator. Remove each Petri dish as needed and use within 15 minutes.

Discs. Use round, white paper discs with a diameter of ¼ inch. The paper used for the discs should be Schleicher and Schuell No. 740-E, No. 470-W, or No. 470, or paper of comparable grade, absorption, performance qualities and purity. No. 740-E

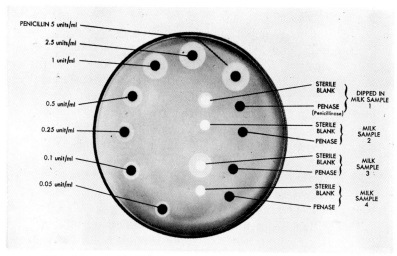

Disk Plate Technique (for Detecting Antibiotics in Milk)
Disk Plate Technique — Bacto — Concentration Disks

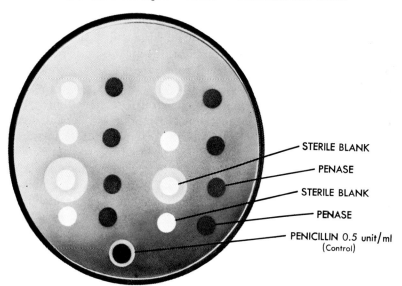

White disks (sterile blanks) dipped in different samples of milk. Zone of inhibition indicates presence of penicillin in milk.
Courtesy Difco Laboratories

is available as discs already punched to the recommended size. The others are available in sheets from which discs may be punched.

Assay procedure. Use a pair of forceps and dip a paper disc completely into the sample to be tested. Shake the disc to remove any excess milk and place on the surface of the agar, touching the disc gently with the tip of the forceps to ensure proper contact of the disc with the agar. Place the discs so that they are at least 20 mm. apart when measured from center to center to avoid overlapping of zones. In this manner many samples may be tested on the same Petri dish. Place a control disc containing 0.05 units of penicillin per ml. on each plate. Incubate at 37° C. for approximately 2½ hours and then examine for any sign of inhibition of the test organism. At this time the test organism will have grown sufficiently so that if penicillin is present in milk at a concentration of at least 0.05 units per ml., a zone of inhibition will be discernible. A definite indication of a zone of inhibition is a positive test. The plate should be held at various angles to the light source until the light is at the proper angle for best observing the zone of inhibition.

Identify Test for Penicillin

If a zone of inhibition is obtained at the end of the incubation period, it may be due to (1) penicillin, (2) an antibiotic other than penicillin, (3) other antibacterial substances. To determine if the activity is due to penicillin, proceed as follows:

Add 0.05 ml. (approximately one drop) of penicillinase concentrate to a 5 ml. aliquot of the positive milk sample, and shake well. Prepare 3 discs from this treated sample, and 3 discs from an untreated aliquot of the sample. Place all discs on one plate. If there is sufficient sample, prepare several replicate plates. Then proceed as described for the test. At the end of the incubation period, observe the plates. A zone around the untreated sample discs, but no zone around the treated sample discs is a positive test for penicillin.

BIBLIOGRAPHY

1. Tobey, James A. Legal Aspects of Milk Sanitation. Second Edition. Published by Milk Industry Foundation (1947).
2. Kirk, J. K. Antibiotics and Our Milk Supply. Proc. Vt. D. Plant Op. and Mgrs. Assn. (1957).
3. Welch, Henry. Problems of Antibiotics in Food as the Food and Drug Administration Sees Them. Am. J. Pub. Health 47,6:701-705 (1957).
4. Levowitz, David. The Problem of Added Water in Milk, and Its Detection. J. Milk and Food Tech. 23,2:40-42 (1960).
5. Dahlberg, A. C. Does the Freezing Point Determination Tell What We Need to Know About Added Water in Milk? Proc. Milk Ind. Foundation Lab. Section (1958).
6. Shipe, W. F. Cryoscopic Techniques for Milk. Proc. XV Int. D. Cong. 3:1844-1850 (1959).
7. Marcus, Theodore. The Control of Watered Milk. Proc. Vt. D. Pl. Op. and Mgrs. Assn. (1957).
8. Robertson, A. H. Cryoscopy of Milk, A 1954-1956 Survey. J. Assn. Off. Ag. Chemists. 40,2:618-662 (1957).
9. American Public Health Association. Standard Methods for the Examination of Dairy Products. Eleventh Edition (1960).
10. Kummerow, F. A. Detection of Foreign Fats in Dairy Products. Proc. Milk Ind. Foundation Lab. Section (1952).
11. Keeney, Mark. Detection of Foreign Fats. Proc. Milk Ind. Foundation Lab. Section (1953).
12. Bhalerao, V. R., and Kummerow, F. A. A Summary of Methods for the Detection of Foreign Fats in Dairy Products. J. D. Sci. 39:956-964 (1956).
13. Robertson, A. H. Some Forms of Adulteration in Dairy Products. J. Milk and Food Tech. 21:154-158 (1958).
14. Keeney, Mark. A Presumptive Crystallization Test for Milk Fat Adulteration. Ice Cream Trade Journal (June, 1953).
15. Vallier, Georges. La Solubilisation De La Caseine Par L'Ammoniague: Son Utilisation Dans La Recherche Des Falsification Du Lait Frais Par Addition de Lait en Poudre. Le Lait 35:372-377 (1955).
16. Choi, R. P., et al. Determination of the Protein Reducing Value of Milk as an Indication of the Presence of Nonfat Dry Milk Solids. J. Milk and Food Tech. 16:241-246 (1953).
17. Clay, C. L., et al. Report on Detection of Reconstituted Milk in Fluid Market Milk. D. Sci. Abst. 17,8:702 (1955).
18. Smith, A. C., and Doan, F. J. Observations on the Effect of Additions of Heat Thickened Protein to Fluid Milk on the Creaming Phenomenon. J. D. Sci. 33:406 (1950).
19. Gnagy, H. J. Report on Gums in Food. A Spectrophotometric Method for the Detection of Certain Stabilizers in Soft Curd Cheeses. D. Sci. Abst. 18,5:446 (1956).
20. Bundesen, H. N., and Martinek, M. J. Procedure for the Separation, Detection, and Identification of the More Common Vegetable Gums in Dairy Products, With Special Reference to Alginates. J. Milk and Food Tech. 17:79-81, 105 (1954).
21. AOAC. Official Methods of Analysis. 8th Ed. (1955).
22. Smitasiri, T., et al. Dyes as Markers for Antibiotic-Contaminated Milks. J. Milk and Food Tech. 21:255-258 (1958).
23. Hargrove, R. E., and Plowman, R. D. Indirect Detection of Antibiotics in Milk. Proc. XV Int. Dairy Congress. 3:1411-1417 (1959).
24. Hargrove, R. E., et al. Use of Markers in Veterinary Preparations for the Detection of Antibiotics in Milk. J. D. Sci. 42,1:202-206 (1959).
25. Furlong, T. E., and Elliker, P. R. An Improved Method of Determining Concontration of Quaternary Ammonium Compounds in Water Solutions and in Milk. J. D. Sci. 36:225-234 (1953).
26. Arret, Bernard, and Kirshbaum, Amiel. A Rapid Disk Assay Method for Detecting Penicillin in Milk. J. Milk and Food Tech. 22:329-331 (1959) and 23:32 (1960).

REVIEW QUESTIONS

1. What are the common types of adulteration of dairy products?
2. How have modern technological advances contributed to the problem of adulteration in the dairy industry?
3. What is a "cryoscope?"
4. What methods have been used to detect the presence of foreign fats in butterfat? Are they equally accurate?
5. What is the principle of the Keeney Rapid Screening Test for Fat Adulteration?
6. What problem has the use of mastitis curatives caused for the dairy industry?
7. Why have certain livestock sprays been prohibited from use on dairy cattle?
8. What factors may be responsible for the zone of inhibition in the test for penicillin?

CHAPTER X

Analysis of Butter and Cheese

I. Butter

Composition of Butter. Butter consists of butterfat, moisture, salt and curd. It varies greatly in composition, but creamery butter would on an average, analyze about as follows:

Fat	80.8%
Moisture	15.9%
Salt	2.4%
Curd	0.9%

Previous to the year 1923 the Internal Revenue Department enforced both a fat and moisture standard for butter, below 80 percent fat and above 16 percent moisture being considered illegal butter. A test case, however, was carried to the United States Supreme Court, and that body ruled that only the 80 percent fat standard could be enforced. Therefore, Congress on March 4, 1923, passed an act defining butter as follows:

"That for the purpose of the Food and Drug Act of June 30, 1906, 'butter' shall be understood to mean the food product usually known as butter, and which is made exclusively from milk or cream, or both, with or without common salt, and with or without additional coloring matter, and containing not less than 80 percentum by weight of milk fat, all tolerances having been allowed for."

Thus, as far as the Federal government is concerned, the moisture content is not considered. This is logical because the food value of butter lies in the fat and not in the other constituents. This standard, therefore, allows the buttermaker to incorporate more moisture in unsalted than in salted butter, as

208   CHEMISTRY AND TESTING OF DAIRY PRODUCTS

long as the fat does not go below the minimum requirement. As far as the consumer is concerned, butter with 18 percent moisture and no salt would have as much food value as butter with 16 percent moisture and two percent salt.

Overrun. It is, however, essential for the creameryman at the time of churning to know the moisture content of the butter so as to obtain a check on the overrun in the plant. By making a moisture and salt test of the butter and allowing around one percent for curd, and then subtracting the sum of the percentages of these three constituents from 100, the percent of fat can be obtained. If the calculations show too high a percent of fat, more water can be worked into the butter to bring the fat content down near the legal limit. Three different methods of calculating overrun are used, namely, the factory, churn and composition overruns.

1. Factory overrun may be defined as the increase in pounds of butter obtained over the pounds of fat purchased. This increase is possible because of the moisture, salt and curd retained in the butter. In a well regulated creamery this overrun is around 22 to 23 percent.

Example: A Creamery purchased 1000 lbs. of cream testing 30 percent, and obtained from it 363 lbs. of butter.

$$1000 \times 0.30 = 300 \text{ lbs. fat purchased}$$
$$363 - 300 = 63 \text{ lbs. gain in butter over fat}$$
$$63 \div 300 = 21\% \text{ overrun}$$

2. Churn overrun is the percentage increase as calculated from the pounds of butter made over the actual pounds of fat that went into the churn, and thus not accounting for any of the losses occurring between the weigh can and the churn. This method checks the efficiency of the buttermaker.

Example: Ten pounds of the cream in problem 1 were unaccounted for, leaving 990 lbs. testing 30 percent that went into the churn. The pounds of butter were 363.

$$990 \times 0.30 = 297 \text{ lbs. fat}$$
$$363 - 297 = 66 \text{ lbs. gain in butter over fat}$$
$$66 \div 297 = 22.22\% \text{ overrun}$$

3. Composition overrun is based on the chemical analysis of the butter. It is the sum of the percentages of moisture, salt and curd, divided by the percentage of fat.

Example: The butter analyzed 80.8 percent fat, 16.2 percent moisture, 2.2 percent salt and 0.8 percent curd.

$100 - 80.8 = 19.2\%$ sum of H_2O, salt and curd
$19.2 \div 80.8\%$ fat $= 23.76\%$ overrun

Moisture Test of Butter

Sampling. The securing of a sample of butter is one of the important steps in testing for moisture, as butter varies greatly in water content, both in the churn and in the final package. When sampling butter from a churn, cut through the butter with a ladle, to free the product of loose water, and then take a small portion from the side of the cut. Repeat this operation in ten to twelve different places, so as to be certain of a representative sample. Manhart[1] found, after working the butter 30 revolutions, the moisture content at the gate end, the middle and gear end, respectively, was 14.30 percent, 14.42 percent and 14.20 percent, a maximum difference of 0.22 percent. With eight-ounce and four-ounce prints, take the whole print for a sample.

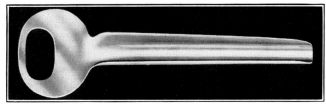

Butter Trier
Courtesy Cherry-Burrell Corp.

Tub butter should be sampled by means of a trier. Insert the instrument diagonally, taking a plug extending from one side at the top, through the center to the opposite side at the bottom. Three such cores should be taken. It is important to have a trier that is a true arc of a circle,[2] otherwise it will make a ragged cut and squeeze out moisture. Some triers give as much as one percent less moisture than the truest one. In sampling pound prints take two entire quarters diagonally opposite each other.

All samples should be placed in a jar that can be sealed tightly so that no moisture can escape. Pint or half-pint fruit jars are excellent receptacles for this purpose.

Preparation of the sample. The preparation of butter samples for analysis is particularly important in order to obtain a homogeneous mass and at the same time not allow any loss of moisture to occur during the operation. Many different procedures are followed, but the more important ones are the two official methods of the Association of Official Agricultural Chemists and the so-called factory method.

The directions for preparing a sample of butter by the official methods are as follows:

Shaking Method[3] — Procedure

Soften entire sample in sample container, by warming in H_2O bath maintained at as low temperature as practicable, not over 39°. Avoid overheating which results in visible separation of curd. Shake at frequent intervals during softening process to reincorporate any separated fat and observe fluidity of sample. Optimum consistency is attained when emulsion is still intact but fluid enough to reveal sample level almost immediately. Remove from bath and shake vigorously at frequent intervals until sample cools to thick, creamy consistency and sample level can no longer readily be seen. Weigh portion for analysis promptly.

Mechanical Stirrer Method[3] — Procedure
Equipment

(a) Cylindrical jar holder. Mounted on shaft of hand centrifuge or similar apparatus for rotation of jar on vertical axis by means of crank and transmission mechanism. Sample container must be cylindrical.

(b) Spatula. Made of steel with wooden handle and fairly stiff blade at least $1\frac{1}{4}''$ wide and $5''$ long, with blade shaped to fit jar.

Preparation

Warm sample in closed sample jar to about 25° C., avoiding melting of any portion to extent that oil and aqueous phases separate. (Most easily managed samples are those that have been warmed to room temperature overnight, but it is not essential that entire contents of jar be thus uniformly tempered, provided all parts are at least in plastic state. In principle, sample should be warmed only enough to permit initiation of mixing.)

Place jar in jar holder and wedge in position with shims of corrugated pasteboard. Remove cover and scrape off any butter adhering thereto with back of spatula blade. Immerse blade in jar with edge against side, holding it at slight angle from vertical and plane of blade at slight angle to radius. Operate crank with other hand, slowly at first, and with increasing speed as lumpiness decreases. In short time lumpiness disappears rather abruptly; from this point tilt spatula occasionally on its horizontal axis, moving back and forth to bring toe of blade little past axis of rotation, then up and down against side of jar (so twisted on its vertical axis as to throw material first to the side and then to center of jar). After lumpiness disappears, mixing 1 minute additional is normally adequate. When mixing has been completed, wipe off blade against mouth of jar and replace cover promptly. Start analysis promptly, in no case later than 1 hour after mixing.

It is suggested that operators make use of either of the following two methods in preparing butter samples for analysis, whether done officially or in the factory.

Method I. Using temperatures as low as possible, soften the entire sample in a closed vessel until sufficiently thin for agitation, avoiding any undue oiling-off or separation of ingredients. Cool and at the same time shake the sample vigorously until a semi-solid mass is obtained. Then open and stir with a spatula or case knife until the entire sample is of the consistency of smooth soft grease. Weigh the portions for analysis at once or, if preferred, hold in air at a temperature that will maintain the sample in proper condition for stirring and then restir immediately before weighing. The proper holding temperature for most butters is between 21°C. and 27°C. though some samples of hard butter may withstand a slightly higher temperature.

Method II. Temper the sample in a closed container set in an incubator or warm room and held at a temperature of 26°C. to 28°C. for a sufficient time to soften the whole butter mass without any appreciable oiling-off. (Water may be used for the tempering if the above conditions are fulfilled and if the difference in temperature of the tempering water and the surrounding air is less than 5°C. so as to prevent any condensation of moisture on the inside of the jar.) Then open and stir with a spatula or case knife until the entire sample is in a uniform condition having the consistency of smooth soft grease. Weigh out the portions for analysis as in Method I.

Laboratory Method. *a. Apparatus.* In the laboratory or factory method of testing butter for moisture, the following pieces of apparatus are necessary: Scales, 10 gram weight, aluminum cup, alcohol lamp, a pair of tongs and air tight jars for the butter sample. Special butter moisture scales, which read the percent of moisture direct, are used. They should be sensitive to 30 milligrams, the same as the cream scales. Two small sliding weights are present on the graduated beams for reading the percent of moisture. There are also coarse and fine tare adjustments for balancing the empty cup; the former is used to nearly balance the cup and the latter, which is very sensitive, to complete the operation. One, new, single pan balance requires no loose weights when samples of 2, 5, or 10 grams are used.

Moisture and Butterfat Balance
Courtesy Macalester Bicknell Co.

Aluminum cups are best for moisture work, because they heat and cool quickly. They should be so constructed as to leave no pockets in which water can collect and they should also be large enough to keep the foam from boiling over.

An alcohol flame is the most satisfactory heat for this method of testing. It gives sufficient heat to drive off the moisture without losing any of the other contents.

b. Operation of the test. The steps in testing the prepared sample are as follows:

(1) Have the scales properly set up. The older balances are leveled in the following way: Move the large tare weight to the extreme right and set all the other weights on the zero mark. If the pointer does not now come to rest at the center, raise or lower the scales by means of the screws (on which the scales rest) until it does.

(2) Balance the empty cup on the right hand pan by means of the tare weights. The cup must be clean and dry. Be sure that the riders for reading the moisture are on zero.

(3) Weigh into balanced cup 10 grams of the prepared butter. Small amounts of butter are added or removed until an exact balance is reached.

(4) Heat the butter over an alcohol flame. Rotate the cup while heating so as to aid the moisture to escape and to prevent the foam from rising too high. If there is danger of the foam running over, remove the cup from the flame until it subsides, or if this is not sufficient, dip the bottom of the cup into cold water. When the moisture is nearly gone, the foam will disappear except for a thin layer remaining on the surface of the fat. At this stage watch the color of the sample and as soon as it has turned brown remove the cup from the flame.

The brown color indicates that all the moisture is driven off. Another factor which indicates the moisture is gone is the pleasant odor, characteristic of butter in cooking, given off at the completion of the test. This odor comes from the fat, butyrin.

(5) Cool the cup and contents to room temperature. During this operation, the cups should be kept covered or put in a desiccator, or otherwise the melted fat would take up moisture from the air as it cools. It will take eight to ten minutes for the cup and its contents to reach room temperature. If care is exercised, the cooling may be done rapidly by holding the cup in cold water, and then wiping it dry before reweighing. Cooling the cup and contents before reweighing is necessary, because the hot air currents arising from the hot cup and fat buoy them up and consequently cause them to weigh less than when cold and no currents are arising. Also the density of the cup and fat is slightly less when hot, because of the expansion due to the heat. With the light aluminum cups ordinarily used this difference in weight is enough to make the moisture test read around 0.3 percent too high, if reweighed while still hot. For rapid work, fairly close results may be secured by rebalancing the cup and contents while hot and deducting 0.3 percent from the reading. For different sized cups a few trials will show what factor to deduct. For the more exact work, however, cool the cup and contents before reweighing.

(6) As soon as the cup and residue are cooled, rebalance and read the percent of moisture. In performing this operation, the 10 gram weight must be on the left hand pan and the cup with

Desiccating Cabinet
Courtesy Fisher Scientific Co.

Desiccator
Courtesy Fisher Scientific Co.

its contents on the right hand pan as before. Then by means of the two weights for reading the moisture, rebalance the cup and contents and read the percent of moisture direct. The best method of rebalancing is to set the lower weight on the 10 percent mark and complete the balancing with the upper weight. The latter is more sensitive than the former and it reads to tenths of one percent, whereas the closest reading with the other is two-tenths of one percent. Thus, if the larger weight is set on the 10 percent mark, and if the smaller one has to be moved over to the 4.8 percent graduation to rebalance the cup and contents, the percent of moisture would be 14.8 percent (10 + 4.8). This means that the moisture in the ten grams of butter weighed 1.48 grams, which is 14.8 percent of the ten gram sample.

Official Method. The sample for the official method is prepared the same as for the laboratory test. Then by means of a chemical balance, between 1.5 and 2.5 grams of butter are weighed into a flat bottomed dish not less than five centimeters in diameter. The dish must previously have been thoroughly cleaned and dried. The sample is weighed to tenths of milligrams. The butter is dried in an oven at the temperature of boiling water, and the samples are weighed at hourly intervals until the weight becomes constant, which indicates that the moisture is gone. The cups and contents must be cooled in a desiccator before each weighing. The percent of moisture is calculated from the loss in

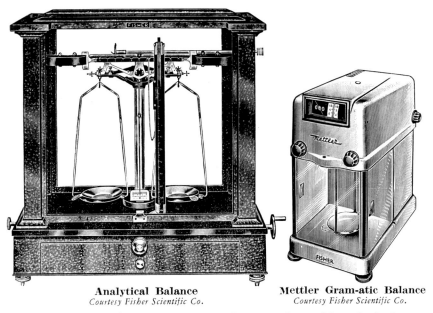

Analytical Balance
Courtesy Fisher Scientific Co.

Mettler Gram-atic Balance
Courtesy Fisher Scientific Co.

weight. The following example will show how this calculation is done:

```
16.5234 grams, wt. of cup and butter
14.2342 grams, wt. of empty cup
─────────
 2.2892 grams, wt. of butter

16.1741 grams, wt. of cup and dried butter

16.5234 grams
16.1741 grams
─────────
 0.3493 grams, wt. of moisture
```

$$\frac{0.3493}{2.2892} \times 100 = 15.26\% \text{ moisture in sample}$$

The writer has found from experience in running many moisture determinations, that it takes close to 2½ hours at the temperature of boiling water to evaporate the moisture from a sample of butter. Thus considerable time is saved by making the first run 2½ hours before removing the sample for the first weighing, and then reheating for 30 minutes as a check. Very few samples require any additional heating beyond these two periods.

Testing Butter for Salt (Cornell Method)

The salt content of butter should be uniform from day to day and of such an amount as the trade desires. It is an advantage to the buttermaker to know the percent of salt in his butter, both from the standpoint of meeting the consumers' demand and in securing as large an overrun as possible.

Principles of the Salt Test. Certain salts form reactions in the same way as do acids and bases. In such cases they follow the same principle of chemistry, namely that equal volumes of acids and bases of the same chemical strength will exactly neutralize each other. Solutions of silver nitrate ($AgNO_3$) and sodium chloride (NaCl) form reactions, and so the former compound is

Thelco Moisture Oven
Courtesy Howe and French

used in determining the quantity of salt (NaCl) in butter. Thus, if the amount of silver nitrate solution of a given strength, that it takes to exactly neutralize all the salt in a definite amount of solution, is known, then the quantity of salt in the sample can be determined. Since butter is in a solid form, it has to be melted in hot water and then a portion of the brine is titrated. By using a definite amount of water for melting a definite amount of butter and then using a definite amount of the brine for titration, the percent of salt in the butter can be read directly from the burette.

Procedure in Testing. The steps in testing butter for its salt content are as follows:

(1) The sample of butter must be taken and prepared as in testing for moisture. Salt varies greatly in butter and so the same pains must be taken to secure a representative sample as in

ANALYSIS OF BUTTER AND CHEESE

the moisture test. The salt test is usually run in connection with the moisture test and so the same sample of butter can be used for both.

(2) Either weigh out 10 grams of the prepared butter or use the residue from the moisture test; as given under the laboratory method.

(3) Measure out exactly 300 ml. of water with a temperature around 140° F. Add a portion of this water to the cup containing the ten grams of butter or the residue from the moisture test, and wash it into a pint milk bottle or some other glass container similar in size. Repeat this operation of adding hot water and rinsing until the moisture cup is cleared of its contents, and then add any of the remaining 300 ml. of water to the bottle. In this test the water must measure exactly 300 ml. for direct reading.

(4) Shake the contents of the bottle thoroughly, so as to distribute the salt evenly throughout the solution. Then let the bottle stand two to three minutes for the fat to rise to the surface.

(5) Into a white cup measure out a Babcock pipetteful (17.6 ml.) of the clear brine. In obtaining this sample see that the tip of the pipette extends through the layer of fat so as to get only the clear salt solution.

(6) Add four or five drops of a 5% potassium chromate solution (K_2CrO_4) made by dissolving 5 grams of potassium chromate crystals in 100 ml. of distilled water. This compound acts as an indicator, showing when all the salt is neutralized by the change in color that takes place. When added to the brine a lemon yellow color is imparted, but when a slight excess of silver nitrate solution over that required for neutralizing the salt has been added to the brine, a light brown color will appear.

(7) Titrate with tenth normal silver nitrate solution. Add the solution, stirring constantly, until the brine turns a light brown color. This is the end point and no more of the silver nitrate solution must be added.

(8) Read to tenths the number of milliliters of silver nitrate solution used in the titration. Each milliliter of tenth normal silver nitrate solution represents one percent of salt. Thus, if it took 2.4 ml. of the solution to neutralize the salt, the salt test of the butter would be 2.4 percent.

A twentieth normal silver nitrate solution may be used, in which case the reading would have to be divided by two to get the percent of salt. The advantage of a weaker solution lies in the fact that if an excess is added the error would not be as great as in case of the stronger solution.

Chemical Reactions in the Salt Test. Two reactions occur in the operation of the salt test as follows:

1. $AgNO_3$ + $NaCl$ = $AgCl$ + $NaNO_3$
   Silver nitrate + Sodium chloride = Silver chloride + Sodium nitrate

2. $2AgNO_3$ + K_2CrO_4 = Ag_2CrO_4 + $2KNO_3$
   Silver nitrate + Potassium chromate = Silver chromate + Potassium nitrate

The first reaction is between the silver nitrate and the sodium chloride, and then, when all the salt has been chemically combined and a slight excess of silver nitrate is added, a reaction occurs between the indicator, potassium chromate, and the silver nitrate as shown in the second equation. The silver chromate formed is the compound that gives the brown color, when all the salt has been chemically combined.

Why the Salt Test Reads Direct. The molecular weight of sodium chloride is 58.5 and that of silver nitrate 170. Thus, a normal solution of sodium chloride may be made by dissolving 58.5 grams of pure salt in a liter (1,000 ml.) of water, and similarly a normal solution of silver nitrate is made by dissolving 170 grams of pure silver nitrate crystals in a liter of water. According to the principle of chemistry, as mentioned on page 216, the following statements may be made:

1,000 ml. $N/1AgNO_3$ sol. will neut. 1,000 ml. N/1NaCl sol. or 58.5 g.
1 ml. $N/1AgNO_3$ sol. will neut. 1 ml. N/1NaCl sol. or .0585 g.
1 ml. $N/10AgNO_3$ sol. will neut. 1 ml. N/10NaCl sol. or .00585 g.

Assume that it took 3 ml. of tenth normal silver nitrate solution to neutralize the salt in a Babcock pipetteful of brine. Since one milliliter of tenth normal nitrate solution will neutralize 0.00585 grams of salt, there would be 3 × 0.00585, or 0.01755 grams of salt in one pipetteful. The pipette delivers 17.5 ml. Then in 300 ml. of solution, the amount used in the test, there would be 300/17.5, or 17 14 pipettefuls. Therefore, the amount of salt in

17.14 pipettefuls, or 300 ml. of the salt solution, would amount to 17.14 × 0.01755, or 0.3 grams. This means that the ten grams of butter contained 0.3 grams of salt. The percent of salt in this butter would be 0.3/10 × 100, or 3 percent. Hence, if ten grams of butter are melted in 300 ml. of water, and one pipetteful of the brine is titrated, the percent of salt may be read direct from the burette without calculation. Each milliliter of tenth normal silver nitrate solution used equals one percent of salt.

If definite amounts are not used the percent of salt in butter can be calculated from the following formula:

$$\frac{\text{ml. N/10 AgNO}_3 \times 0.00585 \times \frac{\text{ml. sol.}}{\text{ml. sample}}}{\text{grams butter}} \times 100 = \% \text{ salt}$$

For example, suppose that ten grams of butter were melted in 200 ml. of water, and that it required 6 ml. of tenth normal silver nitrate solution to neutralize the salt in a 25 ml. sample of the salt solution, then by applying the formula we have

$$\frac{6 \times 0.00585 \times \frac{200}{25}}{10} \times 100 = 2.8\% \text{ salt}$$

Standard Solution Method. Some operators prefer to use the standard silver nitrate solution in place of the deci-normal strength in testing butter for salt. The standard solution consists of 29.06 grams of silver nitrate in a liter of solution. Ten grams of butter or the residue in the moisture test are mixed with 250 ml. of hot water and a 25 ml. sample of the brine titrated with the standard solution. If the 17.6 ml. pipette is used, measure out 176 ml. hot water instead of 250 ml. Each milliliter of silver nitrate solution represents one percent of salt. The principle of the test is as follows:

1. 170 g. $AgNO_3$ combines with 58.5 g. NaCl.
2. 2.906 g. $AgNO_3$ combines with 1.0 g. NaCl, or 10% of the butter, since 10% of the entire amount is titrated (25 ÷ 250).
3. 0.02906 g. $AgNO_3$ represents 0.01 g. NaCl or 1% of salt in the butter (10 g. butter × 10% of butter titrated × 1% = 0.01 g. NaCl).
4. Since 0.02906 g. $AgNO_3$ are contained in 1 ml. of standard solution, 1 ml. represents 1% NaCl.

FAT DETERMINATION OF BUTTER

Kohman Method. In the Kohman method of analyzing butter, gasoline or petroleum ether are used to extract the fat. By running a moisture and salt test at the same time and calculating the curd by difference, a complete analysis of the sample can be obtained. The steps are as follows:

(1). Run a moisture test as usual (see laboratory method, page 211), record the percent of moisture and leave all the weights in the same position as when reading the percent of moisture. Be careful not to move the scales.

(2). Add a portion of gasoline or petroleum ether to the residue, enough to nearly fill the moisture cup (average size). Stir with a glass rod and then let it stand a few minutes until all the solid particles have settled to the bottom.

(3). Pour off as much of the clear solution as possible, being careful not to lose any of the sediment. This clear solution contains the fat.

(4). Repeat steps two and three.

(5). **Slowly** heat until sediment is dry. Do not allow the beaker to rest on the hot plate or flame since rapid heating will cause the solids to pop out of the beaker and thus destroy the accuracy of the test.

(6). Cool and reweigh. The newer balances have beams that read up to 100 percent, and thus the balancing can be done by the movable riders without adding any weights to the pans. The reading will include both the moisture and fat percentages.

If the older balances are used which read only 30 percent, weights must be added. In this case either add 8 grams to the right hand pan with the 10 gram weight still on the left pan, or remove the 10 gram weight and add 2 grams to the right hand pan. This procedure represents 80 percent. Then complete the balancing with the riders and add 80 percent to the reading. The sum will be the total percentage of moisture and fat.

(7). Run a salt test on the residue in the same manner as given on pages 216-217.

(8). The percentage of curd is obtained by subtracting the sum of the percentages of moisture, fat and salt from 100. If the work has been carefully done, the amount of curd will be around 0.5 to 1.0 percent for average butter.

Sediment Test of Cream and Butter

Sediment tests are made of cream and butter to check the conditions under which the butter is manufactured. Stewart[4] discussed the merits of several methods for preparing cream and butter samples for sediment testing. Of the procedures studied in the survey, only the method using N/20 HCl gave satisfactory results for either cream or butter. Directions for the several methods now in common use in the butter industry are given below.

The alkalis or acids used in these tests should be of the specified strengths, as strong acids and strong alkaline solutions will dissolve hair. The principle of the test is based on the action of the acid or alkali used to dissolve the curd in the butter so as to enable the melted butter to pass through the filter. The addition of gasoline or hot water dissolves the last traces of fat left on the disc.

Standards

The Research Committee of the American Butter Institute has tentatively approved[5] a standard chart for use with a two ounce sample of cream. Photographic copies of these standards may be obtained from the American Butter Institute, 110 North Franklin Street, Chicago 6, Illinois; price 35 cents each. These standards have also been approved and made official by many state regulatory agencies.

In these instances, the standard chart is prepared on an individual master negative bearing the regulatory agency name. No official standards for butter have been adopted.[5]

The Sediment Testing of Salt[5]

Procedure: To eight ounces salt add 1,000 ml. hot distilled or filtered water. Stir thoroughly to insure complete solution of the salt. Filter through a 7½ ruled shark-skin filter as in The Sediment Testing of Butter.

Examine filter paper with wide field microscope.

The Sediment Testing of Sweet Cream[5]

(1). Procedure: One quart of the thoroughly mixed sample is placed in a suitable container and warmed to a temperature of 85 to 90° F., with constant stirring. Two ounces of the sample are then added to a 400 ml. beaker or enamel container. Eight ounces of filtered water (distilled preferred), previously heated to 180° F. are then added to the cream and stirred thoroughly with a thermometer or stirring rod.

(2). Operating the sediment tester: The warm mixture of cream and water is poured into the sediment tester which has previously been thoroughly rinsed with filtered water. Force the mixture through the filtering material. The sediment tester is then thoroughly rinsed by forcing through at least four ounces of filtered, hot water (180° F.), before removing the sediment sample.

(3). Handling the soiled filtering material: Withdraw soiled sediment strip and examine — mount and handle as suggested above for milk sediment disc.

The Sediment Testing of Sour Cream[5]

The three methods most commonly used are outlined. This does not preclude any modification which the individual operator may find more suitable to his conditions, provided that such method or modification will not destroy any of the extraneous matter.

Equipment — (The following equipment is required for all three methods.)

(1). Two ounce sample jars

(2). Improved Parsons cream sediment tester

(3). Granite cup, glass beaker or pint fruit jar

(4). Reel of cream filtering material

(5). All water should be free from sediment (filtered or distilled).

(a) Acid Method (Recommended by Dr. B. E. Horall, Purdue University).

Reagent: Hydrochloric acid solution of .03 normal strength. (3.2 ml. concentrated C.P. HCl in 1,000 ml. distilled or filtered water.)

Procedure: Mix two ounces of cream with five ounces of hot acid solution (180° F.) and filter. Proceed as with sweet cream.

(b) Soda Solution Method (Generally known as the Illinois Method).

Reagent: Baking soda.

Procedure: Place six ounces water, free from sediment, in the cup or pint jar. Add one-fourth to one-half teaspoon baking soda (free from sediment), add to this mixture a two ounce representative sample of cream. Heat to 160° F. Do not permit temperature to rise above 165° F. because of chemical changes. This is best accomplished by placing the cup in a kettle or pail of hot water, providing a double boiler effect. Stir the mixture thoroughly while heating. Allow two minutes time to dissolve the curd and filter. Proceed as with sweet cream.

(c) Parsons Method.

Reagent: A dry mixture consisting of 25% powdered sodium hexametaphosphate (Calgon) and 75% sodium bi-carbonate. The two chemicals should be uniformly mixed.

Procedure: Place a two ounce sample of cream into a pint fruit jar. Add two to four ounces hot water (180°-200° F.) and one teaspoonful of the reagent. Stir well and filter. Proceed as with sweet cream.

The Sediment Testing of Butter[5]

Reagent: Hydrochloric acid, 2% solution.

Procedure: Heat to boiling 225 grams (½ pound) butter in a one liter beaker with 200 ml. of the 2% hydrochloric acid solution. Stir during the latter part of the heating. Filter at once through a 7½ cm. ruled shark-skin filter paper supported on a 40 mesh brass section in a Buechner or Hirsch funnel. Do not allow mixture to accumulate on paper. When filtering is impeded, add hot water until the paper is cleared. Then add small amounts of the butter mixture alternately with the hot water treatment.

Samples which foam strongly when heated to boiling usually filter slowly at first. Such samples are frequently, but not invariably, helped by the addition of 10 ml. alcohol, with mixing, just prior to filtering. In some samples which are difficult to filter, continuous heating and stirring for five to ten minutes will be of assistance.

Alternate Method. Filtration on Sediment Pad: Weigh 225 gms. of butter into a liter beaker. Add 250 ml. hot water and 20 ml. of 85% phosphoric acid. Bring to a boil stirring continuously. Filter with suction through 1¼ inch Johnson sediment pad. Wash with hot water.

Examine filter paper with wide field microscope.

FREE WATER IN BUTTER

The recently published United States Standards for grades of butter[6] (April 1, 1960) establishes the final U. S. grade for butter samples according to a flavor classification, subject to disratings for body, color and salt. This final U.S. grade may be lowered if disratings for body, color and salt exceed permitted amounts for any flavor classification.

Free water is frequently encountered when the butter has not been worked sufficiently to ensure complete incorporation of water droplets. Uneven salt distribution may also be a cause of this defect. The test which follows is that recommended in the Laboratory Manual of the American Butter Institute.

Free Water in Butter[5]
Equipment and Reagents:

Fine wire, piano wire .015.

Filter paper, 9 cm. in diameter. (No. 40 Whatman or comparable grade is satisfactory.)

Indicator solution. (To 0.25 gm. brom phenol blue add 100 ml. 95% alcohol and 1 ml. of N/1HCl.)

Petri dishes.

1 ml. pipette.

Preparing Paper for Use:

Place several filter papers in a petri dish and saturate with the indicator solution. Drain excess solution off papers and allow them to dry in air. Since the impregnated papers when exposed to the atmosphere change rather rapidly from the original yellow color to blue, they should be protected by placing in a petri dish with a tightly fitting cover. Ordinarily the papers are prepared just before use.

Procedure:

The butter to be tested should be cold (40 to 50° F.) since poorly dispersed water shows up more readily at lower temperatures. A fresh surface is exposed by cutting with the fine wire and allowing to stand a few minutes. The freshly prepared paper is then placed on the surface and all points brought in contact with the butter by rolling a pipette over it. Remove the paper and allow to dry. Moisture areas on the butter show up as blue spots on the paper. If desired for a permanent record seal the paper in a cellophane envelope.

II. Cheese

Cheddar cheese is the most important of the many kinds of cheese sold on the market. The methods of analysis of this brand of cheese will apply as well to other kinds of hard cheeses.

Composition of Cheese. Thom and Fisk[7] give the following figures as the composition of green Cheddar cheese.

Table 35. Composition of Green Cheddar Cheese

	Minimum %	Maximum %	Average %
Water	33.16	43.89	37.33
Total solids	66.84	56.11	62.67
Fat	30.00	35.89	33.41
Casein	20.80	25.48	23.39
Sugar, ash, etc.	4.86	7.02	5.87

The federal standard for Cheddar cheese is that it shall not contain over 39 percent of water and the fat content of the dry matter shall be not less than 50 percent.

Sampling for Analysis. A trier is used in taking a sample of cheese for analysis. Since a cheese is variable in composition, it is necessary to take several plugs from different parts of the cheese in order to obtain a representative sample. Three vertical plugs are usually taken, one in the center, one near the edge and the third half-way between the other two. Short pieces from each plug should be put back into the opening so as to prevent mold growth. The plugs taken should be kept in sealed jars until ready to prepare them for analysis, as the moisture is quite readily lost. In preparing the sample, chop the plugs of cheese with a case knife or similar instrument until the cheese is cut into particles about the size of a kernel of wheat. It is advisable to leave the cheese in the jar during this process. The sample is then ready for either a moisture or fat test.

Moisture Test. Several methods for determining moisture of the various dairy products were presented in Chapter VI. For official procedures for cheese moisture analyses, the reader is directed to Official Methods of Analysis (AOAC).[3] The laboratory or factory method of moisture testing as used for butter cannot be applied to cheese without some modifications, because the heating process will cause the cheese to swell and sputter badly, thus losing some of the sample. In addition, constituents other than moisture are apt to be driven off due to the high temperature of heat and the fact that the cheese sticks to the bottom of the cup.

a. Troy method. Troy[8] developed a simple test that required much less time than the official methods wherein the samples had to be dried in an oven for several hours. He used a double walled drying cup large enough to hold a small flask. A cover fits over the chamber containing the flask, with a hole in the center just large enough for the neck of the flask to pass through. The cover aids in holding the heat and in maintaining a more uniform temperature. Between the walls is the heating medium, usually tallow or some other hard fat of a high boiling point. The high boiling point oil is necessary so that a sufficiently high temperature will be obtained to dry the cheese sample. A small opening is present at the top of the cup for introducing the oil, and for inserting a thermometer. The thermometer is held in place by a cork through which it passes. An alcohol flame is used for heating the oil.

The steps in operating the test are as follows:

(1). While preparing the sample, have the alcohol lamp lighted so that the oil will be up to temperature by the time the weighing is done.

(2). Weigh into a small flask 5 grams of the finely cut cheese. The butter moisture scales may be used.

(3). Place the flask and contents into the double walled drying cup and put on the cover.

(4). Hold the cheese in the drying cup for 50 minutes, keeping the oil at a temperature of 140° C. to 145° C.

(5). The flask is then removed and cooled in a desiccator.

(6). Reweigh. The loss in weight equals the grams of moisture. To convert to percentage divide the grams of moisture by 5 and multiply by 100. If the butter moisture scales are used, the percent of moisture may be taken direct as in the butter moisture test except the reading must be doubled since only 5 grams of cheese are used in place of the 10 grams as in the case of butter.

b. Olive oil method. Gould[9] has developed a test for cheese moisture that is suitable for the laboratory. A moisture determination by this procedure can be completed in approximately

20 minutes, with only 5 to 7 minutes actually being required for the heating period. The sample of cheese is heated in the presence of olive oil and sodium chloride. These two compounds prevent burning and sputtering. The detailed steps are:

(1). To moisture cup add 20 ml. of olive oil or cotton seed oil and 1 gram NaCl.

(2). Balance the cup and contents on the right hand pan of butter moisture scales.

(3). Add 5 grams of finely cut up cheese.

(4). Heat over alcohol flame, giving the dish a circular motion while heating to prevent cheese from sticking.

(5). The moisture is gone when bubbling ceases.

(6). Cool in desiccator, and reweigh.

(7). Double the percent reading since 5 grams instead of 10 grams were used.

c. Karl Fischer titration method. In recent years, a method using the Karl Fischer solution has become popular in European control laboratories. In a report before the 1956 International Dairy Congress, Raadsveld[10] concluded that the Karl Fischer method given below was the most reliable of four methods studied for the determination of moisture in cheese. Another paper by Peter and Sandor[11] before the same body presented evidence supporting the Karl Fischer method while using a suspending medium of absolute alcohol and xylene. The fact that the test can be run in 5-7 minutes is of obvious practical advantage. This procedure regularly shows results from 0.3 to 1.0 percent higher than those obtained by drying methods. The method as recommended by Raadsveld[10] is as follows:

(1). Karl Fischer solution is as purchased from chemical supply houses. It should be stored in a brown, glass-stoppered bottle and is dispensed from a storage burette which is protected from atmospheric moisture by means of a suitable desiccant.

(2). Use a titration flask provided with a magnetic stirrer.

(3). For each analysis, use between 100-200 milligrams of twice-grated and thoroughly mixed cheese.

(4). Suspend the sample in a mixture of four parts of dry chloroform and one part dry methanol.

(5). Stir reaction mixture continuously during titration and eventually heat to about 50° C.

(6). The end of the titration is marked by a permanent change in the color of the mixture from colorless to yellow to light brown.

(7). Estimate the titer of the Karl Fischer reagent by titrating to the same color a chloroform-methanol mixture containing a known amount of water.

Sediment Test of Cheese

Testing cheese for sediment is a difficult process. Not only is it necessary to dissolve the cheese curd so that it can be made to pass through the standard filter pads, it is also necessary to supplement visual inspection of cheese sediment disks with microscopic examination if the exact quantity and origin of the sediment is to be properly evaluated. Miersch and Price[12] have classified the extraneous material remaining on cheese sediment disks as critical or non-critical. The former includes any insects or substances of animal origin. They also consider vegetable matter or debris as critical under certain circumstances. Non-critical residues would include vegetable matter, soil particles, dust, ash, cloth fibers, and wood or metal particles.

The two procedures described below have received general acceptance in testing cheese samples for extraneous materials.*

Spicer and Price Method[13]
Equipment and Reagents:

Meat Grinder or Waring Blender
Torsion Balance at least 100 gram capacity
Clean 200 ml. beaker
Clean quart bottles
Pressure sediment tester
Wash bottle with distilled water
Parchment paper
Cheese knife
Electric mixer, adjustable speed
Sediment tester disks
Hot plate or burner
Low power microscope

Sodium citrate (dissolve 150 grams in distilled water and dilute to 1000 ml.).

Preparation of Cheese Sample:

Place cheese sample on clean parchment paper, remove the rind with a knife and cut the cheese into strips of suitable size for the food chopper. Care should be taken that particles of paraffin or dust specks do not get into the cheese. Be sure chopper is clean and free from rust. Grind cheese and collect in a clean 200 ml. beaker.

*Neither method appears to have any marked effect on the common types of extraneous matter in cheese. The two solvents appear to have about e ual value although the acid method works somewhat faster on young cheese.

Procedure:

(1). Weigh 100 grams of crushed cheese into a clean quart container.

(2). Add 200 ml. of filtered solvent solution.

(3). Place container and contents in a water bath maintained at 140° F. (Overheating causes protein materials to coagulate, producing a slime which clogs the filter pad.)

(4). Adjust mechanical agitator to the container and stir at approximately 1000 r.p.m. for 15 minutes.

(5). Add 100 ml. distilled water and 100 ml. of solvent solution to the mixture.

(6). Agitate until cheese particles are dissolved (usually 30 minutes or less).

(7). Divide contents of container in two equal parts and filter each part through a single sediment disk (compact surface up).

(8). Since 100 grams of cheese represents about one quart of milk, each disk will have the residue of approximately one pint of milk. Thus it can be compared with standards for whole milk.

Orthophosphoric Acid Method[14]

Apparatus:

(1). One cheese trier

(2). One food grater with 1/8" hole

(3). A 1-liter beaker

(4). One pressure milk sediment tester

(5). Milk sediment pads

(6). Irish poplin cloth pads of the same size as the milk sediment pads

(7). One 500-ml. graduate

(8). One 5-ml. graduate

(9). One glass stirring rod

(10). One hot plate or gas burner and tripod

Reagents:

(1). An aqueous solution containing 1.0 percent by weight of orthophosphoric acid. This is made by adding 7 ml. of orthophosphoric acid to 1000 ml. of water

(2). Acetone

Preparation of Sample:

Take a sample of cheese by plugging the cheese to be examined two or three times with the cheese trier. Grate the cheese plugs on the cheese grater, being careful to exclude all outside contamination.

Determination:

Weigh 50 grams of the grated cheese into a clean 1-liter beaker and add to it 500 ml. of 1.0 percent phosphoric acid. Heat to boiling, stirring occasionally to prevent cheese from sticking to the bottom of the beaker. Boil until all the cheese is dissolved and filter immediately through the pressure milk sediment tester with the poplin cloth pad on top of the milk sediment pad. Wash the beaker and tester with distilled water, forcing the wash water through the filter pads. Remove the pads carefully and examine the extraneous matter on the surface of the poplin cloth pad. The moisture remaining in the sediment pad can be removed by rinsing the pad with 5 ml. of acetone before it is taken from the sediment tester.

Salt Test of Cheese. The testing of cheese for salt is not as simple as testing butter for salt because the soluble proteins resulting from the digestion of the curd interfere with the results. Three tests will be presented here.

Method 1. A.D.S.A. test. The American Dairy Science Association committee on chemical methods for the analysis of milk and dairy products recommends the following procedure[15]:

(1). Weigh accurately approximately 3 grams of ground or chopped cheese into a 300 ml. Erlenmeyer flask and add 10 ml. of 0.1711 N silver nitrate solution, or an amount more than sufficient to combine with all of the chlorine.

(2). Add 15 ml. of halogen-free, chemically pure nitric acid and 50 ml. of water and boil.

(3). As the mixture boils add approximately 15 ml. of saturated potassium permanganate solution in 5 ml. portions. Boil until all cheese particles are digested.

(4). Dilute the solution to about 100 ml., decant off the liquid into a beaker, and wash the precipitate by adding 100 ml. of water and decanting again.

(5). Add 3 ml. of a saturated solution of ferric ammonium sulfate as an indicator and titrate the excess silver nitrate with 0.1711 N potassium or ammonium sulfocyanate.

(6). Run a blank on the reagents used, following the same procedure, except to add sugar to destroy the excess of permanganate.

(7). The number of ml. of silver nitrate used in (1) minus the titration value, divided by the weight of sample equals the percentage of sodium chloride in the sample.

(8). The reagents are made up by standardization against a salt solution containing 10 grams of chemically pure, dry sodium chloride per liter.

Method 2. Marquardt test. Marquardt[16] found that a direct titration method, similar to that used for butter, may be used for cheese less than 5 days old, since no appreciable breakdown of the protein would occur within that time.

The following procedure may be used:

(1). Use 10 grams of cheese and 300 ml. of water. Bring to a boil, remove and let stand for 4 hours.

(2). Titrate 17.6 ml. with $N/20 AgNO_3$, using potassium chromate as indicator. The brown color is the end point.

(3). Each ml. of $N/20 AgNO_3$ used represents 0.5 percent salt.

Method 3. Mercurimetric test. Arbuckle[17] investigated the use of mercuric nitrate in determining the salt content of dairy products. He found that it could be used satisfactorily for the estimation of the salt content of new cheese (approximately five days old or less) and of cottage cheese.

A. Reagents:

(1). Mercuric nitrate solution — 0.1711N. Dissolve 29.31 g. of mercuric nitrate C.P. in a few hundred ml. of water with the addition of 40 ml. of 2N nitric acid and make the solution to one liter volume with distilled water. The correct amount of nitric acid must be used in preparing the standard mercuric nitrate solution since if more or less acid is used the end point in the titration technic will not be as sharp.

(2). Nitric acid — 2N. Dilute approximately 127 ml. of

concentrated nitric acid (about 16N) to one liter volume with distilled water. Use to prepare the mercuric nitrate solution.

(3). s-Diphenylcarbazone indicator solution. Dissolve 100 mg. of s-diphenylcarbazone (Eastman 4459) in 100 ml. of neutral ethyl alcohol and store in a dark place, preferably in a refrigerator. The correct amount of indicator must be used and best results are secured if the indicator is stored not longer than one month.

(4). Standard sodium chloride — 0.1N. Dissolve 5.845 g. of sodium chloride C.P. dried at 120 degrees C., in water and make to one liter volume. Use the standard sodium chloride to standardize each new solution of mercuric nitrate. Standardize each new solution of mercuric nitrate by measuring 25 ml. of the standard sodium chloride solution into a 125 ml. Erlenmeyer flask. Add 0.6 ml. of indicator solution and titrate slowly with the mercuric nitrate solution. The clear solution turns a pale violet at the end point and an intense violet-blue color on the addition of one drop of mercuric nitrate in excess. The normality of the mercuric nitrate equals $\dfrac{25}{\text{ml. of mercuric nitrate used}} \times 0.1$. It should require approximately 14.6 ml. of the mercuric nitrate solution for the titration of 25 ml. of the standard salt solution.

B. Procedure:

(1). Grind the sample of new cheese or cottage cheese finely to reduce to a homogeneous consistency.

(2). Weigh 10 grams of the prepared sample into a 400 ml. beaker. Add 250 ml. of distilled water and heat to 150-160° F.

(3). Mix thoroughly and allow to cool to room temperature.

(4). Pipette a 25 ml. portion of the solution into a 125 ml. Erlenmeyer flask. Add 0.6 ml. of s-diphenylcarbazone indicator.

(5). Titrate slowly with standard mercuric nitrate solution using a burette graduated in 0.1 ml. divisions. Single drops should be approximately 0.05 ml. The clear and colorless solution turns a pale violet color on the addition of one drop of mercuric nitrate at the end point and an intense violet-blue color on the addition of one drop of mercuric nitrate in excess.

(6). Each milliliter of 0.1711N mercuric nitrate used in the titration equals one percent salt in the sample.

In protein-free solutions the color change is as described above, but in solutions containing protein the color of the turbid liquid is salmon red after the addition of the indicator which changes to violet as the titration is started. However, as the titration proceeds the solution becomes clear and pale yellow to colorless and at the end point there is a sharp change to pale violet.

BIBLIOGRAPHY

1. Manhart, V. C. Variability in Composition of Butter from the Same Churning. J. D. Sci. 11, p. 55 (1928).
2. Ellenberger, H. B., and Newlander, J. A. The Trier Method of Sampling Butter for Analysis. Vt. Bul. 265 (1927).
3. Methods of Analysis of the Association of Official Agricultural Chemists. 8th Edit. p. 271 (1955).
4. Stewart, G. F. A National Survey of Methods for the Determination of Sediment in Butter and Cream. J. D. Sci. 20, pp. 509-519 (1937).
5. Amer. Butter Inst. Laboratory Manual. Method of Analysis. Chicago, Illinois.
6. U.S.D.A. United States Standards for Grades of Butter (Reprint from Federal Register of January 28, 1960).
7. Thom, C., and Fisk, W. W. The Book of Cheese, p. 223, The MacMillan Co., New York (1918).
8. Troy, H. C. A Cheese Moisture Test. Cornell Agr. Ext. Bul. 17 (1917).
9. Gould, I. A. A Comparative Study of Methods of Determining the Moisture Content of Cheddar Cheese. J. D. Sci. 20, pp. 625-637 (1937).
10. Raadsveld, C. W. Comparative Experiments on the Determination of the Moisture Content of Cheese. Proc. XIV Int. Dairy Congress. Vol. 2, Part 2:890-901 (Rome, 1956).
11. Peter, A., and Sandor, Z. Volumetric Method for the Rapid Determination of the Water Content of Cheese. Proc. XIV Int. Dairy Congress. Vol. 3, Part 2:364-373 (Rome, 1956).
12. Miersch, R., and Price, W. V. Testing for Extraneous Matter in Cheese. J. D. Sci. 27:881-895 (1944).
13. Spicer, D. W., and Price, W. V. A Test for Extraneous Matter in Cheese. J. D. Sci. 21:1-6 (1938).
14. Turner, A. W., et al. Orthophosphoric Acid as a Cheese Solvent. J. D. Sci. 25:777-778 (1942).
15. Wilster, G. H., et al. Determination of Fat, Moisture and Salt in Hard Cheese. J. D. Sci. 20, p. 29 (1937).
16. Marquardt, J. C. Methods for Determining Salt in Various Cheeses. N. Y. Sta. (Geneva) Tech. Bul. 249 (1938).
17. Arbuckle, W. S. Salt in Butter and New Cheese. Nat'l Butter and Cheese J. Vol. 37, p. 41, May (1946).

REVIEW QUESTIONS

1. How much more moisture can unsalted butter have than salted butter testing 80 percent fat and still meet the Federal requirement for legal butter?
2. What is meant by overrun?
3. Which overrun represents the highest percentage of increase in butter, the factory, churn or composition overrun? Why?
4. Why should the curve in a butter trier be a true arc of a circle?
5. Why should samples of butter be prepared in closed containers?
6. Give two factors that indicate when all the moisture is gone from a sample of butter tested for moisture.
7. Will the moisture test reading be too high or too low if the cup and contents are reweighed while hot? Explain.
8. Write the chemical reactions that take place in the salt test of butter.
9. What is the Federal standard for Cheddar cheese?
10. What purposes do olive oil and salt serve when making a moisture test of cheese by that method?

CHAPTER XI

The Acidity of Milk and Its Products

Apparent Acidity. As soon as it is drawn, milk will show an acid reaction to phenolphthalein as an indicator even though no real acidity has developed. This acidity of fresh milk is due to the presence of casein, phosphates, albumin, carbon dioxide and citrates. It is termed apparent acidity. The approximate percentage of titratable acidity contributed by each of these compounds is[1]:

Casein	0.05—0.08 % acid
Phosphates	0.05—0.07 % acid
Albumin	0.01 % acid
Carbon dioxide	0.01—0.02 % acid
Citrates	0.01 % acid

The apparent acidity will average around 0.13 to 0.17 percent but occasionally shows rather wide range depending on the individuality or breed of cow.

Table 36 shows the apparent acidity results obtained by McInerney[2] and Caulfield and Riddell.[3] Nearly all the tests were in a range of 0.10 to 0.22 percent.

Table 36. Apparent Acidity of Milk from Different Breeds

Breed	(McInerney) Average Acid	(Caulfield and Riddell) Average Acid
	%	%
Holstein	0.136	0.161
Jersey	0.162	0.179
Guernsey	0.139	0.172
Ayrshire	0.123	0.160

In general, those breeds with the higher fat tests will have the higher apparent acidities because of the increased percentage of casein and phosphates that go with the milks richer in fat. Such milk, however, is equal in quality to milk of lower apparent acidity. Grading milk by an acid standard works an injustice in such cases where the apparent acidity is high.

With litmus as an indicator, freshly drawn milk will show both an acid and an alkaline reaction turning blue litmus paper red and red litmus paper blue. This is called an amphoteric reaction. This result is partly due to the proteins, whose amino acids, having in their chemical structure both amino groups (basic) and carboxyl groups (acidic), react with both bases and acids. The phosphates,[4] some of which are acidic and some basic also enter into these reactions.

Real Acidity. Milk, freshly drawn, contains not more than 0.002 percent lactic acid. Any increase in acidity over the apparent is due to the breaking down of the milk sugar or lactose by the action of bacteria, converting it to lactic acid:

$$C_{12}H_{22}O_{11} + H_2O + \text{bacteria} = 4C_3H_6O_3$$
$$\text{Lactose} \quad \text{Water} \quad\quad\quad\quad \text{Lactic acid}$$

This acidity is called real acidity. The organism causing the reaction is *Streptococcus lactis*. The structural formula of lactic acid may be expressed as follows:

$$\begin{array}{c} \text{H} \\ | \\ \text{H} \quad \text{O} \quad \text{O} \\ | \quad\; | \quad\; \| \\ \text{H-C-C-C-O-H} \\ | \quad\; | \\ \text{H} \quad \text{H} \end{array}$$
Lactic acid

In practical work the real acidity is not distinguished from the apparent, both being included in the titration and expressed as real acidity. Fresh milk as received at the factory tests around 0.15 or 0.16 percent acid. Some milk will run higher depending on the amount of apparent acidity. Milk will taste sour to most people when it contains around 0.20 percent of acid, and will curdle when 0.50 to 0.55 percent of acid has developed or a hydrogen ion concentration of pH 4.7. Acid development will cease when 0.80 to 0.90 percent of acid, or pH 4.1 has been reached, as the *Streptococcus lactis* organism is not tolerant to an acidity beyond this point. If part of the acid is neutralized, these bacteria will start working again on the lactose until as much acid is produced as before. This may be repeated until all the milk sugar is gone.

Lactic acid organisms are most active between temperatures of 70° F. to 100° F. They are practically inactive below 40° F. and nearly all of them are killed at pasteurization temperature, 143° F. for 30 minutes. If milk is cooled promptly after milking to 40° F. or below, the milk will retain its sweetness for some time.

Importance of Lactic Acid. Practically all dairy operations depend on knowing the amount of acid in milk or its products. At the receiving stand one method of grading milk and cream is according to their acidity. In general, milk with a high percent of acid contains a large number of bacteria, and since bacteria get into milk through lack of care and cleanliness, an acid test will give an idea of the condition under which the milk was produced. It is important to know the acid content of cream in ripening it for churning. The steps in cheese making are regulated almost entirely by the acid development. In storing cream it is essential to know that it is sweet when put in storage. In shipping milk and cream, it is necessary to know that the acidity is low enough to warrant its arriving in good condition. When pasteurizing milk and cream, the acid content must be sufficiently low so that the heat will not cause curdling. This factor also enters into milk used for condensing where the product is subjected to considerable heat. These examples are sufficient to show the importance of having some method for readily determining the percent of acid in milk.

Testing for Acidity. It is a principle of chemistry that an alkali will neutralize an acid, that is, when these two chemicals are mixed together products will be formed that are neither acids nor bases. This principle is made use of in testing milk for acid. By measuring the amount of alkali of a given strength that it takes to neutralize the acid in a sample of milk, the amount of acid in the sample can be ascertained. The process of determining the amount of acid in a sample is called titration. Tenth normal sodium hydroxide (N/10 NaOH) is the alkali used and phenolphthalein is the indicator for showing when all the acid has been neutralized. Phenolphthalein is colorless in an acid medium, but turns pink in an alkaline solution. Thus, when enough of the alkali has been added to a sample of milk to neutralize the acid and then a slight excess added, a pink color will appear. This is the end point and no more alkali should be added. Phenolphthalein as an indicator, however, does not show the true neutral point, pH 7.0, but gives the color change at a hydrogen ion concentration of pH 8.3 to 8.4 well beyond neutrality. Nevertheless, this indicator is generally used and is acceptable as far as comparative results are concerned. It should be remembered, however, that where neutralization is practiced as in the reduction of acid in highly soured cream for churning, all the calculated amount of neutralizer, using phenolphthalein as an indicator, should not be added or an alkaline condition will result. An alkaline medium is conducive to the growth of putrefactive bacteria.

Apparatus. The apparatus consists of a burette, white cup, pipette, tenth normal sodium hydroxide and phenolphthalein. The burette is used for measuring the alkali. It is graduated

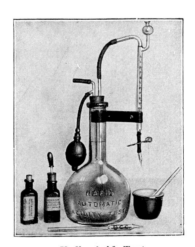

Nafis Acid Test
Courtesy Louis F. Nafis, Inc.

to one-tenth of a milliliter and holds ten milliliters. The burette is connected to a supply bottle and is filled from it by means of a bulb. The surface of the alkali always stops on the zero mark. A white cup for holding the milk is preferable, because the pink color is more readily seen. Either an 8.8 ml. or a 17.6 ml. pipette may be used, preferably the latter because the addition of too much alkali would not make as great an error on a large sample as it would on a smaller one. The phenolphthalein is prepared by dissolving one gram of phenolphthalein in 100 ml. of 95% alcohol, which makes a 1% solution.

Procedure in Testing for Acidity. A committee of the American Dairy Science Association studied procedures for making acidity tests during the period of 1948-1956. As the result of this committee activity, the following procedures have been recommended to the dairy industry as standard tests.[5]

a. Milk, skimmilk, buttermilk, and whey

(1). Measure 9 ml. of the product into a white cup. An 18-g. Babcock pipette may be used for milk, skimmilk, buttermilk, and whey by doubling the amount of indicator and dividing the final result by two.

(2). Add three to four drops of indicator (1% alcohol solution of phenolphthalein).

(3). Slowly add N/10 sodium hydroxide from a burette, stirring the sample with a stirring rod.

(4). When the first definite and relatively permanent shade of pink has been reached, read the burette to determine the ml. of N/10 sodium hydroxide used.

(5). Multiply this reading by 0.1 to express the titer as the so-called percent of lactic acid.

b. Sweet Cream

(1). Measure 9 ml. of cream into a white cup. When products such as cream, condensed milk, and ice cream mix are very viscous, it is more accurate to weigh out 9 g. of the sample than to use a pipette. With the ice cream mix, however, the amount of added water must be adjusted to bring the serum solids down to from 8 to 9%.

(2). Rinse the pipette with one filling of distilled water.

(3). Add six drops of indicator (1% alcoholic solution of phenolphthalein).

(4). Titrate to the first definite and relatively permanent shade of pink, using N/10 sodium hydroxide.

(5). The ml. of N/10 sodium hydroxide are multiplied by 0.1 to express the result as percent acidity.

c. Sour Cream

In creameries where high-acid cream is being tested for the purpose of standardizing the acidity, it is more exact to weigh a 9-g. sample and proceed as for sweet cream.

d. Condensed Milk (Unsweetened)

(1). Measure 9 ml. into a white cup.

(2). Add enough distilled water, used to rinse the pipette, to restore the sample approximately to the concentration of the original milk.

(3). Add indicator (1% phenolphthalein solution) at the rate of three or four drops to every 9 ml. of mixture, and titrate with N/10 sodium hydroxide to the first definite and permanent shade of pink.

(4). Multiply the titer by 0.1 to express it as percent acidity.

e. Ice Cream Mix

(1). The sample should first be warmed to room temperature, to eliminate the high viscosity that may be encountered.

(2). Measure 9 ml. of the warmed mix into a white cup.

(3). Rinse the pipette with one filling of distilled water.

(4). Add six to eight drops of indicator (1% alcoholic phenolphthalein solution) and titrate to the first definite and relatively permanent shade of pink.

(5). The ml. of N/10 sodium hydroxide solution used in this titration are converted into percent acidity by multiplying by 0.1.

f. Chocolate Milk

The modified spot-plate test:

(1). Warm the chocolate milk sample to room temperature.

(2). Use 9 g. of sample, 2 vol. of water.

(3). After each addition of N/10 sodium hydroxide, stir and remove two drops of the mixture with the stirring rod and transfer to each of two depressions in a spot plate. Use one spot for a control and the other for comparison.

(4). To the second spot add one drop of phenolphthalein indicator.

(5). The first appearance of a faint pink color is the end-point for the mixture.

(6). Report the results as percentage of lactic acid by multiplying the ml. of N/10 sodium hydroxide used by 0.1.

g. Butter[6]

Use 18 grams of butter and N/50 solution of sodium hydroxide. Weigh the butter into the titrating cup and add 90 ml. of previously boiled distilled water, cooled to about 150° F. Add 1 ml. of phenolphthalein indicator and titrate. Add the alkali solution fairly rapidly until a reasonably permanent pink color is obtained. This will usually require about one-half minute. Upon continued stirring the end point will fade. Continue the addition of alkali solution in small quantities, and the agitation, until 3 minutes have been consumed in all. Only enough additional alkali solution is added during the final stirring to maintain the light pink color of the mixture fairly uniform. A light definite pink color should then be evident in the watery portion when the fat rises upon discontinuing the agitation. The tendency for the pink color to fade out renders recognition of the final end point somewhat uncertain. It is for this reason that the 3 minute titration period is recommended.

When calculated as lactic acid, each ml. N/50 alkali solution, is equal to 0.01% acid. Normal, sweet-cream butter will usually give a value of .02-.04 percent titratable acidity.

h. Cheese

Weigh 3 grams of cheese into a white cup, add 10 ml. of water a little at a time while working the cheese to a paste with a pestle. Titrate slowly with N/10 sodium hydroxide and complete as under (a). The percent acidity equals the number of milliliters of alkali used times 0.3. For instance, if 8.4 ml. of N/10 NaOH were needed, $8.4 \times 0.3 = 2.52\%$ acid.

Why the Acid Test Reads Direct. The molecular weight of lactic acid is 90, and that of sodium hydroxide is 40. A normal solution of lactic acid contains 90 grams of the pure acid in 1000 ml. of water. Similarly a normal solution of sodium hydroxide is made. Since equal volumes of acids and bases of the same chemical strength will exactly neutralize each other, the following statements may be made:

1,000 ml. N/1 NaOH sol. will neut. 1,000 ml. N/1 lactic acid, or 90 grams.
1 ml. N/1 NaOH sol. will neut. 1 ml. N/1 lactic acid, or 0.09 grams.
1 ml. N/10 NaOH sol. will neut. 1 ml. N/10 lactic acid, or 0.009 grams.

Assume that it took 3 ml. of tenth normal sodium hydroxide to neutralize the acid in 18 grams of milk. Since one milliliter of tenth normal sodium hydroxide will neutralize 0.009 grams of lactic acid, 3 ml. will neutralize 0.027 grams of lactic acid. The percent of acid would then be 0.027/18 times 100, or 0.15 percent. This result is the same as dividing the 3 ml. of alkali used in titrating by 2 and moving the decimal point one place to the left. It must be remembered that this method holds true only when 18 grams of the sample are taken and titrated with tenth normal sodium hydroxide.

If definite amounts of the sample are not used the percent of acidity may be calculated by the following formula:

$$\frac{\text{No. ml. N/10 NaOH} \times 0.009}{\text{grams of sample}} \times 100 = \% \text{ acid}$$

For example, suppose that it took 4.2 ml. of tenth normal sodium hydroxide solution to neutralize the acid in 25 grams of milk, what would the percent of acidity be? Substituting in the formula, we have

$$\frac{4.2 \times 0.009}{25} \times 100 = 0.15\%$$

Rapid Acid Test. The use of the acid test at the receiving stand for accepting or rejecting milk or cream has been used extensively in the past and still has its adherents. If all herds gave milk with the same amount of original acidity, this method of checking the quality of milk would have its advantages. Thus, if the original acidity was 0.15 percent and a standard of 0.17 percent was employed as the upper limit for acceptance, it would show how much acidity had developed due to bacterial action. If the same breed of cows is kept in a locality, and the apparent acidity taken into account, a certain acidity standard might be set up without doing any producer an injustice. It should be thoroughly understood by the operator that a milk might go above the acid standard used and still be a first class product, as the acid was not due to bacterial action but to its composition.

The method usually followed is to put a half ounce dipperful of the milk or cream in a white cup, add to it the same amount of alkali, mix and then note if the pink color remains or disappears. If the color remains, the acidity is below the standard, but if it disappears the acidity is above. In this method of testing the alkali solution must be of a certain strength for each particular standard of acidity selected. Tenth normal sodium hydroxide may be used for this purpose by diluting it with distilled water and then adding phenolphthalein as indicator. The proper strength of alkali to use may be calculated as follows: In using the regular acid test we know that if a sample of milk has 0.15 percent acidity, it will require 3 ml. of tenth normal sodium hydroxide to neutralize it when 17.44 ml. of the

THE ACIDITY OF MILK AND ITS PRODUCTS 243

milk are titrated (the 17.6 ml. pipette delivers 17.44 ml.). In order to avoid the use of decimals we will increase these figures to 1744 ml. of milk and 300 ml. of alkali. Now if equal amounts of milk and alkali are to be used, it will be necessary to dilute the tenth normal sodium hydroxide so as to make the 300 ml. up to 1744 ml. For this amount of alkali it will require approximately 25 ml. of phenolphthalein and so the quantity of water to add should be deducted by this amount. So in this case we have 300 ml. of tenth normal sodium hydroxide solution, 25 ml. of indicator and 1419 ml. of distilled water to be mixed together to make 1744 ml. of alkali solution, which will then measure 0.15 percent acidity in milk when equal amounts of each are used. Similarly any strength solution of alkali for any standard of acidity may be made.

Table 37 gives the relative amounts of tenth normal sodium hydroxide, distilled water, and phenolphthalein to use in making up an alkali solution for several different standards of acidity:

Table 37. Method of Making Acid Standards*

N/10 NaOH		Distilled water		Phenolph- thalein		Total Volume	Acidity Standard
ml.		ml.		ml.		ml.	%
a. 300	+	1419	+	25	=	1744 } measures	0.15
b. 172	+	814	+	14	=	1000	
a. 320	+	1399	+	25	=	1744 } measures	0.16
b. 183	+	803	+	14	=	1000	
a. 340	+	1379	+	25	=	1744 } measures	0.17
b. 195	+	791	+	14	=	1000	
a. 360	+	1359	+	25	=	1744 } measures	0.18
b. 206	+	780	+	14	=	1000	
a. 380	+	1339	+	25	=	1744 } measures	0.19
b. 218	+	768	+	14	=	1000	
a. 400	+	1319	+	25	=	1744 } measures	0.20
b. 229	+	757	+	14	=	1000	
a. 500	+	1219	+	25	=	1744 } measures	0.25
b. 287	+	699	+	14	=	1000	
a. 600	+	1119	+	25	=	1744 } measures	0.30
b. 344	+	642	+	14	=	1000	
a. 700	+	1019	+	25	—	1744 } measures	0.35
b. 401	+	585	+	14	=	1000	
a. 800	+	919	+	25	=	1744 } measures	0.40
b. 459	+	527	+	14	=	1000	
a. 1000	+	719	+	25	=	1744 } measures	0.50
b. 574	+	412	+	14	=	1000	
a. 1200	+	519	+	25	=	1744 } measures	0.60
b. 688	+	298	+	14	=	1000	

* Either the amounts in (a) or (b) may be used. Those in (a) show the student how to arrive at the different amounts of solutions to use for the different standards, while the figures in (b) are for the convenience of the operators who wish to make up an even liter (1000 ml.) of solution.

Relationship of Acidity in Different Products. The acidity of dairy products is in the plasma portion and therefore to compare the acidity of two products, milk and cream for example, the percentages of acidity should be calculated on the plasma basis. Suppose a sample of milk testing 4.0 percent fat had an acidity of 0.16 percent, what would be the comparable acidity of a sample of cream testing 40 percent fat? The percent plasma in the milk would be 96 percent (100−4) and in the cream 60 percent (100−40). Therefore the comparable acidity of the cream to that of the milk would be 0.16:X : : 96:60 or 0.10 percent. In other words, this milk with 0.16 percent acidity has actually no more acidity than the cream with 0.10 percent.

The relative acidities of the cream churned and the butter obtained from it may be calculated in the same way.

Factors Affecting Titratable Acidity. *a. Action of calcium phosphate.* The percent acidity of milk as determined by titration is affected by any conditions that cause a change in its calcium phosphate. This compound is present as di-calcium phosphate and upon the addition of sodium hydroxide, part of it will immediately change to tri-calcium phosphate and finally to phosphoric acid. This, of course, increases the percent of acidity. Some of the factors affecting this reaction are, speed of titration, dilution of sample, amount of indicator, and temperature of the milk.

An operator who titrates quickly will obtain a lower percent of acidity than a slow moving person since any delay allows more tri-calcium phosphate to be precipitated, with the ultimate result of the formation of phosphoric acid.

Diluting the sample with water before titrating will lower the results, since the addition of water checks the rate of the precipitation of the tri-calcium phosphate. Equal volumes[7] of milk and water will show around 0.02 percent less acidity than the undiluted milk. Nine volumes of water to one of milk will lower it 0.06 to 0.07 percent. Viscous samples as cream, however, are diluted because it is necessary to wash out the cream adhering to the pipette and to allow a clearer observation of the change in color of indicator. Sommer[7] compared three methods of titrating cream with calculated acidities:

(1). The sample was measured with a 9 ml. pipette and not rinsed with water, but rinsed with the titrated cream while it was being used as a stirring rod.

(2). The sample was measured with a 9 ml. pipette and the pipette rinsed by filling once with distilled water.

(3). Nine grams of the sample were weighed and titrated without dilution.

The results were, respectively, 0.096, 0.0847 and 0.1073 as compared to the calculated acidity of 0.0811 percent. Thus, the second method gave the closest results even though the cream was diluted. This is probably due to the fact that the high fat content prevents the color from being noticed as promptly as in whole milk or skimmilk.

The greater the amount of indicator used the sooner will the pink color show when titrating, and therefore a lower reading be obtained. Around 1/20 of the phenolphthalein must be changed before the pink color appears and the more indicator that is used the greater will be the degree of color and thus noted sooner. Caulfield and Riddell[3] titrated 18 grams of whole milk and found that 3 drops of phenolphthalein indicator gave a color change at 0.172; 0.5 ml., 0.159; 1.0 ml., 0.154; 1.5 ml., 0.151; and 2.0 ml., 0.149 percent acidity.

The temperature of the samples will affect the titratable results, the higher temperature giving the greater results. If a sample of milk titrates 0.13 percent acidity at 40° F., it will be approximately 0.15 percent at 75° F. A more rapid change to tri-calcium phosphate occurs at the higher temperatures.

Some investigators have studied the effect of removing the disturbing effect of calcium by the addition of potassium oxalate to the milk before titration. This procedure precipitates the calcium as an oxalate. Van Slyke and Bosworth[8] reduced the average acidity of 21 milk samples from an original of 0.164 to 0.082 percent acid. Thus the difference between the two results was considered to be the result of some chemical reaction involving the calcium. Ling,[9] however, claims the oxalate introduces another error by causing some alkalinity which would lower the titratable acidity below the true acid content.

b. Stage of lactation. The colostrum or first milk after calving is high in percentage of titratable acidity due to its high protein and mineral content. Table 38 shows its course during the first month.[3] Very little change in acidity occurs during the remainder of the lactation period except a slight drop the last month.

c. *Mastitis.* Mastitis lowers the acidity of milk due to the blood that gets through the weakened udder walls into the milk. Blood being alkaline will lower the acidity of freshly drawn milk to around 0.10 to 0.12 percent. The blood also increases the sodium chloride content of milk, and therefore one of the tests for mastitis is the so-called chloride test. Any amount of chlorine

Table 38. Acidity and pH Values of Milk During First Month of Lactation

Days		Titratable Acidity	pH
		%	
	First milking	0.44	6.25
1	Second milking	0.34	6.33
	Third milking	0.26	6.35
2	Fourth milking	0.24	6.38
	Fifth milking	0.24	6.36
3	Sixth milking	0.21	6.46
5		0.20	6.47
10		0.19	6.54
15		0.17	6.57
20		0.16	6.58
25		0.16	6.58
30		0.16	6.58

over 0.14 percent is considered abnormal. As the chloride increases, the percent acid decreases and vice versa.

d. *Enzyme activity.* Enzymes will cause fatty acids to be liberated from the fat and thereby increase the titratable acidity. This hydrolysis of fat may be augmented by homogenizing raw milk, by shaking milk vigorously or by cooling and warming milk through certain temperatures. Cooling to 40° F., then warming to 86° F., and finally cooling to 40° F. again will activate the enzyme. It is naturally active in late lactation and milk at that time often has a bitter flavor.

e. *Effect of feed.* Experiments have shown that feeds such as silage or certain concentrates as corn gluten feed or even the feeding of inorganic acids have no effect on the acid content of milk. Turner and Beach[10] have shown that corn silage in the ration of dairy cows did not increase the titratable acidity of milk. Sommer and Hart[11] fed up to 120 ml. sulphuric acid per day for six days with no change in milk acidity. Anderson et al.[12] studied the effects of corn gluten feed, wherein sulphuric acid is used in the production process of this concentrate on its possible effect on milk acidity and found it had no influence.

Hydrogen Ion Concentration or pH

Actual acidity cannot be measured by titration. For example, a normal solution of hydrochloric acid is equal in strength to a normal solution of acetic acid as measured by titration, yet the former is a strong acid and the latter a weak acid. The exact method is to measure the hydrogen ion concentration or pH.

Pure water dissociates slightly into hydrogen [H^+] and hydroxyl [OH^-] ions, both in equal number, the former being characteristic of acids and the latter of bases. The ionization constant of pure water has been calculated to be 10^{-14} at 20° C. The product of the hydrogen ions and the hydroxyl ions [H^+] $\times$ [OH^-] is equal to this constant. Thus, in a liter of pure water at 20° C., the hydrogen and hydroxyl ions, being equal in number, would each have the value 10^{-7}, since $10^{-7} \times 10^{-7} = 10^{-14}$. Absolutely pure water, therefore, is neutral since it yields the two kinds of ions in equal numbers upon dissociation, and this neutral point is indicated as 10^{-7} or pH 7.0. Any solutions that yield hydrogen ion concentrations greater or hydroxyl ion concentrations less than 10^{-7} are acid, and conversely any solutions that yield hydroxyl ion concentrations greater or hydrogen ion concentrations less than 10^{-7} are alkaline. In other words below pH 7.0 indicates an acid, and above pH 7.0 indicates a base. The pH scale ranges from 0 to 14, the zero representing complete dissociation of an acid yielding hydrogen ions to the extent of one gram per liter of solution, and the figure 14 representing complete dissociation of a base yielding hydroxyl ions to the extent of one gram per liter of solution. If the pH of a solution is known, the pOH can be calculated by subtracting the pH from 14, as for example, a pH 3.0 would mean a pOH 11.0 (14-3). Therefore, whether the substance is acid or alkaline, it is necessary only to give, as is customarily done, the pH value.

pH may be defined as the log of the reciprocal of the hydrogen ion concentration expressed in grams per liter of solution. A normal acid solution contains one gram of **ionizable** hydrogen in a liter of solution but it is not all **ionized**. A 0.1 N solution of hydrochloric acid comes very near being completely ionized and consequently is a strong acid. Around 91.4 percent of the molecules are ionized. The pH may be calculated from this percentage figure as follows: Since a 0.1 N acid solution contains 0.1 gram of ionizable hydrogen, then $0.1 \times .914 = 0.0914$ g. of ionized hydrogen. The normality of this solution with respect to H ions is $\frac{N}{1}$ or $\frac{N}{10.94}$. The pH value of 0.1 N hydrochloric

$\overline{0.0914}$

acid is, therefore, the logarithm of 10.94 or 1.04.

Other examples, as in the two following questions, may be solved in the same way: (1) What is the pH of a normal acid solution if it is 1% ionized? (2) What is the pH of a normal alkaline solution if it is 1% ionized?

Since one gram of ionizable hydrogen ions is present in a liter of a normal solution but only one percent in problem 1 is ionized, there would be $1 \times .01$ or 0.01 g. ionized hydrogen in a liter. The normality of this solution with respect to H ions would be $\frac{N}{1}$ or $\frac{N}{100}$. The pH value of this acid would there-

$\overline{0.01}$

fore be the logarithm of 100 or 2.0. In the second problem the pH is determined by first calculating the OH ion concentration or pOH. This calculation is made in the same manner as in problem 1. This base is dissociated to the extent of one percent and therefore the OH concentration in a liter is $1 \times .01$ or 0.01 g. The normality of this solution with respect to OH ion concentration would be $\frac{N}{1}$ or $\frac{N}{100}$ or pOH 2.0. The pH value would there-

$\overline{0.01}$

Table 39. Relationship of Hydrogen Ion Concentration and pH Ions per Liter of Solution

pH	Hydrogen ions	Hydroxyl ions
	g.	g.
0.0	1.0	10^{-14}
1.0	0.1	10^{-13}
2.0	0.01	10^{-12}
3.0	0.001	10^{-11}
4.0	0.0001	10^{-10}
5.0	10^{-5}	10^{-9}
6.0	10^{-6}	10^{-8}
7.0	10^{-7}	10^{-7}
8.0	10^{-8}	10^{-6}
9.0	10^{-9}	10^{-5}
10.0	10^{-10}	0.0001
11.0	10^{-11}	0.001
12.0	10^{-12}	0.01
13.0	10^{-13}	0.1
14.0	10^{-14}	1.0

fore be 12.0 (14—2), since the sum of the pH and pOH values equal the constant 14.

Table 39 shows the relationship of hydrogen ion concentration and pH.

In order to clarify the meaning of the figures in Table 39, let us consider for example pH 4.0. The hydrogen ions in this case amount to 0.0001 gram per liter of solution. Written in fractional form this would be $\frac{1}{10000}$ or $\frac{1}{10^4}$. The pH is 4.0, which is the fourth power of 10.

Next consider the negative exponents. These become positive as soon as the 10 becomes a denominator. Thus 10^{-5} means $\frac{1}{10^5}$ or pH 5.0.

Whenever the pH is not a whole number, as for example, pH 6.8, it is necessary to use logarithmic tables to obtain the concentration. Thus pH 6.8 equals $\frac{1}{10^{6.8}}$ or $\frac{1}{6,310,000}$ or $\frac{1.59}{10,000,000}$ grams hydrogen ions in a liter of solution.

Buffer Systems in Milk

Substances which, by their presence, cause a solution to resist changes in reaction when acids or bases are added are called buffers. Mixtures of weak acids or weak bases with their corresponding salts make the most efficient buffer systems. Buffering action is extremely important in biological systems since the animal body must be protected from sudden changes in hydrogen-ion concentration.[13]

Titration curves of milk are relatively smooth in the pH range of 5.5 to 8.0. The abrupt change in the curve between pH 8.0 and 8.5 makes phenolphthalein a valuable indicator in the determination of milk acidity. The principal buffers in milk are the proteins, citrates, phosphates, and bicarbonates. All are active when the pH of the milk is between 5.5 and 8.0. Above pH 8.5, the buffering effect of the citrates, phosphates, and bicarbonates is minimal, leaving the protein constituents as the only effective influence. At pH 5.0 or below, only the phosphates show buffering action.[14]

More basic details concerning the relationship of acidity, pH, and buffering action are discussed in the recent text by Jenness and Patton[15] and in the previously mentioned text of the Gortners.[13]

Methods of Determining pH Values. Hydrogen ion concentration may be determined either (a) colorimetrically or (b) by means of a potentiometer.

a. Colorimetric method. In the colorimetric method several different indicators that cover various ranges of pH are used. The unknown substance with indicator added is matched against color solutions of known pH values containing the same indicator, and the pH value noted at that point. This method is generally used for such purposes as checking the reaction of media for bacteriological purposes, and for the detection of mastitis milk. Its accuracy may be increased by the use of a comparator block, which compensates for the turbidity of the solution under test. Table 40 gives the pH values of some of the more common indicators used.

b. Potentiometer method. For precise work in measuring pH values of substances, the potentiometer should be used. This method for determining the pH values of a substance, whether acid, base or salt, consists of measuring the difference in potential or voltage of two electrodes in a sample of the solution. One of the electrodes is a reference electrode, or half cell, with a potential independent of the pH of the solution tested. The other electrode is affected by the solution and the difference in potential of the two electrodes causes a voltage or current that can be measured by the potentiometer and the pH value may be read directly on the instrument scale.

Table 40. Color Changes of Indicators Used in Determining pH

Indicators	Full acid color	Full alkaline color	Sensitive range pH values
Thymol blue	red	yellow	1.2—2.8
Brom phenol blue	yellow	blue	3.0—4.6
Brom cresol green	yellow	blue	3.8—5.4
Methyl red	red	yellow	4.4—6.0
Chlor phenol red	yellow	red	5.0—6.6
Brom cresol purple	yellow	purple	5.4—7.0
Brom thymol blue	yellow	blue	6.0—7.6
Phenol red	yellow	red	6.6—8.2
Cresol red	yellow	red	7.2—8.8
Meta cresol purple	yellow	red	7.4—9.0
Thymol blue	yellow	blue	8.2—9.8
Phenolphthalein	colorless	red	8.0—9.6
Cresolphthalein	colorless	red	8.2—9.8

The reference electrode is usually a calomel electrode. It consists of mercury, calomel (mercurous chloride) and a saturated solution of potassium chloride, all contained in a small glass tube inside a larger one. In the larger tube is also potassium chloride solution, sufficient of which escapes through a glass stopper to form a "salt bridge" for electrical connection with the solution to be tested, and without allowing any outside contamination. The inner glass tube is wired to the potentiometer The measuring electrode may be either the quinhydrone or the glass electrode.

The quinhydrone electrode consists of a gold electrode plus small amount of quinhydrone crystals added to the test solution. It is not applicable throughout the entire pH range, but it is excellent for acid solutions and in strongly buffered alkaline solutions up to pH 9.0.

The glass electrode is now used in preference to the quinhydrone electrode because it may be used over the entire pH range. It functions in unbuffered solutions, and no chemical need be added to the unknown solution. Once these modern pH meters are adjusted for room temperature and standardized against a buffer solution of known pH., the reading of a sample can be made in a few seconds. Once standardization has been completed, a series of samples may be checked for pH in rapid sequence by rinsing the electrodes thoroughly with distilled water between samples. A neutral position on the range switch permits the indicating meter to be disconnected from the electrical circuit during the preliminary warm-up period and when changing test samples.

Beckman Zeromatic pH Meter
Courtesy Macalester Bicknell Co.

Operation of the Test.[16] (1) Set switch on start and connect the power cord. During a five-minute warm-up period, verify needle adjustment at 7.00.

(2) When handling electrodes after start, always have switch in the neutral position.

(3) Rinse and wipe electrodes before measurements. Use fine tissue for this as glass electrodes are very fragile.

(4) Set the temperature dial at electrode temperature.

(5) Standardization: Immerse electrodes in buffer solution and switch to proper range (temperature of buffer solution should be within 10° C. of sample temperature). Adjust standardization control until meter needle indicates exact pH of buffer. Then, switch to neutral and mark needle position with the auxiliary dial pointer.

(6) Measurement: Switch to neutral. Readjust standardization control until needle rests at pointer position. Immerse electrodes in sample, then switch to proper pH range and read pH value directly on the dial. Repeat buffer standardization occasionally during extended series of measurements. Readings may be made with ease to ±0.1 pH unit and can be estimated to ±0.03 pH.

The pH of Milk. Johnston and Doan[17] found that the pH values of herd samples of milk ranged from 6.34 to 6.92 with 97 percent of the samples occurring in the range of 6.45 to 6.80. Seventy-eight percent of the samples showed a pH value in the range of 6.54 to 6.71. These workers found that 7 of 10 samples having pH values above 6.8 showed evidence of mastitis while samples giving readings below 6.4 showed apparent bacterial activity.

Milk will curdle when the pH is about 4.7. The *Streptococcus lactis* organism ceases to convert lactose to lactic acid when the pH is down to 4.1. Other acid formers such as *Lactobacillus bulgaricus* will continue the production of acid until a pH 3.5 is reached. Mastitis milk is alkaline and has a pH around 7.6. The pH of colostrum[3] ranges from 6.25 on the first day to 6.46 on the third.

The figures in Table 41 show the approximate relationship of titratable acidity of milk to the pH value.

Table 41. Relation of Titratable Acidity of Milk to the pH Value

Titratable acidity	pH	Titratable acidity	pH
%		%	
0.50	6.0	0.145	6.7
0.43	6.1	0.125	6.8
0.36	6.2	0.115	6.9
0.30	6.3	0.105	7.0
0.25	6.4	0.095	7.1
0.205	6.5	0.090	7.2
0.165	6.6	0.085	7.3

BIBLIOGRAPHY

1. Rice, F. E. and Markley, A. L. The Relation of Natural Acidity in Milk to Composition and Physical Properties. J. D. Sci. 7, p. 482 (1924).
2. McInerney, T. J. A Note on the Acidity of Fresh Milk. J. D. Sci. 3, p. 228 (1920).
3. Caulfield, W. J., and Riddell, W. H. Some Factors Influencing the Acidity of Freshly Drawn Cows' Milk. J. D. Sci. 19, pp. 235-42 (1936).
4. Barthel, C. Milk and Dairy Products. p. 2, MacMillan and Co. London (1910).
5. Doan, F. J., et al. Procedures for Making Acidity Tests of Fluid Milk Products. J. D. Sci. 40:1643-1644 (1957).
6. Hunziker, O. F. The Butter Industry, 3rd Edit., p. 763 (1940).
7. Sommer, H. H. The Acidity of Milk and Dairy Products. Wis. Rsh. Bul. 127, p. 6 (1935).
8. Van Slyke, L. L. and Bosworth, A. W. The Cause of Acidity of Fresh Milk of Cows and a Method for the Determination of Acidity. N. Y. (Geneva) Ag. Exp. Sta. Tech. Bul. 37 (1914).
9. Ling, E. R. The Titration of Milk and Whey as a Means of Estimating the Colloidal Calcium Phosphate of Milk. J. Dairy Rsh. 7, pp. 145-155 (1936).
10. Turner, B. B., and Beach, C. L. The Effect of Silage on the Acidity of Milk. Storrs Agr. Exp. Sta., 16th An. Rpt., 150 (1904).
11. Sommer, H. H. and Hart, E. B. Influence of Acids in the Ration on the Acidity of Milk. J. D. Sci. 4, p. 7 (1921).
12. Anderson, E. O., White, G. C., and Johnson, R. E. Corn Gluten Feeding and the Titratable Acidity of Milk. J. D. Sci. 19, pp. 317-322 (1936).
13. Gortner, R. A., Gortner, R. A. Jr., and Gortner, W. A. Outlines of Biochemistry. 3rd Ed. John Wiley and Sons, Inc., New York (1949).
14. Davis, J. G. A Dictionary of Dairying, p. 130. Leonard Hill, Ltd. (London, 1955).
15. Jenness, Robert and Patton, Stuart. Principles of Dairy Chemistry. John Wiley and Sons, Inc. (New York, 1959).
16. Beckman Instruments, Inc. Operating Instructions for Model H-2 pH Meter.
17. Johnston, H. K., and Doan, F. J. The Use of a Direct Reading pH Meter for Routine Examination of Milk at the Dairy Plant Intake. J. D. Sci. 26:271-276 (1943).

REVIEW QUESTIONS

1. Distinguish between apparent and real acidity of milk.
2. Explain an amphoteric reaction.
3. By what agency and from what compound is real acidity formed?
4. Is phenolphthalein a good indicator in the acid test of milk? Why?
5. What portion of milk or cream contains lactic acid?
6. Why does diluting a milk sample with water lower the titratable acidity?
7. What is colostrum?
8. Why does milk from cows having mastitis test lower in acidity?
9. Define pH.
10. If the pH of a solution is 3, what would be the pOH?
11. Which has the higher pH value, an acid or a base?
12. What is a buffer? What are the principal buffers in milk?
13. Describe the calomel electrode.
14. Can the acid content of milk be changed by feeding cows such feeds as silage or other high acid feeds?
15. Write the structural formula of lactic acid.

CHAPTER XII

Dairy Chemistry Problems

1. Solutions.

Normal Solution. A normal solution is one that contains the molecular weight in grams of dissolved substance, divided by the hydrogen equivalent of the substance, per liter of solution. For example, a normal solution of hydrochloric acid, HCl, would consist of 36.465 ÷ 1, or 36.465 grams of acid per liter of solution, there being but one replaceable hydrogen atom. For sulphuric acid, H_2SO_4, it would be 98.076 ÷ 2, or 49.038 grams acid per liter of solution, there being two replaceable hydrogen atoms. For sodium chloride, NaCl, it would require 58.454 ÷ 1, or 58.454 grams salt per liter of solution, since sodium would have the equivalent of one hydrogen atom, that is, it unites with one chlorine atom as hydrogen does in HCl.

Fractions of normal solutions are also used as one-tenth normal, written N/10 or 0.1N, and twice normal, 2N, etc. From the definition of normal solution we know that solutions of different compounds thus made are placed on the same basis, that is, one ml. of normal NaOH will exactly neutralize one ml. of normal H_2SO_4, or one ml. of 0.1 normal $AgNO_3$ solution will react completely with one ml. of 0.1 normal NaCl.

The following problems will aid the student in understanding the principles involved:

Problem 1. How many grams of pure phosphoric acid, H_3PO_4, are needed to make 1 liter of normal solution?
Ans. 32.668 g.

Problem 2. How many grams of sodium hydroxide, NaOH, are needed to make 0.1 N solution?
Ans. 4.000 g.

Problem 3. How many grams of silver nitrate, $AgNO_3$, are needed to make 2 N solution?
Ans. 339.776 g.

Problem 4. How many grams of ammonium oxalate, $(NH_4)_2C_2O_4 \cdot 2H_2O$, are needed to make 1 liter of normal solution?
Ans. 80.066 g.

Problem 5. How many grams of ferric chloride, $FeCl_3$, are needed to make 500 ml. of a normal solution?
Ans. 27.037 g.

Problem 6. How many grams of nitric acid, HNO_3, are needed to make 2 liters of 0.1 N solution?
Ans. 12.603 g.

Problem 7. How many grams of aluminum sulphate, $Al_2(SO_4)_3$, are needed to make 2 liters of 0.1 N solution?
Ans. 11.404 g.

Problem 8. How many grams of calcium hydroxide, $Ca(OH)_2$, are needed to make 1 liter of 0.01 N solution?
Ans. 0.370 g.

Problem 9. How many grams of sodium carbonate, Na_2CO_3, are needed to make 1 liter of N solution?
Ans. 53.002 g.

Problem 10. How many grams of sodium thiosulphate, $Na_2S_2O_3 \cdot 5H_2O$, are needed to make .5 liter of N solution?
Ans. 124.097 g.

Molar Solutions. A molar solution is one which contains the molecular weight in grams of dissolved substance per liter of solution. Note how this differs from a normal solution. The entire weight in grams of a molecule of the substance is used, and not divided by the hydrogen equivalent. In many cases both the normal and molar solutions would be similar, as for example, hydrochloric acid there being but one replaceable hydrogen atom. Sulphuric acid, however, would not be similar because it contains two replaceable hydrogen atoms and therefore the molar would be twice as strong as the normal. As in the case of normal solutions we have a molar solution, M/1, a 0.1 molar, M/10, and twice molar, 2M, etc., depending on whether the entire molecular weight in grams of the substance, one-tenth or twice the amount, respectively, was dissolved in sufficient water to make a liter of solution.

Percentage Solutions. The solutions heretofore discussed have involved the molecular weight of the substance dissolved, but when we speak of a certain percentage solution, the molecular weight does not enter into the calculation. For example, suppose a 5 percent solution of sodium chloride is desired, either by volume or weight. In the first case, weigh out 5 grams of chemically pure NaCl, and add sufficient water to make up to 100 ml. of solution. The weight of salt is here equal to 5 percent of the total volume of brine. It is not, however, equal to 5 percent of the total weight since the combined weight of the salt and water is practically 105 grams, and this divided into 5 grams would give less than 5 percent. In the second instance, weigh out 5 grams NaCl and dissolve in 95 grams of water. Here the total weight is 100 grams and the 5 grams salt would constitute 5 percent of the total by weight. By volume it would represent more than this percentage. It should always be stated whether a certain percentage solution is by volume or weight. Usually, when it is not so stated, volume is meant.

Another percentage solution problem may be represented by the following: Change a 95 percent alcohol to one of 75 percent strength. The simplest solution to this problem is to add 20 ml. of water to 75 ml. of the 95 percent alcohol. It is readily seen that the amount of water to add was obtained by taking the difference between the strength of the two alcohols (95-75). This kind of problem may also be solved by the Pearson Square.

Standardizing Acid and Alkali Solutions. In order to make standard acid or alkali solutions, it is necessary to have a starting point. Oxalic acid may be used for this purpose because its crystals are practically free from impurities.

a. N/10 Oxalic Acid, $C_2H_2O_4 \cdot 2H_2O$. Dissolve 1.5754 grams of oxalic acid crystals in a 200 ml. beaker of hot distilled water. Cool and transfer this quantity to a 250 ml. volumetric flask. Make up to 250 ml. with CO_2 free water. (Freshly boiled distilled water will be practically CO_2 free.) Mix well and transfer to a clean stoppered bottle. (Before pouring any standard solution into a wet bottle, rinse the bottle with a little of the solution and discard.) This is your standard solution for checking the sodium hydroxide that follows.

b. N/10 Sodium Hydroxide, NaOH. Add 6 ml. of saturated sodium hydroxide solution to a liter of CO_2 free distilled water. This mixture will approximate a 0.1 normal solution. Titrate this solution against the standard oxalic acid, using phenolphthalein as indicator. Adjust the alkali solution until it is exactly 0.1 normal. For example, suppose 10 ml. of the oxalic acid neutralized 8.8 ml. of the sodium hydroxide, then the latter is too strong and must be diluted. Add 1.2 ml. (10.0 — 8.8) of CO_2 free distilled water for each 10 ml. of the sodium hydroxide solution or 120 ml. for each liter. Check again to see if this amount made exactly an N/10 normal solution of sodium hydroxide.

c. **N/10 Hydrochloric Acid, HCl.** Add 12 ml. of concentrated HCl to a liter of CO_2 free distilled water. Titrate this mixture against the above standard sodium hydroxide and adjust to exactly 0.1 normal strength. Use phenolphthalein as indicator.

2. Moisture and Solids.

Problems concerning comparisons of products with varying moisture contents may at times be puzzling. This type of dairy problem may be illustrated by the sponge. For example, suppose a sponge contained 70% water and then some of the water was squeezed out, after which it analyzed 60% water. What percent of the original water was removed? This type of problem may be solved in two ways:

Solution 1. Assume the weight of the sponge, when it contained 70% water, to be 100 grams, in which case there would be 70 grams of water and 30 grams of dry matter. After squeezing the sponge it would still contain the 30 grams of dry matter but X grams of water. Thus we have:

(1). 70% water and 30% dry matter or 70 grams of water and 30 grams of dry matter, which equals 100 grams, the total weight.

(2). 60% water and 40% dry matter or X grams of water and 30 grams of dry matter, which together equals ? grams, the new total weight.

(3). Since the grams of dry matter do not change then in (2) 40% = 30 grams and 100% = 75 grams, the new total weight. Then the water = 75 — 30 or 45 grams.

(4). 70 grams water in sponge originally.
 45 " " " " after squeezing.
 ―――
 25 " " squeezed out.
 25 ÷ 70 = 35 5/7% of original water removed.

Solution 2. This type of problem may also be solved on the percentage basis by using the water equivalent or parts water per 100 parts of dry matter. Thus:

(1). 70% ÷ 30% = 233⅓ parts H_2O per 100 parts of dry matter.
(2). 60% ÷ 40% = 150 parts H_2O per 100 parts of dry matter.
(3). 233⅓ — 150 = 83⅓, 83⅓ ÷ 233⅓ = 35 5/7% water removed.

Problem 1. In manufacturing evaporated milk what percentage of the water in normal milk, containing 87 percent water, must be removed so that the percentage of solids in the evaporated milk will be twice that in the normal?
                                        Ans. 57.47%

Problem 2. If normal milk, containing 87 percent water, is made into evaporated milk at a ratio of 3 to 1, what is the percent of total solids in the evaporated milk? What percent of the water in the normal milk was removed?

Ans. 39.00% T. S.
76.63% Water.

Problem 3. If normal milk contains 86 percent water and the powdered milk from it contains 4 percent moisture, what percent of the original water was removed in the drying process?

Ans. 99.32%

Problem 4. If grass contains 75 percent water when cut for hay, what percent of this water must be removed to lower the percentage of moisture in the cured hay to 20 percent?

Ans. 91.67%

3. Standardization Problems.

Standardization is the raising or lowering of the percent of fat in milk or cream to a desired standard. This is done by the addition of milk or cream which tests either higher or lower, as the case requires.

There are two kinds of standardization problems. One problem consists of making up a definite number of pounds of milk or cream testing a certain percent. The other method consists in using up a certain amount of milk or cream and adding enough of another product to bring it to the desired percentage. This will result in an indefinite amount of the standardized product.

Problems in standardization may be readily solved by the rectangular method, which is as follows: Draw a rectangle and in the center place the percentage desired. At the left hand corners place the percentages of fat in the materials to be used. Then subtract the figure in the center of the rectangle from the larger one at the left and write the result at the right hand corner diagonally opposite. Next subtract the smaller figure at the left from the one in the center and write the result at the right hand corner diagonally opposite it. These figures at the right hand side of the rectangle indicate the relative amounts of the materials to be used, the fat percentage of which is indicated by the figure opposite to the left. An example of each of these problems will make the directions clear.

Problem 1. Definite Amount. Suppose it is desired to make up 200 pounds of 40 percent cream from whole milk testing 3.5 percent and cream testing 45 percent. How many pounds of each material would be used?

Solution.

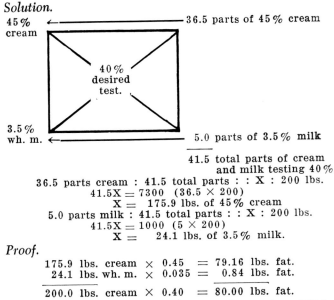

$$36.5 \text{ parts cream} : 41.5 \text{ total parts} :: X : 200 \text{ lbs.}$$
$$41.5X = 7300 \quad (36.5 \times 200)$$
$$X = 175.9 \text{ lbs. of } 45\% \text{ cream}$$
$$5.0 \text{ parts milk} : 41.5 \text{ total parts} :: X : 200 \text{ lbs.}$$
$$41.5X = 1000 \quad (5 \times 200)$$
$$X = 24.1 \text{ lbs. of } 3.5\% \text{ milk.}$$

Proof.

175.9 lbs. cream	× 0.45	=	79.16 lbs. fat.
24.1 lbs. wh. m.	× 0.035	=	0.84 lbs. fat.
200.0 lbs. cream	× 0.40	=	80.00 lbs. fat.

In the above problem it is seen that it requires 175.9 pounds of 45 percent cream and 24.1 pounds of 3.5 percent milk to make 200 pounds of 40 percent cream. These figures are correct because the sum of the pounds of fat in the cream and whole milk used equal the pounds of fat in the standardized cream.

Problem 2. Indefinite Amount. Suppose a creamery man has on hand 500 pounds of 30 percent cream and he wishes to use all of it in making 18 percent cream. He has on hand plenty of skimmilk with which to reduce the test. How much skimmilk must be add to the 500 pounds of cream and how many pounds of 18 percent cream will there be?

Solution.

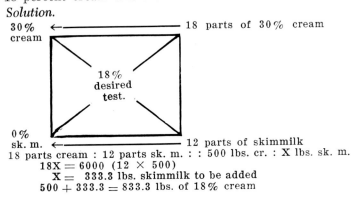

$$18 \text{ parts cream} : 12 \text{ parts sk. m.} :: 500 \text{ lbs. cr.} : X \text{ lbs. sk. m.}$$
$$18X = 6000 \quad (12 \times 500)$$
$$X = 333.3 \text{ lbs. skimmilk to be added}$$
$$500 + 333.3 = 833.3 \text{ lbs. of } 18\% \text{ cream}$$

DAIRY CHEMISTRY PROBLEMS

Proof.

$$500 \text{ lbs. cream} \times 0.30 = 150 \text{ lbs. fat.}$$
$$333.3 \text{ lbs. sk. m.} \times 0. = 0 \text{ lbs. fat.}$$
$$833.3 \text{ lbs. cream} \times 0.18 = 150 \text{ lbs. fat.}$$

Problem 1. How many pounds of each of 4.5% whole milk and skimmilk 0% must be used in making up 250 lbs. of 3.7% milk?

Ans. 205.6 lbs. wh. m.
44.4 lbs. sk. m.

Problem 2. How many pounds each of 3.2% milk and 40% cream must be used in making up 250 lbs. of 3.7% milk?

Ans. 246.6 lbs. wh. m.
3.4 lbs. cream.

Problem 3. Show how to make up 150 lbs. of 35% cream from 40% cream and skimmilk.

Ans. 131.25 lbs. cream.
18.75 lbs. sk. m.

Problem 4. Repeat (3) except use 3.5% whole milk in place of the skimmilk.

Ans. 129.5 lbs. cream.
20.5 lbs. wh. m.

Problem 5. How many pounds of skimmilk must be added to 75 lbs. of 40% cream to make whole milk testing 4%?

Ans. 675 lbs. sk. m.

Problem 6. How many pounds of whole milk testing 3.0% must be added to 100 lbs. of 35% cream to make 4.5% milk?

Ans. 2033⅓ lbs. wh. m.

Problem 7. How many pounds of skimmilk must be added to 400 lbs. of 4.0% milk to make 3.0% milk?

Ans. 133⅓ lbs. sk. m.

Problem 8. How many pounds of 45% cream must be added to 50 lbs. of 5.0% milk to make a 35% cream?

Ans. 150 lbs. cream.

Problem 9. Make up 120 lbs. of 18% cream for an ice cream mix. Materials on hand 40% cream and 4.0% milk.

Ans. 73⅓ lbs. wh. m.
46⅔ lbs. cream.

Problem 10. For cream cheese make up 260 lbs. of 8.0% milk from 30% cream and 4.0% milk.

Ans. 220 lbs. wh. m.
40 lbs. cream.

4. Percentage Problems.

Percentage means parts per 100. Thus if a sample of milk tests 4.0 percent fat, it means that 4 pounds of fat are present in every 100 pounds of such milk. In percentage problems it sometimes aids in solving them by bearing in mind a simple statement as $2 \times 4 = 8$, wherein, if any two of the figures are given, the third may readily be obtained by calculation. For example, how many pounds of milk testing 5.0 percent fat would be required to produce 10 pounds of fat? Substituting in the above statement $2 \times 4 = 8$, the unknown pounds of milk correspond to 2, the 5.0 percent fat to 4 and the 10 pounds of fat to 8. Therefore

(1) $8 \div 4 = 2$
(2) $10 \div .05 = 200$ pounds of milk.

Problem 1. How many pounds of fat in 300 lbs. of milk testing 3.5%?
Ans. 10.5 lbs.

Problem 2. How many total pounds of fat in 250 lbs. of milk testing 3.7% and 125 lbs. of milk testing 4.2%?
Ans. 14.5 lbs.

Problem 3. What would be the test of a mixture of the following:

275 lbs. of milk testing 3.0%
357 lbs. " " " 3.3%
102 lbs. " " " 5.6%
140 lbs. " " " 4.8%

Ans. 3.71%

Problem 4. If 450 lbs. of milk testing 3.7% were mixed with 550 lbs. of other milk, and the mixture tested 4.0%, what was the test of the 550 lb. lot?
Ans. 4.245%

Problem 5. If 790 lbs. of whole milk were separated and gave 90 lbs. of 30% cream and 700 lbs. of skimmilk testing 0.04%, what did the whole milk test?
Ans. 3.45%

Problem 6. If a quantity of 3.8% milk was separated and gave 250 lbs. of 38% cream and the skimmilk contained 1.5 lbs. of fat, how many pounds of whole milk were separated?
Ans. 2539.5 lbs.

Problem 7. How many pounds of cream testing 25% fat can be obtained from 10,550 lbs. of milk testing 3.8%, if 2.0 lbs. of fat were lost in the skimmilk?
Ans. 1595.6 lbs.

Problem 8. Find the percent of fat in the cream when 45,000 lbs. of milk testing 4% were separated and gave 6,500 lbs. cream. The loss in the skimmilk was 1% of the total fat.
Ans. 27.4%

CHAPTER XIII

Bacteriology of Milk

Bacteria are small one celled organisms belonging to the plant kingdom. Each bacterium possesses a firm and well differentiated cell wall which resembles that of a plant rather than the protoplasmic cell wall of an animal. This cell wall is a protective membrane and all food is taken in by the organism through this membrane. Most bacteria are of one of three shapes: spheres, straight rods or spiral rods. A spherical organism is termed a *coccus*, the straight rod a *bacillus* and the spiral rod a *spirillum*.

Size of Bacteria. Bacteria are too small to be seen with the naked eye. It would take approximately 25,000 organisms placed side by side to measure an inch. The unit of measurement is the micron, abbreviated as the Greek letter μ (mu). The micron is 0.001 of a millimeter or approximately 1/25,000 of an inch. Most bacteria encountered in dairy products are 0.5 to 1.0 micron in width and 1.0 to 2.0 microns in length. Some are much longer, such as the bulgaricus and acidophilus organisms which measure around 8 to 10 microns in length. Thus, in order to see and study these minute cells it is necessary to use a high powered microscope, one which magnifies around 800 times.

Reproduction. Bacteria reproduce by fission or a transverse splitting of the cell into two parts, each section soon developing into a mature organism. This division may take place as often as once every 30 minutes under favorable conditions. Therefore the increase is very rapid being a geometric progression. One organism would in 10 hours, theoretically, produce over a million individuals. Milk held at a warm temperature would soon have enough organisms to make it taste sour and the count would be around 100,000,000 organisms per milliliter, and when curdled as high as one billion, as measured by the plate count. This increase would not go beyond a certain limit as the toxic products formed and the change in reaction of the medium would inhibit the growth of the bacteria.

Reproduction is also carried on by the production of spores in some organisms. This usually occurs in the older individuals and under adverse circumstances. These spores are very resistant and remain dormant until conditions are favorable again for growth and reproduction of the species. None of the cocci produces spores but many of the bacilli and possibly a few of the spirilla are spore-forming.

Requirements for Growth. For growth bacteria require food, moisture, suitable temperature and a medium with the proper reaction. Milk is an excellent food for bacteria as it contains proteins, fat and carbohydrates and meets the above requirements of moisture and suitable reaction. Fresh milk has an average pH of 6.6 which is nearly ideal for most organisms and if the milk is held at the proper temperature for best growth, conditions are perfect for a rapid increase in numbers. The optimum temperature for development of most bacteria lies between 70° F. and 100° F.

The number of bacteria in milk decrease during the first five to six hours after milking. This interval of time is called the *germicidal period.* Two reasons are given for this decrease in numbers. First, an inhibiting agent, lactenin, is present in milk, which kills some of the organisms, and second, a dying out of certain species that do not find milk a suitable medium for development, occurs. This germicidal period is influenced by temperature; it continues longer at low temperatures, is almost destroyed at 143° F. and is entirely so at 163° F. It is customary at some milk plants to take advantage of the germicidal action and allow the patron to bring the morning's milk without cooling it, if delivered before a certain time. However, some organisms are not checked in their development by this germicidal action of milk, among which are the lactic acid bacteria which sour milk, and so it is not advisable to place too much dependence on this germicidal action to hold down the bacterial count. Furthermore, Frayer[1] has shown that delayed cooling affects the future quality of the milk, even though the bacterial count may be low when delivered at the plant. Cooling milk immediately to and holding at a temperature of 40° F. or below seems to produce the best results in respect to its quality both present and future.

Sources of Milk Bacteria. *a) Internal sources.* It was once thought that milk in the udder was entirely free of bacteria. In 1900 Ward[2] cut up udder tissue and found bacteria in the gland and in the milk ducts, and in 1913 Harding and Wilson[3] studied aseptically drawn milk from udders of 78 dairy cows and found an average of 428 bacteria per milliliter of milk. Therefore no matter what precautions are taken in obtaining low count milk from a cow, there will be at least 400 to 500 bacteria present. These organisms are mostly cocci and are harmless. Chromogens or color producing bacteria are also present, the colonies being mostly yellow and orange.

b. External sources. Most bacteria in milk come from external sources such as the utensils, stable air, coat of animal and the milker. The utensils are the greatest source if not properly cleaned. *Streptococcus lactis* is the predominating organism in dirty utensils. It grows prolifically in improperly cleaned milking machines and pails, feeding and developing on any droplets of milky water or dried milk left in the containers and tubes. Unclean utensils and lack of cooling are the greatest factors in causing high bacterial counts in milk. Other organisms from the utensils are rod-shaped bacteria, chromogens, clumps of micrococci, molds and yeasts.

The stable air produces no great contamination because light and desiccation kill most of the organisms. It is advisable, however, to give the animals the dusty feeds, such as hay, after milking and to regulate the time of sweeping and bedding so as not to raise any unnecessary dust at milking time. The principal organisms from this source are spore formers such as *Bacillus subtilis* and some yeasts and molds.

The coat of the animal may be a source of very undesirable organisms, since the cow may be covered with manure, bedding and soil, and particles of these materials might fall into the milking pail. The species of organisms from this source are mainly the Escherichia-Aerobacter or coli-aerogenes group which produce gas and bitter flavors in dairy products.

Number of Bacteria in Milk. Bacterial standards for milk vary according to the city and/or state where it will be marketed. A study of sanitary milk and ice cream legislation in the

United States by the National Research Council[4] indicated that 37 of the 48 states had defined bacterial limits for raw milk sold at retail but that most cities had prohibited the retail sale of raw milk. Over 80 percent of the states and cities studied had established bacterial limits lower than 30,000 per milliliter. The following maximum numbers of bacteria in several dairy products are sometimes used:

Raw Milk
    Certified. Usually 10,000 or less bacteria per ml., and must conform to the requirements of the American Medical Commission.
    Best Grade. 50,000 or less per ml.

Pasteurized Milk
    Certified Milk (Pasteurized). Bacteria content not to exceed 10,000 per ml. at any time before pasteurization and not to exceed 500 per ml. following pasteurization and until delivery to the consumer.
    Best Grade. Bacteria count not to exceed 200,000 per ml. prior to pasteurization and not more than 30,000 per ml. following pasteurization and until delivered to the consumer.
    Non-fat Milk.[5] Must be pasteurized and bacteria count not to exceed 200,000 per ml. prior to pasteurization, not to exceed 10,000 per ml. total bacteria and no more than 3 coliforms per ml. following pasteurization.
    Ice cream, ice cream mix, or ice milk mix.[5] Must be pasteurized. Bacteria count must not exceed 200,000 colonies per gram before pasteurization and must not exceed 50,000 colonies per gram or 10 coliforms per gram at the time of examination.

The number of organisms required to produce certain changes in milk has been studied by Hammer.[6] He inoculated sterile milk with different bacteria and made plate counts as the defects occurred. When sufficient acidity in milk just perceptible to the taste had been produced by *Streptococcus lactis*, the count was from 30 to 90 million per ml., and when sour, the count was from 53 to 166 million. A slightly ropy condition produced by *Alcaligines viscosus* in sterile milk required from 15 to 44 million per ml., while a very ropy condition showed a count of 450 to 786 million. A bitter flavor was produced by an organism when the count was 10 to 66 million per ml. A sweet curdling organism produced slight coagulation with only a count of 1.25 to 4.9 million per ml., and when the coagulation was firm, the count was 16 to 56 million.

Bacteriological Examination of Milk.

Plate Method. The estimation of the number of bacteria in milk by the plating procedure consists in mixing a definite volume of milk with melted agar, allowing the agar to solidify and then counting the colonies that appear upon incubation. A colony consists of the progeny of one cell or group of cells, and the results are reported as so many colonies per milliliter of milk.

a. Preparation of agar. The agar used must conform to the formula described in Standard Methods[7] or give equivalent results. Dehydrated base stock is recommended. The medium has the following composition:

Pancreatic digest of casein (U.S.P.)	5	g
Yeast extract (natural, enzymatically converted amino-acid product derived from primary yeast. On termination of digestion, pH of liquified yeast is adjusted, the material filtered, and resulting filtrate (containing all B-complex vitamins and growth factors natural to brewer's yeast) is dried to a powder, containing approximately 11.0% total nitrogen	2.5	g
Glucose	1	g
Agar, bacteriological grade	15	g
Distilled water	1	liter

These media are commonly obtained in the powdered form from the supply houses and are prepared by simply adding a specified quantity of the dry mixture to water in a flask, heating the mixture to boiling to dissolve the solid ingredients and then pouring in tubes or flasks in the desired quantities. These are then sterilized in the autoclave at 15 pounds pressure for 20 minutes. It is best to check the reaction produced in each lot of dry ingredients and adjust the reaction to pH 7.0 if necessary. Laboratories can make up the media from the ingredients if desired. However, when this is done, the analyst must assume responsibility for any errors attributable to the preparation of media from ingredients.[7]

b. Adjustment of reaction. Place 5 ml. of distilled water in a test tube, then 5 ml. of the agar and finally 5 drops of brom thymol blue indicator. Mix and compare the color with standards and if too acid, add N/10 NaOH to the agar in the tube until the color corresponds to that of the desired standard, preferably pH 7.0. Knowing the amount of alkali used for 5 ml. of agar, calculate the amount of normal sodium hydroxide to add to one liter of agar.

c. Plating the milk. The first step in plating milk is to dilute the sample to a suitable concentration so that the colonies of bacteria will be properly spaced for convenient counting. For

268   CHEMISTRY AND TESTING OF DAIRY PRODUCTS

Milk Dilution Bottle
Courtesy Kimball Glass Co.

samples of milk of unknown quality use dilutions of 100, 1,000 and 10,000, and for pasteurized milk or raw milk of high quality a dilution of 1:100 should be sufficient. The accompanying chart shows clearly how to make the dilutions. By means of pipettes graduated to 1.0 and 1.1 ml. and the use of 99 ml. phosphate

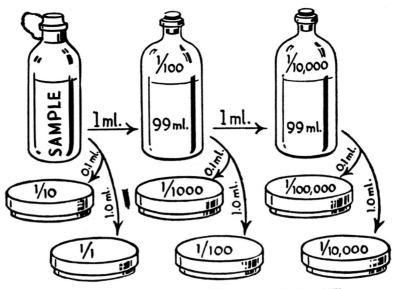

Showing Method of Making Dilutions in Plating Milk

buffered water blanks, any desired dilution can be made. All the glassware must have been sterilized by heating in an oven for at least one hour at not less than 170° C., and the dilution blanks in the autoclave at 15 lbs. pressure for not less than 20 minutes. The sample of milk and all dilutions must be well mixed before making any transfers by shaking the bottle at least 25 times in an up and down motion of about one foot. This mixing not only distributes the bacteria but also aids in breaking up clumps and chains of organisms.

Melt the prepared nutrient agar in boiling water and then temper to 45 degrees C. Transfer the proper amount of diluted milk to the Petri dishes and then pour in 10 to 12 ml. of tempered agar. All pouring lips of tubes or bottles must be sterilized by flaming them just before pouring. Mix the agar and sample thoroughly by rotating and tilting the dish carefully without splashing the media onto the cover or on the edge of the dish. Next set the plates on a level surface and allow the media to solidify. Then invert the plates and place in the incubator, leaving at least one inch space between each pile. It is advisable to not stack higher than four in a pile. Incubate for 48 hours either at 35 degrees C. or 32 degrees C., since both temperatures of incubation are now recognized as standard. However, the 32° C. incubation is preferred since many bacteria which grow at 32° C. will not grow at 35° C. A tolerance in time is allowed of ±3 hours.

d. Counting plates. Select for counting those plates that have between 30 and 300 colonies, preferably the latter. If a

Quebec Colony Counter
Courtesy Will Corp.

plate has more than 300, it is overcrowded and the number of organisms is less than it should be, while those plates with less than 30 colonies would be too low to be representative. When counting, place a guide plate, ruled into square centimeters, beneath the Petri dish and then by the use of a hand lens that magnifies at least 1½ times count all colonies, following the lines on the guide plate so that no colony is missed. The use of a hand tally for recording mechanically the number of colonies facilitates the enumeration and tends to be more accurate. After the plate is counted, multiply the number obtained by the dilution used and report this figure as the "Standard Plate Count" per milliliter. For example, if the number of colonies on the 100 dilution plate was 250, the estimated number of bacteria would be 25,000 (250 × 100) per ml.

Direct Microscopic Method. The direct microscopic or Breed method of counting bacteria in milk consists of examining a stained film of milk under a compound microscope. It is much quicker and less expensive than the plate method and individual organisms and leucocytes can be observed and studied. The plate method, however, is better for low count milk and for the study and propagation of pure cultures. The apparatus required consists of a microscope, pipettes, slides and stains.

a. Microscope. A compound microscope with an oil-immersion lens is used. It must be standardized so that a factor can be applied to determine the number of bacteria per milliliter of milk from the average number of organisms counted in several fields. The steps in this operation are as follows: First determine the diameter of a field, using a micrometer. This instrument is a glass slide with fine graduations measuring in one-hundredths millimeters. Suppose the diameter of a field was 0.16 mm., then the area of it would be πr^2 or 3.1416×0.0064 which equals 0.02 sq. mm. The next step is to determine the number of fields in the entire film of milk examined. The size of this film is 0.01 milliliter spread over one square centimeter or 100 square millimeters, and therefore 100 sq. mm. ÷ 0.02 sq. mm. = 5000 fields. This figure represents 0.01 milliliter of milk and the figure for one milliliter would be 5000 × 100 or 500,000. This last number is the microscopic factor, and the average number of bacteria counted in several fields must be multiplied by this factor to obtain the number of organisms in one milli-

BACTERIOLOGY OF MILK

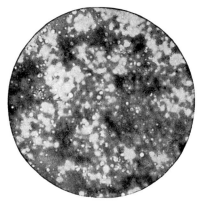

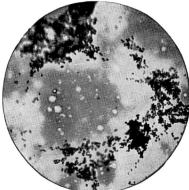

When a drop of clean, high grade milk is dried on a glass slide and then colored by immersion in a blue stain, it has the appearance shown above when seen under the microscope. No bacteria can be found. The background shows white circles where the fat drops have been dissolved out of the dried milk solids. A few white blood corpuscles may occur in milk of this type, though none are to be seen in the picture. ×600.

Where high grade milk is placed in improperly cleaned utensils it takes up masses of bacteria such as are shown in the above picture. When the milk is stained for microscopic examination these become evident as deep blue specks of various shapes. Bacteria in utensils produce putrid, unpleasant odors as they cause the decay of the remnants of milk left in rinse water or in open seams. ×600.

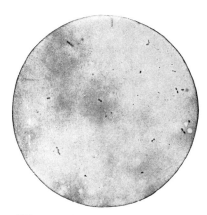

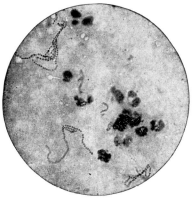

Where high grade milk is allowed to stand without adequate cooling, the bacteria that cause the normal souring of milk grow rapidly. These show in the above picture as pairs or double pairs of minute specks. Bacteria of this type produce a sour, but not putrid odor. ×600.

Where milk is drawn from a cow suffering from mastitis (garget), the milk contains enormous members of white blood corpuscles and bead-like chains of bacteria that cause the inflammation. These are so characteristic that they are readily found in milk even where it is diluted with large quantities of good quality milk. ×600.

Photomicrographs of Milk Samples

From Circ. 93 N. Y. State Agricultural Exp. Sta.

Microscope
Courtesy Fisher Scientific Co.

liter of milk. Thus, if 50 fields were counted and the average number of bacteria found was 0.6, the number per milliliter would be 500,000 × 0.6 or 300,000.

b. Pipette. The pipette used in measuring the milk is calibrated to deliver 0.01 ml. It should hold 0.1395 grams of mercury at 20 degrees C. The tip should be blunt and of such a form that it will discharge the milk cleanly without running back on the side of the tip. A metal syringe which delivers 0.01 ml. has recently been approved in Standard Methods.[7] Sometimes loops, made of such a diameter as to deliver 0.01 ml. of milk, are used in place of pipettes. These are satisfactory for use in quality control laboratories for the milk industry or for teaching purposes but are not permitted to be employed for official control work.

c. Slides. Any ordinary glass slide may be used for the milk film. Glass or cardboard guides ruled into square centimeters

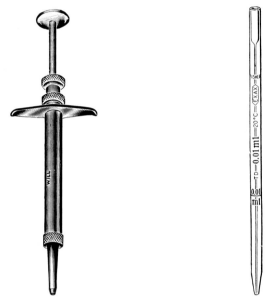

0.01 ml. Transfer Syringe
Courtesy Will Corp.

Breed Pipette
Courtesy Fisher Scientific Co.

must be placed beneath the slide so that the milk may be spread over a space of exactly one square centimeter. Several films may be placed on a slide.

d. *Stains.* Three stains have been approved for official preparation of slides for microscopic examination. The slide is prepared in three steps. These include fixing the smear, defatting the milk film, and finally staining the film. The reader is directed to the description of these stains in Standard Methods.[7] The stain most commonly used in unofficial analyses is the Newman-Lampert formula. It combines in one operation the three steps required in official procedures. The formula and method of making this stain is as follows:

Newman-Lampert Formula

Methylene blue powder, certified	1 g.
Ethyl alcohol, 95%	54 ml.
Tetrachloroethane, technical	40 ml.
Acetic acid, glacial	6 ml.

Add the alcohol to the tetrachloroethane in a flask and heat to not over 70 degrees C. to mix these two liquids. Methyl alcohol may be used in place of ethyl, in which case do not exceed 60° C. Add the warm mixture to the methylene blue powder and shake vigorously until the dye is completely dissolved. Cool and then slowly add the acetic acid and mix. Filter the mixture through filter paper and store in a tightly stoppered bottle. The purpose of this stain is three-fold: to fix the film of milk on the glass slide, to dissolve the fat and to stain the bacteria and leucocytes.

e. Preparation of the slides.

(1). Mix the sample and then draw the milk into the pipette above the graduation mark.

(2). Wipe the outside of the pipette with a clean piece of cheesecloth, and then touch the tip with the cloth to bring the milk down to the mark.

(3). Blow the drop of milk onto the slide and by means of a bent needle spread it over exactly one square centimeter.

(4). Dry the film in a warm place and on a level surface. The drying should not extend beyond five minutes, otherwise bacterial growth might occur. On the other hand the drying should not be too rapid as the film will become cracked and peel off during the staining and washing operations.

(5). Using a dropper, place enough stain on the slide to cover the smear and let dry in a slanting position.

(6). When thoroughly dried, wash the slide in water to remove excess stain.

(7). Dry the slide, after which it is ready for examination.

f. Counting the bacteria. Place a drop of cedar oil on the stained film of milk and then use the oil immersion lens for locating the bacteria. If the milk has a low number of organisms, count more fields than if the sample has numerous bacteria. Standard Methods[7] give the following number of fields to count depending on the microscopic factor and the number of bacteria.

Range of individual microscopic counts	Number of fields to be examined if the field diameter measures	
	0.206 mm. (300,000 factor)	0.146 mm. (600,000 factor)
30,000 to 300,000	30	60
300,000 to 3,000,000	20	30
Over 3,000,000	10	20

Methylene Blue Reduction Test. The methylene blue reduction test is based on the fact that the color imparted to milk by the addition of a dye such as methylene blue will disappear more or less quickly. The removal of the oxygen from milk and the formation of reducing substances during bacterial metabolism causes the color to disappear. The agencies responsible for the oxygen consumption are the bacteria. Though certain species of bacteria have considerably more influence than others, it is generally assumed that the greater the number of bacteria in milk, the quicker will the oxygen be consumed, and in turn the sooner will the color disappear. Thus, the time of reduction is taken as a measure of the number of organisms in milk although actually it is likely that it is more truly a measure of the total metabolic reactions proceeding the cell surface of the bacteria. Hobbs[8] has given a detailed study of the mechanics of dye reduction to which the interested reader is directed.

a. Apparatus. The necessary equipment consists of test tubes with rubber stoppers, a pipette or dipper graduated to deliver 10 ml. of milk and a water bath for maintaining the samples at 35°-37° C. The bath should contain a volume of water sufficient

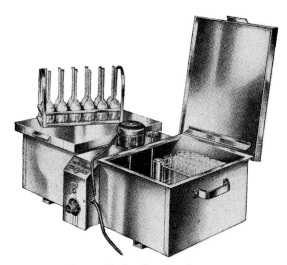

Losee Reduction Incubator
Courtesy Macalaster Bicknell Co.

to heat the samples to 35° C. within 10 minutes after the tubes enter the water and should have some means of protecting the samples from light during the incubation period. If a hot-air chamber is used, the samples should be heated to 35° C. in a water bath since warm air would heat the milk too slowly.

The dry tablets contain methylene blue thiocyanate and may be obtained from any of the usual laboratory supply houses. They should be certified by the Commission on Standardization of Biological Stains. The solution is prepared by autoclaving or momentarily boiling 200 ml. of distilled water in a light resistant (amber) stoppered flask and then adding one methylene blue tablet to the flask of hot water. The tablet should be completely dissolved before the solution is cooled. The solution may be stored in the stoppered, amber flask or an amber bottle in the dark. Fresh solution should be prepared weekly.

b. Procedure in testing.

(1). Sterilize all glassware and rubber stoppers either in an autoclave or in boiling water.

(2). Measure 1 ml. of the methylene blue thiocyanate solution into a test tube.

(3). Add 10 ml. of milk and stopper.

(4). Tubes may be placed in the water bath immediately or may be stored in the refrigerator at 32-40° F. for a more convenient time of incubation. When ready to perform the test, the temperature of the samples should be brought to 35° C. within 10 minutes.

(5). When temperature reaches 35° C., slowly invert tubes a few times to assure uniform creaming. Do not shake tubes. Record this time as the beginning of the incubation period. Cover to keep out light.

(6). Check samples for decolorization after 30 minutes of incubation. Make subsequent readings at hourly intervals thereafter.

(7). After each reading, remove decolorized tubes and then slowly make one complete inversion of remaining tubes.

(8). Record reduction time in whole hours between last inversion and decolorization. For example, if the sample was still blue after 1.5 hours but was decolorized (white) at the 2.5 hour reading, the methylene blue reduction time would be recorded as 2 hours. Decolorization is considered complete when four-fifths of the color has disappeared.

c. Suggested classification by Standard Methods:

Class 1. Excellent, not decolorized in 8 hours.

Class 2. Good, decolorized in less than 8 hours but not less than 6 hours.

Class 3. Fair, decolorized in less than 6 hours but not less than 2 hours.

Class 4. Poor, decolorized in less than 2 hours.

d. Factors affecting the test. Many factors affect the methylene blue reduction test and therefore the steps of operation should be uniform. Since the oxygen content must be used up before the color disappears, any manipulation that increases the oxygen affects the test. Cold milk holds more oxygen than warm milk; pouring milk back and forth from one container to another increases the amount, and at milking time much oxygen may be absorbed.

The kind of organisms affect the rate of reduction. The coliforms appear to be the most rapidly reducing organisms, closely followed by *Streptococcus lactis,* some of the faecal *Streptococci,* and certain micrococci.[8] Thermoduric and psychrophilic bacteria reduce methylene blue very slowly if at all.[8] A large number of leucocytes affect the reduction time materially.

Light hastens reduction and therefore the tests should be kept covered. The concentration of the dye should be uniform as an increased concentration lengthens the time of reduction. Increasing the incubation temperature augments the activity of the bacteria and therefore shortens the reduction time.

The creaming of the test samples causes a number of organisms to be removed from the body of the milk and brought to the surface with the rising fat. This factor causes variations in the reduction time, since the bacteria are not evenly distributed. Johns[9] found the accuracy of the test was increased, reduction time shortened and decolorization more uniform if the samples were periodically inverted during incubation.

The Resazurin Test. The resazurin test is conducted similar to the methylene blue reduction test with the judgement of quality based either on the color produced after a stated period of incubation or on the time required to reduce the dye to a given end-point. Numerous modifications have been proposed. The two most commonly used are the "one-hour test" and the "triple-reading test" taken after one, two, and three hours of incubation. Other modifications have value in specific applications.

The procedure for making the resazurin test is as follows: Prepare resazurin solution by dissolving one resazurin tablet (dye content/tablet, approximately 11 mg., certified by Biological Stain Commission) in 200 ml. of hot distilled water as was done in the methylene blue test. Place one ml. of dye solution in a sterile test tube, then add 10 ml. of sample. Stopper the tube, place in the incubator and, when the temperature reaches 35° C., invert to mix the milk and dye. Incubate at 35° C. Tubes are examined and classified at the end of an hour in the "one-hour test" or at the end of three successive hourly intervals in the "triple-reading test." The following relationships of color and quality are generally accepted:

Color of Sample	Quality of Milk
1. Blue (no color change)	Excellent
2. Blue to deep mauve	Good
3. Deep mauve to deep pink	Fair
4. Deep pink to whitish pink	Poor
5. White	Bad

Standard Methods[7] accepts as an optional procedure the use of vials containing the proper amount of dye which has been prepared by evaporating to dryness the 1 ml. portion of dye. These may be prepared and stored for later use as needed, simply adding 10 ml. of milk to each tube when ready to complete the test.

Frayer[10] made a comprehensive study of the resazurin test and concluded that this test may be made a valuable time saving tool if properly conducted and intelligently interpreted, but should be supplemented by microscopic examination.

Morgan[11] compared the resazurin test with the Breed microscopic method on 235 samples of milk and found the test reliable. On the other hand, Thornton et al.[12] state that the test is an unreliable index of bacteriological quality in milk. Atherton et al.[13] found the resazurin reduction time of refrigerated bottled milk at either 20° or 37° C. was much too long to be of any value in evaluating bacteriological spoilage of stored milk.

Economic Importance of Milk Bacteria. *a. Streptococcus lactis.* This is the most rapid producer of acid in milk. It grows rapidly between temperatures of 70° F. and 100° F. and converts the lactose in milk to lactic acid. In this respect it is harmful to the market milk industry as any development of acid is objectionable. Its growth, however, can be controlled by having all utensils clean and by promptly cooling freshly drawn milk to 40° F.

Streptococcus lactis is useful in many manufacturing processes such as cheese and butter making, in cultured buttermilk and in starters. In pure butter cultures or starters it has an associative action with two other organisms, *Leuconostoc dextranicum* and *Leuconostoc citrovorum*. These latter organisms are also known as *Streptococcus paracitrovorus* and *Streptococcus citrovorus*, respectively.[14] When sufficient lactic acid has been produced by S. *lactis* to make a suitable reaction, *L. dextranicum* and *L. citrovorum* become active and convert the citric acid in milk to acetic acid and then to acetylmethyl-carbinol and diacetyl. The first compound is odorless but its importance lies in its being the source of the second which gives a fine flavor and aroma to the culture. Thus in a butter culture or starter, S. *lactis* provides the acid flavor and *L. dextranicum* and *L. citrovorum* the flavor and aroma. Starters are used to ripen cream for churning, and the resulting butter is imbued with the delicate flavors and aroma produced by these organisms. The main chemical reactions are shown by the following formulae:

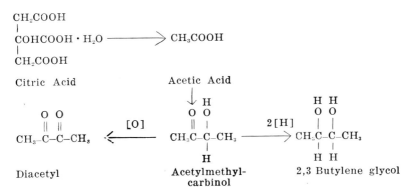

If considerable acid is present acetylmethylcarbinol is oxidized to diacetyl, but if a limited amount of acid is present the carbinol will be reduced to butylene glycol. The addition of citric acid to milk to supplement that normally present will cause an increased production of the carbinol and thus more diacetyl.

b. *Lactobacillus bulgaricus* and *Lactobacillus acidophilus*. These organisms are used in milk drinks. Metchnikoff, a Russian, pioneered in advocating the use of sour milk in the prolongation of life, the idea being that these acid producing organisms would develop in the intestinal tract and replace the putrefactive bacteria. It has been discovered, however, that only the acidophilus organism will develop in the intestines as the bulgaricus bacterium can not tolerate the lowered surface tension in the tract. Therefore more acidophilus than bulgaricus milk is used. Both organisms produce a high acid, up to 3.0 to 4.0 percent, this acid giving a sharp but pleasant flavor to the milk. Those consumers who do not like high acid flavors can still obtain the advantages of acidophilus milk by the use of a very concentrated culture of these organisms. A small amount of a special preparation is added to the milk, which supplies sufficient organisms so that the milk can be consumed while still sweet without waiting for fermentation to take place.

c. *Escherichia-Aerobacter organisms*. The *Escherichia-Aerobacter* group of bacteria, as represented by *Escherichia coli* and *Aerobacter aerogenes*, are a detriment to the dairy industry. They are gram-negative, non-spore forming bacteria which ferment lactose to acid and gas and give milk and its products a very undesirable flavor. Milk that has any appreciable number of these organisms cannot be used for cheese making, because of the gas produced which in some cases is great enough to cause a cheese press to buckle under the excessive pressure. The flavor of the cheese is also very inferior.

Pasteurization of milk or cream destroys the coliform group and a test for the presence of this group in pasteurized milk is frequently used to determine whether the milk in question has been contaminated by improper handling following pasteurization. Normally for routine or official control of pasteurized samples, testing is limited to presumptive tests on one of three approved[7] media, one of which is liquid and two are solid. The production of gas in the liquid medium or the development of dark red colonies on the solid media are accepted as presumptive evidence of the presence of coliform organisms. In general, quality control laboratories prefer the use of solid media since these procedures are similar to other bacteriological techniques in common use. However, if the operator is interested in a more exact estimate of coliforms at low levels (<10/ml.), the liquid medium is more rewarding. Completed tests require an additional transfer following positive presumption test so are seldom used unless a differentiation of coliform types or some other specific information is desired.

d. *Psychrophiles.* In recent years, those responsible for quality control in the fluid milk industry have had to consider a group of organisms known as *psychrophilic* ("cold-loving") bacteria. They are a poorly defined group of organisms which show maximum activity at temperatures slightly above freezing — those temperatures associated with refrigerated storage. The need for longer storage life of raw milk, intensified by the changing pattern of milk plant operations following World War II, brought these organisms from the point of mere laboratory interest to a state of genuine industry concern.

The *psychrophilic* species found in milk will grow over a wide temperature range.[13] Growth slows noticeably at the extremes. Counts made following incubation temperatures of 3.3° C. or 35° C. are significantly lower than those made following incubation at 5° C. or 32° C., respectively. The psychrophiles are generally Gram-negative, non-spore-forming rods. *Pseudomonas, Achromobacter, Alcaligenes,* and *Flavobacterium* are the predominating genera. The majority of these *psychrophilic* species are said to be inert, that is, their metabolic products do not produce marked changes in the milk. However, there are still many which are either *proteolytic* or *lipolytic* and some show pigmentation or fluorescence. They are not disease-producing bacteria but may ruin the market quality of milk through the production of undesirable flavors or extremely high counts during a normal storage period. While pasteurization will destroy *psychrophiles,* it cannot overcome the off-flavors or other pre-pasteurization damage to milk constituents.

Psychrophiles usually originate in soil or water. Their presence in milk is generally associated with improper sanitary care of milk handling equipment. *Psychrophiles* have been reported[15] to survive both heat and chemical sterilization when protected by milkstone deposits. Untreated farm water supplies frequently contain large numbers of these low temperature species. This may be a definite factor in milk quality since many strongly *proteolytic* and *lipolytic* bacteria are derived from contaminated water.

Because many *psychrophilic* species are inert, few of the usual tests for milk quality have any value in estimating their numbers or activity. The usual method for determining *psychrophilic* populations in water or milk is to prepare agar plates as in the Standard Plate Count and then make the count after a period of incubation at 5° C. for seven days or longer.

An excellent review article on the *psychrophilic* microorganisms in milk and dairy products was prepared by Thomas and presented in two parts in Dairy Science Abstracts.[16]

e. Sweet curdling. Some organisms have the property of producing a rennin-like enzyme which will curdle milk without the production of any acid, at least in the initial stage of coagulation. The causative organisms may be spore formers such as *Bacillus subtilis* and *Bacillus cereus,* or non-spore producing rods belonging to the genus Proteus. Various species of cocci also sweet curdle milk, the most important organism being *Streptococcus liquefaciens.* Sweet curdling occurs most frequently during the warm season. An incident of this type of trouble occurred on a farm near the Vermont station where the milk kept for home use curdled, but not that taken to the milk plant. The source of the difficulty was discovered when it was learned that the milk retained for home use was always given an extra straining through a cloth. By discarding this cloth strainer the source of the sweet curdlers was removed and the trouble was eliminated.

f. Ropy milk. Milk sometimes attains such a slimy consistency that it can be drawn out in threads several feet long. This condition is known as ropy milk. It should not be confused with gargety milk, which is more on the lumpy order and is due to an infection of the udder. Also gargety milk shows its abnormal condition as soon as it is drawn, whereas ropy milk does not develop until 12 to 24 hours afterward at refrigeration temperature. Several organisms may cause ropy milk, the most common one being *Alcaligenes viscosus.* It produces a capsule which is the cause of the sliminess. This organism grows well at fairly low temperatures and therefore ropy milk develops while the product is held in the refrigerator. *Aerobacter aerogenes* has also been known to produce ropy milk.

Ropy milk does not occur often, but at the time of its occurrence it usually develops into an epidemic. Drastic measures are necessary to bring it under control. The milk is not harmful, but its nature is objectionable. In combating the outbreak, sterilize all utensils and equipment with which the milk comes in contact. Since the water in the cooling tank is often the source of the trouble, the tank should be drained and then cleaned with a sterilizing solution. The original source may be the stagnant pools in the pasture through which the cows wade and contaminate their coats, then carrying the bacteria to the stable and eventually to the milk pail. It may be necessary to fence off these water holes in the pasture fields. The epidemics can be controlled if particular attention is paid to all details and not neglecting any single factor that might be the cause.

Mastitis. Mastitis is an inflammation of the udder or mammary gland. In the acute form, the condition of the cow is obvious. The udder is swollen and inflamed, little or no milk is available, and prostration of the animal is common. In the more common form known as "chronic" or "latent" mastitis, the animal may appear perfectly normal even though laboratory tests indicate the presence of mastitis organisms or abnormal milk.

The immediate cause of mastitis is in dispute. One school of thought places the blame on certain predisposing causes (injury to the udder, poor milking practices, etc.) which reduce the animal's resistance and permits organisms present to produce the disease. Another group believes specific organisms may be transmitted from cow to cow and cause the disease to be spread in this manner. The first group believes the best control for mastitis is the elimination of predisposing causes, the second believes the disease can be checked by control measures against the causative organism. While *Streptococcus agalactiae* was the organism most commonly associated with mastitis twenty years ago, today a score of different species have been cultured from mastitic milk with *Staphlococci* predominating. When herds have been routinely treated with antibiotics, Streptococcal mastitis has nearly disappeared. However, despite the change in bacterial flora associated with mastitis and even though tremendous sums of money have been spent for antibiotic treatment in the past few years, there has been little change in the total incidence of mastitis during the period.[17]

Mastitis is a serious problem for the dairy industry. In addition to an estimated[18] loss of animals and milk amounting to

$225,000,000, the presence of certain organisms causing human diseases and of residual antibiotics which may impair human health is a cause for considerable alarm. Residual antibiotics in the milk supply resulting from mastitis treatments also have far-reaching effects in the cheese industry. Available information on preventive measures, diagnosis, and control measures have been well summarized by Plastridge,[18] Murphy,[19] and others.

Several tests may be applied to milk to detect mastitis in dairy cows. Physical (barn) tests would include the Whiteside Test, the California Mastitis Test (CMT), strip cup method, and others. Chemical methods for detecting mastitis would include the brom thymol blue test papers or solutions, catalase test, etc. Laboratory tests would include the examination of smears under the microscope, blood agar tests, Hotis test, and others. None of these tests is able to completely characterize mastitis but several tests, taken together, may have significance. The more common tests are presented below:

(1). *Microscopic method.* Incubate the sample overnight at 32-37° C., and then examine under the microscope for the typical long chains of *Streptococcus agalactiae,* for leucocytes (white blood corpuscles), or other causative organisms such as staphlococci, coliforms, or yeasts. The presence of leucocytes in milk indicates an abnormal condition of the udder, since these agents act as scavengers to destroy bacteria. A large number of leucocytes plus long chains of bacteria, 10 or more organisms in a chain, indicate mastitis. In general 500,000 leucocytes per ml. may be considered the limit for normal milk.

(2). *Brom thymol blue test.* The principle of this test is to measure the reaction of the milk. In case of mastitis the udder tissues are attacked by the mastitis organisms and blood exudes into the udder and milk. Since blood is alkaline it changes the milk from a slightly acid to an alkaline reaction, and therefore an acid reaction indicates normal milk and an alkaline reaction indicates mastitis milk.

The test is made by adding one ml. of brom thymol blue indicator to 5 ml. of milk in a test tube, and then noting the color produced. Normal milk will be a yellow green, and abnormal milk green to dark green or blue. The degree of color corresponds closely to the amount of infection. The test cannot be used on milk from recently freshened cows usually up to 5 days, or on that milk from nearly dry cows, as these milks are apt to be alkaline in any case.

The brom thymol blue solution is made by dissolving 0.2 grams of the powder in 25 ml. of 95 percent alcohol and then adding 75 ml. of water. Finally, sufficient N/10 NaOH is added to give the solution a blue color.

Prepared brom thymol blue test papers are available which are preferred for barn use. These absorbent papers contain the proper amount of dyestuff to give the proper color reaction when a drop of milk from the udder is placed on the test paper.

(3). *Chloride test.* The principle of this test is to measure the chloride content of milk. A percentage of 0.09 to 0.14 percent chloride in milk is normal but 0.14 percent or above is abnormal. Blood is high in sodium chloride, and since some blood enters into mastitis milk it increases the chloride content above normal.

The test consists of mixing in a test tube 1 ml. of milk, 4 drops of indicator, potassium chromate, 5 ml. of silver nitrate solution, and then noting the color produced. A brown color indicates normal milk and a lemon yellow color mastitis milk. The silver nitrate solution is made by dissolving 1.3415 grams of the crystals in one liter of water. This strength solution, when used in the above portion, will show the yellow color if 0.14 percent or more of chloride is present in the milk.

(4). *Hotis test.*[20] This method of testing for mastitis is done by mixing 0.5 ml. of sterile 0.5 percent aqueous solution of bromocresol-purple with 9.5 ml. of milk in a test tube and incubating at 37° C. for 24 hours. If the milk has the mastitis streptococci, the color will change from a purple to a yellow shade during the incubation period as a result of acid produced by these bacteria. In addition small flakes or balls of growth from 0.5 to 4 millimeters in diameter will usually form on the side of the tube.

(5). *Modified Whiteside test.* The principle[21] of the Whiteside test seems to depend directly or indirectly on the presence of leucocytes which causes a precipitate to form upon the addition of sodium hydroxide.

The test is as follows: Place 5 drops of milk on a plate of glass, add one drop of normal sodium hydroxide to the milk, stir 20 seconds and watch for a precipitate. The degree of infection corresponds to the amount of precipitate, if no precipitate the milk is normal.

(6). *Efficiency of the different tests.* Fay[22] evaluated some of the tests for mastitis as based on their value in detecting not only positive but also negative cases. The percentage efficiencies on this basis for the following tests were:

1.	Microscopic, incubated sample	87%
2.	Hotis	63%
3.	Leucocyte count	55%
4.	Manipulating udder for detecting fibrous tissue	40%
5.	Catalase	39%
6.	Chloride	38%
7.	Brom thymol blue	36%
8.	Strip cup	26%
9.	Microscopic, non-incubated sample	14%
10.	Abnormal appearance of milk	8%

Pathogenic Bacteria. The pathogenic bacteria in milk are of two kinds, those originating from the milkman and other contacts and those from the cow. The diseases that are due to direct human infection are the more serious and include scarlet fever, diphtheria, septic sore throat, typhoid fever and diarrhea or gastroenteritus. Diseases traceable to the cow are tuberculosis and undulant fever. It is fortunate that these disease organisms do not produce spores and can therefore be destroyed by pasteurization. This operation provides a safe product up to that point but contamination may occur afterward. However, nearly all milk-borne epidemics are due to the consumption of raw milk and rarely to the use of pasteurized. It should be emphasized that Public Health Service reports have shown a steady decline in the number of outbreaks traced to milk and milk products in recent years so that today such problems are rare. So, while continued attention is necessary to maintain the present status, disease outbreaks traceable to milk are not a major Public Health concern today. The recent textbook by Hammer and Babel[23] presents a comprehensive discussion on the pathogenic bacteria in dairy products as does a review of milk-borne diseases which appeared in the Journal of Dairy Research.[24]

Scarlet fever is caused by *Streptococcus pyogenes* var. *scarlatinae*. In the majority of cases of milk-borne epidemics, the organism was introduced from cases or carriers on the farms or in the plants. During the past 20 years very few outbreaks of the fever have been traced to milk.

Diphtheria is due to the organism *Corynebacterium diphtheriae*. Like scarlet fever direct human infection is the source of the spread of this disease as far as milk is concerned.

Septic sore throat is caused by *Streptococcus pyogenes* var. *epidemicus*. The disease may be spread by milk infected either directly by man or indirectly from the udder of the cow. In the latter case the organism would have been transferred from an infected throat of an individual to his hands and then to the udder by way of the teat canal. This fact has often led consumers to believe that the mastitis streptococcus is the cause of septic sore throat, but such is not the case.

Typhoid fever takes the lead in milk-borne epidemics. Milk carries the organism, *Bacterium typhosum* more readily than does water. A bad outbreak of typhoid fever occurred in Montreal in 1927 and lasted over four months, involving around 5,000 cases. The source of the trouble was traced to a certain milk supply.

Diarrhea mainly affects infants from one to two years of age. It is not the bacteria alone but also the toxic products produced by the organisms that cause the irritation. Nearly all the cases occur with bottle fed and not with breast fed babies. The causative organism is *Bacterium dysenteriae*.

Tuberculosis in humans may be caused by the bovine type of organism although it differs somewhat from the human strain. Cows may have tuberculosis in the neck glands, intestinal tract, in the lungs and in the udder. When the location is in the udder, the milk is most dangerous. However, only around three percent of all cases are in the udder tissues. Children under 5 years of age are most subject to infection from the bovine type, from 5 to 16 less so, and for adults the percentage incidence is small. The eradication of tubercular cattle from the herds has progressed rapidly during the last few years and the chances of tubercular infection from the bovine type of organism is very much lessened. The causative organism of the bovine type is *Mycobacterium tuberculosis* var. *bovis*. It exists in milk but does not multiply. It is inactive at 84° F. or below, active at 96° F. to 105° F. and is killed by pasteurization.

Undulant or malta fever has been known for an indefinite period of time, but only within the past few years has it been associated with milk from abortive cows. It has been established that the same organism, *Brucella abortus,* that causes abortion in cows can produce undulant fever in humans. Steps are being taken to eradicate the disease germ from the dairy herds, but in the meantime the only safe procedure is to pasteurize this milk when used for human consumption.

Yeasts. Yeasts like bacteria are unicellular, but differ from bacteria in size, cell structure and methods of reproduction. In shape many are elliptical and some cylindrical. The smallest yeasts may be no larger than bacteria but most of them are much larger, measuring 3μ to 10μ in diameter and 3μ to 100μ in length. The cell structure is somewhat advanced over that of bacteria, since the protoplasm can be differentiated into cytoplasm and a well defined nucleus. Reproduction by budding consists of an outgrowth from the side of the wall of the parent cell, and then later breaking loose as another individual cell. In some rare cases reproduction occurs also by fission. Spores are produced by the true yeasts but not by the pseudo or false yeasts.

It is the false yeasts that mainly concern the dairy industry because they ferment lactose with the production of alcohol and carbon dioxide. They can tolerate high acid conditions in which other organisms cannot develop and therefore in warm weather may produce undesirable changes in dairy products. Cream, that is shipped some distance and is not kept cold, may become yeasty in odor and taste and in some cases so gassy that the lids are forced off the cans. The two common species of yeasts that produce these results are *Torula cremoris* and *Torula sphaerica.*

Some yeasts are present in most butters but the counts are not high if the cream was of good quality and the churn was kept in a sanitary condition. Yeasts may readily be controlled by pasteurization.

Molds. Molds differ from yeasts and bacteria in that they are multicellular. They grow in long threads or hyphae, collectively called mycelium. They reproduce mainly by spore formation. Warm temperatures and high humidity are ideal conditions for growth. As the temperature is lowered, growth decreases until at 32° F. little if any growth occurs. *Oidium (Oospora) lactis* is the most common mold found in dairy products. It is a white appearing mold, and therefore at certain stages of growth is not very noticeable. Some common molds[25] found in butter and cheese are *Aspergillus flavus,* showing as a white felt on the surface with yellow or yellowish green conidia, *Aspergillus niger,* white felt appearance with chocolate brown or black conidia, *Penicillium roqueforti,* a blue green mold, especially useful in ripening Roquefort cheese, *Rhizopus nigricans,* a black mold, and *Mucor sylvaticus,* having extensive growth of mycelium with grayish or black sporangia.

Mold growth in butter originates largely from the cream and improperly cleaned equipment such as churns, vats, pumps and piping and refrigerators. The vegetative growth and spores in milk and cream are readily destroyed by pasteurization, and any growth in the equipment can be destroyed by hot water, 165° F. or higher. An application of formalin to such places as shelves in a curing room is also an effective means of checking mold growth.

Method of Counting Yeasts and Molds in Butter. Yeast and mold counts of butter may be made by plating, provided the agar is of sufficiently high acid content to prevent bacterial growth. One procedure is as follows: To standard agar add a solution of tartaric acid 2.5 percent and dextrose 50 percent, using 8 ml. to 10 ml. of agar to one ml. of the acid dextrose mix-

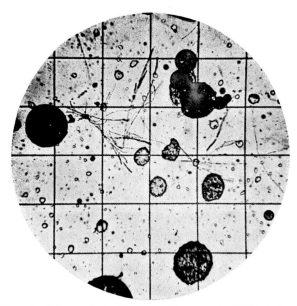

Mold Mycelia Filaments in Butter (Microscopic Field 1.5 sq. cm.)
Courtesy American Butter Institute
Microphotograph by G. W. Shadwick

ture. Use dilutions 1, 10 and 100 and incubate at 70° F. to 85° F. for 2 to 5 days. Count the colonies before mold growth becomes too heavy. The yeasts and molds may be counted together or separately as they are easily distinguishable.

The interpretation of results may be expressed by the following classification:

Production Conditions	Mold and Yeast Count
Good	-50 per gram
Fair	51-100 " "
Poor	101-500 " "
Very poor	500 + " "

Production of High Quality Milk. Most of the bacteria in milk come from the body of the cow and from unclean utensils. To produce a milk of low bacterial count, it is necessary to keep the body of the cow clean, use small-top milk pails, use clean utensils and cool the milk promptly after milking and keep it cool. If these conditions are properly met, the count need not exceed 5,000 bacteria per ml.

a. Clean cows. The body of the cow, especially those parts of the belly, flanks and udder that are immediately above the milk pail, may be a source of bacterial contamination, because the loose hairs, particles of manure and bedding may fall into the milk, carrying with them many organisms. Harding and Prucha[26] found that when cows were unusually dirty and were milked into open pails, the dirt in the unstrained milk weighed 10.8 mg., while in case of cows cared for in the better dairies the amount of dirt was 2.5 mg. per quart. The amount of dirt in the first case would cause an increase of about 17,000 bacteria per ml. It is therefore evident that the coat of the animals should be kept reasonably clean. This involves keeping the stables clean, using plenty of bedding, brushing the animals and, just before milking, wiping the udder and flanks.

b. Strainers. Straining milk does not remove any appreciable number of bacteria, since the organisms are so small that they pass through the mesh of the strainer, but the practice is necessary in order to remove as much as possible of the insoluble particles of dirt and thus improve the appearance of the milk The importance of sediment in milk and dairy products was discussed in Chapter VIII.

c. Utensils. One of the greatest factors affecting the bacterial content of milk is the condition of the utensils. The milking pails and milking machines with all their parts must be properly cleaned and sanitized after each milking in order to produce a milk with low bacterial count. The introduction of farm bulk milk coolers, pipeline milkers, and portable milk dumping stations have created new problems for farmers and quality control workers. Recent studies[27] at Vermont have indicated that there are many suitable products and procedures for cleaning and sanitizing equipment on farms but that the dairy and chemical industries have failed in their efforts to teach proper sanitary practices. Equipment surfaces should preferably be of seamless, stainless-steel construction to minimize their effect on the desirable flavor and bacteriological quality of milk. Faulty cleaning procedures result in serious lowering of milk quality through the action of the psychrophilic bacteria during our present extended storage of raw milk prior to pasteurization and of pasteurized milk before it is consumed in the home.

d. Cooling the milk. The greatest single factor in the dairy industry today affecting the bacterial count of milk is the temperature to which milk is cooled and held. As soon as the milk is drawn and strained, it should be placed in the cooling tank and cooled down to 40°F. or below. When milk is handled in properly cleaned utensils and stored in farm bulk tanks or cans which are free from milkstone buildup, very little increase in bacterial growth occurs at this temperature. Ayers, Cook and Clemmer,[28] in Table 42, give figures showing the effect of three different temperatures on the growth of bacteria in milk.

Table 42 — Effect of Temperature of Milk on Bacterial Development.

Temperature	Fresh	24 hrs.	48 hrs.	72 hrs.	96 hrs.
		Bacteria per ml. of milk			
40° F.	4,295	4,138	4,566	8,427	19,693
50° F.	4,295	13,961	127,727	5,725,277	39,490,625
60° F.	4,295	1,587,333	33,011,111	326,500,000	962,785,714

In summary it might be stated that there is very little excuse for a high bacterial count in milk, if proper and regular methods are followed in caring for the animals and their product.

BIBLIOGRAPHY

1. Frayer, J. M. The Production of High Quality Milk. Vt. Bul. 343 (1932).
2. Ward, A. R. The Invasion of the Udder by Bacteria. N. Y. (Cornell) Bul. 178 (1900).
3. Harding, H. A., and Wilson, J. K. A Study of the Udder Flora of Cows. N. Y. (Geneva) Tech. Bul. 27 (1913).
4. Dahlberg, A. C., and Adams, H. S. Sanitary Milk and Ice Cream Legislation in the United States. Nat. Res. Council Bul. 121 (July, 1950).
5. Mass. Dept. of Pub. Health. Rules and Regulations for Milk and Milk Products (1956).
6. Hammer, B. W. Studies on the Numbers of Bacteria in Milk, Showing Various Changes. Ia. Res. Bul. 29 (1916).
7. APHA. Standard Methods for the Examination of Dairy Products. Eleventh Edition (1960).
8. Hobbs, Betty C. The Part Played by Bacteria in the Reduction of Methylene Blue in Milk. J. D. Res. 10:35-38 (1939).
9. Johns, C. K. Concerning the Accuracy of the Methylene Blue Reduction Test. J. D. Sci. 21:227-239 (1938).
10. Frayer, J. M. The Resazurin Test — A Preliminary Study. Vt. Bul. 435 (1938).
11. Morgan, G. F. V. The Resazurin Test in Factory Practice. D. S. Abs. 4:85 (1942).
12. Thornton, H. R., et. al. The Reduction of Resazurin in Milk and Aqueous Solutions. Reprint, Can. J. of Res. 19 (1941).
13. Atherton, H. V., et al. Changes in Bacterial Population and Characteristics of Bottled Market Milk During Refrigerated Holding. Penn. Ag. Exp. Sta. Bul. 575 (1954).
14. Breed, R. S., et al. Bergey's Manual of Determinative Bacteriology, 6th Ed. Williams and Wilkins Co., Baltimore (1948).
15. Parker, R. B., et al. Psychrophilic Bacteria — A Sanitation Problem. J. Milk and Food Tech. 16,136-139,152 (1953).
16. Thomas, S. B. Psychrophilic Micro-organisms in Milk and Dairy Products Part I and II. D. Sci. Abst. 20,5:355-370 and 20, 6:447-468 (1958).
17. Penn. Ext. Service. The Mastitis Complex and the Antibiotic Problem. (1957).
18. Plastridge, W. N. Bovine Mastitis: A Review. J. D. Sci. 41:1141-1181 (1958).
19. Murphy, J. M. Mastitis — The Struggle for Understanding. J. D. Sci. 39:1768-1773 (1956).
20. Hotis, R. P. A Simple Method for Detecting Mastitis Streptococci in Milk. U.S.D.A. Circ. 400 (1936).
21. Dunn, H. O., et al. Nature of the Material in Mastitic Milk Responsible for the Whiteside Reaction. J. D. Sci. 25:712 (1942).
22. Fay, A. C. An Evaluation of Some of the Tests Used in Detecting Mastitis in Dairy Cattle. J. Milk Tech. 1:38-42 (1938).
23. Hammer, B. W., and Babel, F. J. Dairy Bacteriology. John Wiley and Sons, Inc., New York (1957).
24. Reviews of the Progress of Dairy Science, Section F. Milk Borne Diseases. J. D. Res. 20:115-126 (1953).
25. Macy, H. Some of the Factors Influencing the Growth of Molds in Butter. Minn. Tech. Bul. 64 (1929).
26. Harding, H. A., and Prucha, M. J. Germ Content of Milk. Illinois Bul. 236 (1921).
27. Atherton, H. V. A Field Study of the Sanitary Care of Milking Equipment on Dairy Farms. J. Milk and Food Tech. 22:273-278 (1959).
28. Ayers, S. H., et al. The Four Essential Factors in the Production of Milk of Low Bacterial Content. U. S. D. A. Bul. 642 (1918).

REVIEW QUESTIONS

1. What are bacteria?
2. How do bacteria reproduce?
3. What conditions are necessary for bacterial growth?
4. Explain what is meant by the germicidal period.
5. How many bacteria are usually present when milk turns sour?
6. What is the advantage of the Newman-Lampert formula over the three official stains used for the DMC?
7. What is the principle of the methylene blue reduction test?
8. Why should the test be kept covered during incubation period?
9. Name some advantages and disadvantages of the organism Streptococcus lactis to the dairy industry.
10. What organisms give the flavor and aroma to butter cultures?
11. How do the acid producing organisms, Lactobacillus, bulgaricus and acidophilus differ from Streptococcus lactis?
12. What kind of organisms curdle milk without acid formation?
13. How would you distinguish between gargety milk and ropy milk?
14. Which is the best test for detecting mastitis?
15. What is the principle of the Whiteside test for mastitis?
16. What is meant by pathogenic bacteria?
17. What is a cause of undulant fever?
18. How do yeasts differ from bacteria?
19. How must the agar used for growing bacteria be modified for the growth of yeasts and mold?
20. What are the two greatest factors affecting the bacterial content of milk?

CHAPTER XIV

Nutritive Value of Milk and Its Products

Food Value of Milk. *a. Nutrients in milk and their functions.* Milk is the most unique and ideal food for man. It meets the nutritional needs of the body better than any other single food. It contains the food constituents, (1) protein, (2) carbohydrates, (3) fat, (4) minerals, and (5) vitamins, and furthermore contains these dietary essentials in fairly suitable proportions.

(1). *Protein.* The primary purpose of protein is for the formation of new tissue (growth) and repair of broken down tissues. The muscles, glands, other organs, many hormones and enzymes, all require protein. It can also serve as a source of heat and energy, but these particular functions are more efficiently served by the carbohydrates and fat. All proteins contain the elements, carbon, hydrogen, oxygen, and nitrogen, and in addition some contain sulphur and phosphorus. Proteins are very complex compounds made up of a number of amino acids. At least 23 individual amino acids are now recognized. Feeding experiments have shown that daily supplies of at least eight of these amino acids are essential for maintenance of normal body functions in an adult man.[1] The remaining ones are either non-essential or can be synthesized by the body.

Milk contains several different proteins and the amino acid content of each of these is shown in Table 43. It will be noted that these proteins contain all the essential amino acids and in suitable proportions. This is very important since protein foods, to be effective, must contain all the essential amino acids and, equally important, these must be ingested simultaneously.[1] In an excellent discussion of the nutritive value of milk proteins, Henry[2] reported that "compared with an 'ideal' protein, milk proteins are deficient only in the sulfur-amino acids, cystine and methionine."

Table 43. Amino Acid Analysis of Bovine Milk Proteins[a]*

Grams amino acid per 100 g. of protein

	Casein proteins			Serum (whey) proteins		
	α-Casein	β-Casein	γ-Casein	β-Lacto-globulin	α-Lactal-bumin	Immune globulins (euglobulin and pseudo-globulin)	Serum albumin
Approximate percent of total milk proteins	45-63	19-28	3-7	7-12	2-5	1.4-3.1	0.7-1.3
Glycine	2.8	2.4	1.5	1.4	3.2	5.2[c]	1.8
Alanine	3.7	1.7	2.3	6.2	2.1	4.8[c]	6.3
Valine	6.3	10.2	10.5	5.8	4.7	9.6	5.9
Leucine	7.9	11.6	12.0	15.6	11.5	9.6	12.3
Isoleucine	6.4	5.5	4.4	8.4	6.8	3.0	2.6
Serine	6.3	6.8	5.5	5.0	4.8	11.5[c]	4.2
Threonine	4.9	5.1	4.4	5.9	5.5	10.5	5.8
Lysine	8.9	6.5	6.2	11.4	11.5	6.8	12.8
Arginine	4.3	3.4	1.9	2.9	1.2	4.1	5.9
Methionine	2.5	3.4	4.1	3.2	1.0	0.9	0.8
Cysteine+half-cysteine	0.43	0.1	0.0	3.4	6.4	3.2	6.5
Aspartic acid	8.4	4.9	4.0	11.4	18.7	9.4[b]	10.9
Glutamic acid	22.5	23.2	22.9	19.5	12.9	12.3[b]	16.5
Tyrosine	8.1	3.2	3.7	3.8	5.4	6.7[b]	5.1
Phenylalanine	4.6	5.8	5.8	3.5	4.5	3.9	6.6
Tryptophan	1.6	0.65	1.2	1.9	7.0	2.7	0.6
Histidine	2.9	3.1	3.7	1.6	2.9	2.1	4.0
Proline	8.2	16.0	17.0	4.1	1.5	10.0[b]	4.8
Total[d]	115.7	117.4	113.3	113.9	111.6	116.3	113.4

a. The eight proteins in this table account for about 97% of the total proteins and about 90% of the total nitrogen present in normal milk.
b. Figures taken from the amino acid analysis of the immune pseudoglobulin of colostrum which is closely related to, or the same as, that of milk.
c. Unpublished data on the immune globulins of colostrum.
d. In some cases values for phosphate and amide N (as ammonia) are included in the totals.
The authors are indebted to Dr. B. L. Larson for preparation of this table.
*From material presented by Hansen and Carlson.[1]

Such foods as bread and cereals are lacking in some of the essential amino acids, but when fed in combination with milk, the less adequate proteins of the cereal grains are well supplemented by those in the milk. Therefore, a meal of bread and milk is a fair ration in itself. Foods of animal origin — milk, eggs, glandular organs and lean meat — are well supplied with the essential amino acids and, therefore, it is advisable to include a liberal supply of these foods in the diet, especially for growing children as they are producing new tissues at a rapid rate.

(2). *Carbohydrates.* Carbohydrates serve the body as providers of heat and energy, any excess of this nutrient being stored as body fat. They are compounds containing the elements carbon, hydrogen and oxygen, the hydrogen and oxygen being in the proportion of two hydrogen atoms to one oxygen atom, as in water.

The carbohydrate in milk is the milk sugar, or lactose. It is an excellent food for babies, providing a much firmer flesh than other kinds of sugar. It has also a favorable effect in the intestinal tract. It requires a longer time than other sugars for complete digestion, and this delayed action provides a suitable medium for the growth of such beneficial organisms as the species *Lactobacillus acidophilus,* which crowd out the putrefactive bacteria. Lactose has been shown to have a favorable influence on the utilization of calcium and phosphorus in the immature animal.

(3). *Fat.* Like the carbohydrates, fats provide the body with heat and energy. They also aid in the proper metabolism of the sugars, and serve as a carrier of the fat soluble vitamins, A, D, E and K. The contribution of milk fat to the flavor of dairy products has much to do with consumer acceptance of these products. Fats contain the elements carbon, hydrogen and oxygen, but differ from the carbohydrates in having a relatively small amount of oxygen as compared to the amounts of carbon and hydrogen. This high amount of carbon and hydrogen in fats gives this food nutrient, upon oxidation in the body, a fuel value 2.25 times as high as that of either the proteins or the carbohydrates.

Milk fat is an excellent food as it is highly digestible, and contains a large number of short chained fatty acids which are now believed to have special food functions. Experiments[3] have shown that butterfat has certain nutritive properties not possessed by other fats, and this difference is not due entirely to the presence of the vitamins normally present in butterfat, but to the kinds or quantities of the fatty acids constituting the fat itself. The unsaturated fatty acids in milk fat have also been found to be essential for nutrition. Burr and Burr[4] fed rats a diet nearly free of fat, and found that the animals developed a scaly condition of the skin, necrosis of the tail, a degeneration of the kidneys, a failure of growth and finally death. The addition to the diet of an unsaturated fat or of pure linoleic acid corrected the trouble in a very effective manner, while the feeding of saturated fatty acids had no effect.

Boutwell et al.[5] have shown that butterfat together with lactose as a diet for young rats promoted the synthesis of B-complex vitamins through the action of intestinal bacteria. When the butterfat was replaced with a vegetable fat, the organisms were handicapped in building B-complex vitamins, and the growth of the rats was retarded to a large degree. These findings serve as a warning against the use of "filled milk" in which the butterfat is replaced by cheaper fats. Babies living entirely or largely on milk and, hence, dependent on lactose as a carbohydrate, might have their health seriously jeopardized by feeding this type of milk.

(4). *Minerals*. The inorganic mineral elements necessary for normal nutrition include calcium, phosphorus, iron, copper, manganese, magnesium, sodium, potassium, chlorine, iodine, cobalt and zinc.

Milk is a good source of calcium and phosphorus which are so important in the formation of bone and teeth. Approximately 98.5 percent of the calcium and 86 percent of the phosphorus of the body are present in the skeleton and teeth. The remaining part of the calcium circulates in the blood and permeates the soft tissues, performing useful functions such as aiding in the contraction and relaxation of the heart muscles, maintaining the proper relative proportions of calcium salts to other salts, and inducing clotting of the blood to prevent fatal bleeding from breaks in the blood vessels. The phosphorus in addition to its bone and teeth formation, is an essential element of all cells, and functions in the activation of enzymes and serves as a buffer in the blood. The interrelationship of calcium, lactose, and vitamin D as they occur in milk assures favorable conditions for calcium utilization in the body.

A child needs for best results[6] 1.0 gram of calcium and 1.32 grams of phosphorus per day. A quart of milk contains around 1.20 grams of calcium and 0.9 gram of phosphorus. These figures show why a growing child needs a quart of milk daily since this amount will furnish approximately 84 percent of the daily recommended allowance of calcium for growing children during the period of greatest need. A quart of milk will also furnish full calcum needs for most other age groups. It is interesting to note that the ratio of calcium to phosphorus in milk is 1.3:1, regarded as ideal by nutritionists for proper utilization in the body.[1]

Iron and copper are necessary for the formation of hemoglobin, the coloring matter of the red blood cells. The iron enters into the composition of the hemoglobin, but the copper appears to serve only as a catalyst. Iron also aids in the transportation of oxygen through all the arteries and capillaries to the innermost cells of all the organs of the body. Diets of milk alone after the nursing period do not contain sufficient iron and copper to prevent nutritional anemia. Manganese is also lacking in milk, this element being necessary to support normal growth and reproduction.

Experiments[3] at Wisconsin have shown that the addition of iron, copper and manganese to milk (plus vitamin D for children) will make it a complete food, so that it can be used as the sole article of diet. Rats have been fed this "mineralized milk" up to the fifth generation with good results. Several men students for three months lived entirely on mineralized milk and orange juice, maintained their weight and felt well and satisfied. The three minerals were added as follows: 8 grams of soluble ferric pyrophosphate, 0.4 gram copper sulphate and 0.4 gram manganese sulphate were dissolved in 250 ml. of water and 1 teaspoonful of this mixture was added to the milk each day.

The remaining minerals, magnesium, sodium, chlorine, cobalt and zinc, are adequately supplied by the average diet. Iodine in some localities may be deficient.

Milk, however, is a good source of magnesium which is important for muscle activity and nerve stability; of sodium and chlorine which occur to a large extent in the blood, acting as a regulator of neutrality and osmotic pressure; and of potassium, found largely within the body cells and aiding in muscular action.

Cobalt occurs in milk as cyanocobalamin or Vitamin B_{12}. However, supplemental cobalt or vitamin B_{12} in the diet of cows failed to increase the vitamin B_{12} content of the milk.[7] Cobalt is found in extremely small amounts in most of the organs of the human body, and is known to increase the number of red corpuscles in the blood. Some cases of anemia have not responded to the usual iron treatment, except when cobalt was present as an impurity in the iron compound. Cobalt has also been found to be effective in correcting failing appetite and wood chewing habits of dairy cattle, indicating that this element is necessary in the diet even though the amounts needed are minute.

A small amount of zinc is present in milk, appearing in greatest quantity in the colostrum. Zinc is necessary for normal growth as it is always present in human tissues, especially in the bones, hair, muscles and such organs as the liver, pancreas and kidneys. Stirn et al.[8] have shown that a zinc-free diet will retard growth in rats, and will interfere with the proper development of the fur.

Iodine is necessary for the production of the hormone, thyroxin, which regulates the rate of metabolism in the body. If insufficient iodine is provided, the thyroid gland, in its endeavor to provide sufficient of the hormone, becomes enlarged and goiter results. Milk has only a small amount of iodine, but plenty of it can be supplied in the diet by the use of iodized salt.

The role in human nutrition of the chlorides and citrates found in milk has not been defined.[1]

(5). *Vitamins.* The chemistry and some of the properties of the vitamins have been considered in Chapter I, and only their nutritive values will be discussed here.

Vitamin A is a very important food accessory. It occurs as carotene in plants, a yellow colored material which is transformed to the colorless vitamin A in the animal body. Milk is a valuable source of this vitamin. It is fat soluble and therefore found in the milk fat. Vitamin A is important in the diet as it is necessary for growth, health and reproduction. It keeps the epithelial tissues healthy, and thus aids in preventing infection. A lack of this vitamin will cause night blindness, or the inability to see in a dim light that is still sufficient for normal vision. In extreme cases xerothalmia or total blindness will result. It also aids in maintaining normal glandular functions, and in fact increases the life span of an individual by several years. A child needs from 2,000 to 2,500 International Units daily. A quart of 4.0 percent milk will supply from 1100 to 3000 units depending on the type of feed. Green feeds in the ration of dairy cows increase the amount of carotene or vitamin A in the milk. Cows turned to pasture will give milk containing double the content of vitamin A over that on barn feeding, and it requires but two weeks of pasture feeding to produce this change in amount.

Vitamin B_1, or thiamin, is widely distributed in eggs, meat, cereals and vegetables, and is therefore no problem in the ordinary diet. Milk is a good source of this vitamin. It is necessary in the diet for growth, appetite, muscular co-ordination, intestinal motility, and prevention of nervous diseases as beri-beri in man and polyneuritis in animals. An average adult requires at least 1.5 milligrams daily while a child needs 1.0-1.5 milligrams daily.[9] A quart of milk contains 0.40 milligrams.[1]

Vitamin C or ascorbic acid is abundant in citrus fruits and berries, such as strawberries and raspberries. Pasteurized milk does not provide at all times a sufficient amount for the needs of children. This vitamin is necessary in the diet to prevent scurvy, a disease in which the gums become swollen, bleed easily and the teeth loosen. It is also important as a stimulant to the appetite and for growth and health. A child needs a minimum of 50 milligrams and an adult requires 75 milligrams daily. A quart of milk will contain approximately 15 milligrams,[1] however the method of pasteurization employed has considerable influence on the vitamin C content of bottled milk. The holding process of pasteurization destroys around 18 percent of the vitamin, but high temperature[10] pasteurization has been shown to cause no loss.

Vitamin D is necessary in the diet to aid in calcium and phosphorus retention so as to make strong bones and prevent rickets. Our natural diets do not contain sufficient amounts for human needs. The largest sources are the fish oils, yeast and the action of sunlight. Milk contains around 40 International Units per quart, whereas a child needs from 300 to 800 units. Fortunately, the amount of vitamin D can be increased to adequate quantities by either irradiation of the milk, feeding irradiated yeast to dairy cows, or by the addition of a vitamin D concentrate direct to the milk. Irradiated milk is produced by exposing milk in a thin film to the action of ultra-violet light. This method will provide 400 units per quart, which is the standard usually followed. Formerly it was 135 units per quart. Yeast being high in sterols, will, when irradiated, provide a large amount of vitamin D and upon feeding it to dairy cows produces a milk containing in excess of 400 units per quart, the present standard being 430. Fortified milk or the direct addition of a vitamin D concentrate will also produce the desired number of units. These methods of increasing vitamin D in milk were controlled by patents and were licensed, but after much litigation they were relinquished for general use.

Vitamin B_2, G, or riboflavin is a very essential vitamin. Its occurrence in milk makes that product a very important food. This vitamin is necessary for growth, proper nutrition of the eyes, tissues and linings of the mouth and lips and prevention of premature aging of the skin. It has been shown to be essential in the formation of red blood cells and hemoglobin. Milk is a good source of this vitamin. It shows up in whey as a greenish yellow pigment. The content of this vitamin in milk is not affected to any appreciable degree by the feed the cow receives. The different breeds of dairy cows yield in their milk approximately the same amount of this vitamin, when considering the total yield for the period. A quart of milk[11] contains 1.5 milligrams, and man requires from 1.5 to 2 milligrams daily.

Vitamin E has been found to aid in reproduction as regards rats, mice and poultry. It is not known if it is essential to human reproduction, but it is necessary for the nutrition of muscles in man and for better utilization of vitamin A. Milk has a fair amount of this vitamin but data on the requirements of man for vitamin E and on the tocopherol content of dairy products are limited. Narayanan and co-workers[12] showed the tocopherol content of cows' milk fat averaged 34.8 ug. per gram of fat and 837 ug. per pound of milk. They found no differences between breeds. Colostrum contains more tocopherols than milk and more is present in summer than in winter milk.[7] Many workers have noted a relationship between tocopherol content of milk and development of oxidized flavors.

Niacin, the pellagra-preventive factor, is present to the extent of about 1.5 milligrams per quart.[13] The estimated requirement of man is 10 to 15 milligrams daily. Available evidence indicates that pellagra is not strictly due to niacin deficiency alone but rather to the results of low quality protein in the diet. However, the usual daily intake of milk, meat and vegetables prevents pellagra. The inclusion of milk in the diet alters the intestinal flora, and therefore more niacin may be synthesized by these organisms. People who are afflicted with pellagra usually have diets deficient in other vitamins as well. This vitamin also prevents nervous and digestive disorders, sore mouth and tongue.

Pantothenic acid is essential for the development of the central nervous system, for growth, and for proper utilization of riboflavin and certain other vitamins. In turn, its action depends on the presence of biotin and folic acid. This vitamin may be synthesized by the microflora in the digestive tract of the cow. Milk contains[11] an average of 4.0 milligrams per liter. The amounts needed in human nutrition have not been established at the present time.

Pyridoxine (B_6). Available evidence indicates approximately 1-2 milligrams of the vitamin B_6 group should be consumed daily and this quantity is readily available in ordinary diets. Milk has been shown[7] to contain 400-600 micrograms per liter. Sunlight or ultraviolet light have a marked destructive effect on pyridoxine.

Biotin is one of the newer vitamins belonging to the vitamin B complex. As yet, its functions are not clear but it is known that in humans it increases resistance to disease, and in small animals cures many cases of paralysis and loss of hair coat. Milk is rich in biotin,[11] containing about 50 micrograms per liter. Human requirements are not known but in one case of biotin deficiency, it was found that 150 micrograms daily cured the condition. Therefore, if this amount is needed a quart of milk per day would supply one-third of the biotin requirement.

Inositol and folic acid are also connected with the B vitamins. They appear to work together in the prevention of tumor growths. Milk is a good source of inositol; it contains around 180 milligrams per liter.[11] Man's requirements have not been established, but a good diet supplies around 980 milligrams per day. Thus a quart of milk daily would supply one-fifth of the day's requirement. Milk is not a good source of folic acid, but it is possible that bacteria in the intestinal tract may form folic acid from the inositol in the milk.

Choline is one of the latest nutritional factors in the B complex. Milk is a good source containing 149 milligrams per liter.[14] It appears to have some connection with the utilization of fatty acids in the body. The amounts for humans are not known. It is necessary for normal growth in rats, chicks and dogs.

A grass juice factor has been shown to be present in milk as indicated by the better growth of rats fed on milk from pasture fed cows. As yet, it has not been isolated.

Vitamin B_{12} or cyanocobalamin is essential for the normal development of red blood cells and aids in the growth of children. Milk is a good source for vitamin B_{12}, containing 3-4 micrograms per liter. While human requirements for this vitamin have not been stated, it appears that small amounts of milk, meat, and eggs will provide sufficient quantities for maintaining proper body functions.

b. Energy value. The energy value of a food is measured in terms of calories. The small calorie is the amount of heat required to raise 1 gram of water 1° C. This unit is rather small for ordinary usage and therefore the large calorie is generally employed. A large calorie is the amount of heat required to raise 1,000 grams of water 1° C., or 1 pound of water 4° F.

Proteins and sugars when oxidized in the body will each yield 4 calories per gram, but fats yield 4×2.25 or 9 calories on account of their high content of carbon and hydrogen. In calculating the caloric value of many foods, the above physiological fuel values are commonly used. Thus, the caloric value of a quart of milk, which weighs 2.15 pounds or 975 grams, and has an average test of 3.3 percent protein, 5.0 percent lactose and 4.0 percent fat, may be calculated as follows:

Example of caloric value of one quart of milk.

```
975 grams × 3.3% = 32.18 g. protein, × 4 = 129 cal.
975 grams × 5.0% = 48.75 g. lactose, × 4 = 195 cal.
975 grams × 4.0% = 39.00 g. fat,     × 9 = 351 cal.

     Total                                 675 cal.
```

The food value of different milks should be compared on their content of calories, and not merely on their relative fat percentages. The proteins and sugars vary with the fat content, and must, therefore, be considered in the calculation. When only the fat percentage is known, the simple formula, $E = 12 (F + 2)$, where E equals the calories in 100 grams of milk and F equals the fat test, may be used in comparing the relative food values and prices of two different milks. For example, suppose a milk testing 3.5 percent fat sells for 26 cents a quart, how much would a milk testing 5.0 percent fat be worth per quart on the same food value basis?

Solution:
1. $E = 12 \times (3.5 + 2)$ or 66 calories
2. $E = 12 \times (5.0 + 2)$ or 84 calories
3. $66 : 84 : : 26 : X$
    $X = 33$ cents per quart.

The nutrients in milk are highly digestible. Morrison[15] gives the coefficients of digestibility of the different nutrients as follows: protein 94 percent, nitrogen-free extract (lactose) 98 percent and fat 97 percent. Nevens and Shaw,[16] working with rats, report digestibilities of 92 percent of the protein, 100 percent of the lactose and 99 percent of the fat.

c. Milk for infants. The most important characteristics of human milk, as compared to cows' milk, are its low protein, low ash and high lactose contents. Table 44 gives a comparison of the average composition of the two milks.

Table 44. Average Composition of Human and Cows' Milk

	Water %	Total Solids %	Ash %	Protein %	Lactose %	Fat %
Human Milk	87.43	12.57	0.21	1.63	6.98	3.75
Cows' Milk	87.00	13.00	0.70	3.30	5.00	4.00

Mother's milk is the best food for infants as indicated by the thrifty condition of the breast fed baby. However, it is not always possible for the mother to nurse the infant, and therefore milk from another source must be provided. Fresh cows' milk is the common substitute. Since it differs in composition from human milk, it must be modified so as to resemble the analysis of breast milk as closely as possible. The usual method of modification is to add water to lower the protein and to add some form of sugar, preferably lactose, to increase the sugar content.

(1). *Normal and processed milks.* Washburn and Jones[17] and Newlander and Jones[18] made extensive studies of the values of different grades of milk in infant feeding. The experiments covered the usage of normal cows' milk, sweetened condensed milk, evaporated milk, milk of different fat contents, homogenized milk and remade milk. Baby pigs from two days to four weeks old were used in these trials. It is generally held by nutritionists that the results attained with young pigs may be applied to the feeding of the human infant because of the physiological similarity of their digestive systems. It is recognized that a pig at birth is more fully developed than is the baby and that it matures more rapidly, but in general it can be assumed that each week of early life of a pig approximately corresponds to two to three months of an infant's life. The feeding values of the different milks were compared by the evidences of thrift and vigor while on trial, and by the chemical analyses of the carcasses.

The results showed that normal cows' milk with a fat percentage around 3.5 percent was the best. The animals did not do as well on the lower or higher testing milks. Both the evaporated milk and the powdered whole milk, when diluted to normal proportions, were nearly equal in feeding value to that of the normal fresh cows' milk. The sweetened condensed milk, however, was undesirable as it produced a dangerously fat body, and the bones were but two-thirds as strong as the bones of those animals fed the normal or evaporated milk. Homogenization made the curd in the stomach much more flocculent and friable, and it would appear that this process has some benefits in increasing the digestibility of milk. Remade milk, which was made by emulsifying skimmilk and unsalted butter in the proper proportions, was not equal in feeding value to the normal, evaporated, or powdered milks, but nevertheless proved to be a good food. The conclusions drawn from these trials were that these milks, when properly diluted, would have the same relative values for human babies. A number of recent works[19] simply confirm previous conclusions that many modified milks are, for all practical purposes, nutritionally similar to human milk.

(2). *Soft curd milk.* Milk from individual cows, when coagulated with rennin or pepsin, will give a coagulum that varies widely with different animals. It is known that the milks with the softer curds are more like the characteristics of human milk and are more desirable as substitutes for mother's milk than the harder curds. Interest in soft curd milk was given an impetus when Hill[20] proposed his test for determining the character of the curd of different milks as an index of its suitability for infant feeding.

Briefly, this method consists in coagulating the milk with a calcium chloride pepsin reagent at 35° C., allowing a 10-minute interval for completion of the coagulation, and then measuring the curd tension by the pull required to draw a 10-pronged curd knife through the coagulant. A sensitive spring balance is used for this measurement. A committee of the American Dairy Science Association has drawn up a tentative method[21] for this test so as to insure a uniform procedure among different workers. The main change is the recommendation to use N/10 hydrochloric acid containing 0.45 percent pepsin in place of the calcium chloride pepsin reagent. The other changes are minor.

Natural soft curd milk is characteristic of some individual cows. Milk will vary in curd tension from about 15 to 150 grams by the Hill test and any milk under 30 grams tension is considered as a soft curd. Natural soft curd milk is found in all breeds of dairy cattle, but predominates in the Holstein breed, followed in order by the Ayrshires, Brown Swiss, Guernseys and Jerseys. Soft curd milk is not a different kind of product, but its softer curd is due to the lower content of solids especially casein. Milk high in fat is also higher in casein, and therefore has a greater curd tension.

A soft curd milk, being an individual characteristic of certain cows, will not vary appreciably. Some changes are noted at the beginning and end of the lactation period, when the tension increases to its highest point. Any change that affects the composition of the milk such as sudden weather changes and atmospheric conditions will also affect the curd tension.

Mastitis produces a soft curd and a cow that becomes afflicted with this disease may give milk that would be placed in the soft curd classification. This is primarily due to the decrease in amount of casein, and to some extent to the higher pH values of this kind of milk. Thus, it is essential to know that a soft curd milk comes from a mastitis free herd before advertising it as such. In general, the use of a natural soft curd milk from selected animals, on account of its variability, is less desirable than would be a milk carefully adjusted under controlled conditions.

The curd tension of milk may be reduced by a method known as the base-exchange process. Milk acidified with 0.3 percent citric acid is percolated through a zeolite bed (sodium-aluminum silicate containing potassium) where it gives up about 20 percent of its calcium and phosphorus in exchange for sodium and potassium. By proper regulation of the procedure a milk can be obtained which is normal in respect to pH and the usual ratios of calcium to phosphorus and sodium to potassium. This procedure will cause an appreciable drop in the curd tension. This milk is marketed in several cities as "Sof-Kurd" milk. This method is most widely used for the commercial production of soft curd milk. The process is patented. One objection raised to this kind of soft curd milk is the removal of some of the most important minerals, calcium and phosphorus, which are so necessary for bone development in the young. However, the results of metabolism studies indicate that the remaining minerals are more efficiently utilized. This is attributed to a readjustment of the calcium and the character of the curd of the milk.

Homogenizing milk will lower the curd tension by at least one-half of its original strength. It disperses the fat which introduces more points of weakness in the coagulum, and it may remove some of the phospholipids from the fat globule surface which would also affect the tension of the curd. This method of providing a soft curd milk is based on a sound procedure and will become more universally used. This milk is more readily digestible but not more completely so than raw milk. It has the objection of producing a product with considerable sediment, largely leucocytes, but this adverse effect can be overcome by clarification.

Heating milk affects its curd tension, depending on the temperature and length of time of the heating. Pasteurization has practically no effect, but boiling milk for 5 minutes will decrease the tension sufficiently to place an average milk (50 to 60 grams) down to the soft curd class of under 30 grams. Evaporated milk has a very low tension and a high digestibility due to the high sterilization temperatures used, and consequently the use of evaporated milk in infant feeding has been increasing rapidly. A milk with soft curd properties is essential in infant feeding. Average market milk has a high curd tension and, unless modified in some manner, will react adversely in an infant's stomach. Since the commercial soft curd milks are not everywhere available, and natural soft curd milks require considerable trouble to locate and obtain, parents have turned to the use of evaporated milk. This product is used widely for this purpose, and will continue to be thus used until the dairy industry provides a natural soft curd milk that is readily obtainable.

d. Pasteurized milk. The objection has been raised to pasteurization of milk, that the process lowers the nutritive value of the product. Many experiments have been conducted in comparing the nutritive properties of raw and pasteurized milks. Some changes are effected but for the most part they are insignificant from a nutritional standpoint. Henry and Kon[22] experimenting with rats found that the calcium and phosphorus were equally assimilated from the two milks, and the proteins were of equal digestibility. Vitamin A is not affected by pasteurization. Vitamin B_1 losses depend somewhat on the manner of heat treatment, holder pasteurization causing about 10% loss while HTST appears to have little effect. Destruction of vitamin C by heating also depends on the time and temperature of exposure. A loss of 12% of the vitamin C content and a 10% loss of niacin in holder pasteurization has been reported[23] while King and Waugh[10] indicate that HTST has no effect on ascorbic acid values. Boiling for 30 minutes destroys one-third of the vitamin C while autoclaving produces a 50% reduction.[23] The same review article[23] quoted work showing that contamination with zinc, iron, or lead caused a greater loss of vitamin C at the boiling point than at 5° C. but copper and brass were just the reverse. Storage of pasteurized milk at 10° C. for four days caused pasteurized milk to lose 77% of the vitamin C present in the raw sample and similarly stored raw milk lost 64% of its vitamin C content. Exposure of milk to light causes serious losses of riboflavin (50-60 percent), and vitamin C (50-60 percent)[23] as well as to vitamin A (40-50 percent) and pyridoxine.[7]

Since milk does not provide sufficient quantities of the B vitamins and ascorbic acid even in the raw supply, any reduction due to the almost universal use of pasteurization is not a problem since supplementary sources (citrus fruits and vegetables) must be provided to supply an adequate supply of these vitamins.

It should be kept in mind when considering the nutritive value of pasteurized milk, that pasteurization serves two primary purposes, the destruction of harmful bacteria and the improvement of the keeping quality of the milk.

Food Value of Butter. Butter is a very nutritious and energetic food as it is high in fat and highly digestible. Its composi-

NUTRITIVE VALUE OF MILK AND ITS PRODUCTS 309

tion varies somewhat and therefore the food value has a corresponding variation. Legal butter must contain at least 80 percent fat, and since it contains about one percent protein, its caloric value would be at least 3284 calories per pound. $(80 \times 9) + (1 \times 4) \times 4.536 = 3284$. A higher fat content of course would furnish still more calories.

Butter is a valuable source of vitamin A, especially when the cows are on pasture or fed other carotene-rich feeds. The natural yellow color of butterfat is directly proportional to the carotene content, which indicates a high vitamin A potency. Butterfat of a light yellow or nearly white color is low in carotene.

In 1941, a comprehensive research project was started to study the vitamin A values of butter at different seasons of the year. Nineteen states participated in this work. The results[24] from 14 of these states are given in Table 45.

Table 45. Vitamin A Potency of Creamery Butter

State	Average Yearly Butter Production Million Pounds	Average Vitamin A Potency (In International Units)		
		Winter Butter	Summer Butter	Annual Average
Minnesota	304.4	10,808	17,946	14,855
Wisconsin	151.2	10,663	18,884	16,039
Ohio	69.5	9,698	15,459	13,650
Iowa	241.0	10,946	17,434	15,010
Nebraska	93.0	11,287	20,667	17,030
Kansas	74.7	11,606	15,030	13,768
South Carolina	1.8	9,674	16,253	14,312
Louisiana	0.8	12,283	18,068	16,379
Mississippi	4.0	12,041	17,868	16,533
Washington	32.0	12,162	19,827	18,467
Oregon	28.2	12,889	18,464	16,535
California	43.2	15,140	18,968	18,241
Idaho	37.1	13,312	18,197	16,384
Montana	12.7	18,636	22,018	21,014
Total	1093.9			
Average		11,160	17,955	15,529

It should be noted that winter butter makes up 36 percent and summer butter 64 percent of all the creamery butter in the United States. The winter butter averaged over 11,000 units of vitamin A and the summer butter nearly 18,000 units per pound, while the weighted average vitamin A potency for all the butter is between 15,000 and 16,000 units per pound.

Storing butter under proper conditions does not decrease the vitamin A content, according to Baumann and Steenbock.[25] They found that butter stored for six months at 32° F. showed no decrease in vitamin A content. It was only when butter was exposed to air, light and heat to such an extent and for so long a period as to cause bleaching and to become tallowy, that the vitamin content decreased. Since most butter is stored during the flush season while the cows are on pasture, storage butter may be a much better source of vitamin A than freshly made butter in the winter months.

Food Value of Cheese. Cheese is a food of high nutritive value, rich in protein, fat and the minerals calcium and phosphorus. Various cheeses differ considerably in their composition and therefore in their energy values. Table 46 from several sources[26-27] shows the composition and caloric value of some of the more common varieties:

Table 46. Composition and Energy Value of Various Cheeses.

	Water %	Fat %	Protein %	Lactose and Ash %	Calcium %	Phosphorus %	Calories in 1 lb.
Cheddar	37.33	33.41	23.39	5.87	0.99	0.68	1860
Cottage	74.71	0.92	17.71	6.66	0.99	0.37	445
Cream	51.37	33.00	12.31	3.32	0.64	0.53	1600
Neufchatel	57.76	22.72	15.04	4.48	0.64	0.53	1250
Roquefort	40.16	32.31	21.39	6.14	0.72	0.52	1780
Swiss	34.16	30.61	29.22	6.01	0.99	0.84	1860

It will be noted that all the various kinds of cheese are high in protein, and, with the exception of cottage cheese which lacks the fat, high in energy value. They can be used to advantage as substitutes for such high protein foods as meat and eggs. One pound of American Cheddar cheese contains as much protein as:[26]

> 1.57 pounds of sirloin steak
> 1.35 pounds of round steak
> 1.89 pounds of fowl
> 1.79 pounds of smoked ham
> 1.81 pounds of fresh ham

Cheese is also high in the bone building minerals, calcium and phosphorus and is an excellent source of vitamin A and a fair source of vitamin G. Cottage cheese, of course, is low in vitamin A, but nevertheless is an excellent food as a source of protein and minerals.

Food Value of Ice Cream. The food value of ice cream depends to a large extent on its composition, and since it is purchased by volume, the weight of a unit, such as the quart or gallon, must be considered. If the composition and weight of a quart of ice cream are known, the caloric value can be readily computed. Standards vary in different states, the fat requirement for plain ice cream ranging from 8 to 14 percent and the total solids may run from 28 to 38 percent. Therefore, an ice cream testing 8 percent fat and 28 percent total solids would have only two-thirds the energy value of one testing 14 percent fat and 38 percent total solids. The caloric value per quart of these ice creams may be computed as follows: Assume both samples to weigh the same, an average weight of 1.2 pounds per quart.

1. 8% fat   × 9 cal. = 72 cal. per 100 grams.
   20% solids not fat × 4 cal. = 80 cal. per 100 grams.

   Total   152 cal. per 100 grams.
   Total in 1.2 lbs. or 544 g. = 152 × 5.44 or 826 cal. per qt.

2. 14% fat   × 9 cal. = 126 cal. per 100 grams.
   24% solids not fat × 4 cal. = 96 cal. per 100 grams.

   Total   222 cal. per 100 grams.
   Total in 1.2 lbs. or 544 g. = 222 × 5.44 or 1208 cal. per qt.

Aside from its energy value, ice cream is a valuable source of high quality proteins, which are complete and more readily assimilated than some other proteins. Calcium and phosphorus are present in generous quantities. Since ice cream is high in fat, it is an excellent source of vitamin A, and contains a fair amount of D and E, all three being fat soluble. It also is a good source of vitamin B_1 and provides some vitamin C.

Ice cream is very palatable and highly digestible. It is not only an ideal and nutritious food for people in good health, but also for invalids and convalescents. The fact that it is usually included in the diets of patients indicates that it has special merits as a food.

Food Value of Buttermilk and Skim Milk. Buttermilk and skim milk have similar compositions, both belonging to the high protein class of foods. They contain very little fat but nevertheless are good foods because nearly all the other nutrients of whole milk are present. They are highly digestible and their consumption for household use could be increased to advantage. These milks contain around 350 calories per quart as compared to 675 calories for 4 percent milk, thus having at least one-half the energy value of whole milk. They have all the minerals and vitamins of whole milk except vitamin A, as this food accessory accompanies the fat. In addition to its food value, buttermilk has a beneficial effect on the digestive tract in that it prevents certain putrefactive bacteria from developing.

Chocolate Milk. Chocolate milk is a popular beverage to many people. The addition of a cocoa syrup to milk has been one means of increasing the consumption of milk, especially among school children. Many individuals that do not like plain milk will drink chocolate milk. It would appear that the addition of the syrup with its additional nutrients would increase the nutritive value of the milk, but experiments[28] with rats have shown that added cocoa to the extent of one percent lowers the digestibility of the milk slightly, and the net result is that the cocoa neither harms nor enhances the nutritional value of the milk.

More recent studies[29] have confirmed that normal chocolate milk has about the same nutritive value as unflavored whole milk and that cocoa, contrary to some opinion, does not interfere with the uptake of the nutrients in milk or with the calcium balance and retention in the human body. While experimental evidence indicates the theobromine (0.7 to 2.7 per cent) and the tannins (2-12 per cent) in cocoa do not affect human health, the problem is not completely settled. Cocoa extracts free from theobromine or tannins cause no depressive effect on growth in young rats.

A pleasing flavored chocolate milk may be made as follows: Mix together, by weight, 6 parts of sugar, 1 part of cocoa and 5 parts of water. Heat in a double boiler for a few minutes at 180° F., or if precautions are observed, the heating may be done over an open flame, stirring rapidly to prevent burning, until the mixture comes to a boil. Restore the loss of water. Mix one part of this syrup with 7 parts of milk. To prevent settling it will be necessary to add a stabilizer such as cocoloid. However, if one takes the precaution of mixing the milk just before use, the milk will have a more pleasing taste without the stabilizer.

Table 47 gives a brief summary of the quantities of some of the nutrients in dairy products in relation to human requirements.

NUTRITIVE VALUE OF MILK AND ITS PRODUCTS

Table 47. Summary of Food Nutrients in Dairy Products and Human Requirements (Average)

Nutritional factor	Daily requirements of adults	Nutrients in 1 quart of milk*	Nutrients in 1 pound of butter*	Nutrients in 1 pound of cheese	Nutrients in 1 quart of ice cream
Protein (g.)	70	34	2.7	110	22
Calories	3000	675	3300	1860	1000
Calcium (mg.)	800	1150	91	8300	625
Phosphorus (mg.)	1320	907	73	2250	500
Iron (mg.)	12	0.9	0	4	0.5
Vitamin A (I.U.)	5000	1560	15000	6350	2650
Vitamin B_1 (mg.)	1.5	0.4	Trace	0.1	0.2
Vitamin C (mg.)	75	10	0	0	5
Vitamin D (I.U.) (child)	400	40	450	150	100
Vitamin G (mg.)	1.8	1.7	0.05	1.9	1.0
Niacin (mg.)	20	1.0	0.4	Trace	0.5

*Based on values in USDA Agr. Handbook No. 8 Composition of Foods (1950).[30]

BIBLIOGRAPHY

1. Hansen, R. G., and Carlson, D. M. An Evaluation of the Balance of Nutrients in Milk. J. D. Sci. 39:663-673 (1956).
2. Henry, Kathleen M. The Nutritive Value of Milk Proteins. Parts I and II. D. Sci. Abst. Review Article 63. 19:603-616 and 691-704 (1957).
3. Irwin, M. H. Milk as a Food Throughout Life. Wis. Bul. 447 (1939).
4. Burr, G. O., and Burr, M. N. A New Deficiency Disease Produced by the Rigid Exclusion of Fat from the Diet J. Biol. Chem. 82, pp. 345-367 (1929).
5. Boutwell, R. K., Geyer, R., Elvehjem, C. A., and Hart, E. B. Butterfat Can Promote Building of B-Vitamins. Wis. Bul 465 (1944).
6. Sherman H. C. Mineral Needs of Man. Food and Life, Year Book of Agriculture, U.S.D.A. (1939).
7. Kon, S. K., and Henry, K. M. Reviews of the Progress of Dairy Science, Section D: Nutritive Value of Milk and Milk Products J. D. Res. 21:245-298 (1954).
8. Stirn, F. E., Elvehjem, C. A., and Hart, E. B. The Indispensability of Zinc in the Nutrition of the Rat. J. Biol. Chem. 109, pp. 347-359 (1935).
9. National Research Council, Food and Nutrition Board. Recommended Daily Dietary Allowances (Revised 1958).
10. King, C. G., and Waugh, W. A. The Effect of Pasteurization Upon the Vitamin C Content of Milk. J. D. Sci. 17, p. 489 (1934).
11. Holmes, Julia O. New Developments in Nutrition Research. Milk Plant Mo. 24, No. 2, p. 22 (1945).
12. Narayanan, K. M. et al. Co-vitamin Studies. Part I. Variations in Tocopherol, Carotene and Vitamin A Contents in Milk and Butterfat of Cows and Buffalos. Indian Jour. D. Sci. 9:44-51 (1956). D. Sci. Abst. 18:679 (1956).
13. Bailey, Jr., E. A., et al. A Method for the Estimation of Nicotinic Acid in Milk. J. D. Sci. 24, p. A 269 (1941).
14. Hodson, A. Z. The Nicotinic Acid, Pantothenic Acid, Choline and Biotin Content of Fresh, Irradiated, Evaporated and Dry Milk. J. D. Sci. 28: p. A84 (1945).
15. Morrison, F. B. Feeds and Feeding. 20th edit. The Morrison Pub. Co., Ithaca, N. Y.
16. Nevens, W. B., and Shaw, D. D. J. of Nutrition 6, p. 139 (1933).
17. Washburn, R. M., and Jones, C. H. Studies of the Values of Different Grades of Milk in Infant Feeding. Vt. Bul. 195 (1916).
18. Newlander, J. A., and Jones, C. H. Studies of the Values of Different Grades of Milk in Infant Feeding Vt. Bul. 389 (1935).
19. McGillivray, W. A., and Porter, J. W. G. Reviews of the Progress of Dairy Science, Section D. Nutritive Values of Milk and Milk Products. J. D. Res. 25:344-363 (1958).
20. Hill, R. L. The Physical Curd Character of Milk and Its Relationship to the Digestibility and Food Value of Milk for Infants. Utah Bul. 207 (1928).
21. Doan, F. J. Soft Curd Milk. J. D. Sci. 21, pp. 729-753 (1938).
22. Henry, K. M., and Kon, S. K. Raw and Commercially Pasteurized Milk as Sources of Calcium and Phosphorus for the Growing Rat. J. D. Sci. 20, abs. p. 171 (1937).
23. Kon, S. K., and Henry, K. M. Reviews of the Progress of Dairy Science, Section D. Nutritive Value of Milk and Milk Products. J. D. Res. 18:317-359 (1951).
24. Harmon, E. M. Greater Butter Values. The Dairy World. 24:2, p. 10 (1945).
25. Baumann, C. A., and Steenbock, H. Fat Soluble Vitamins. The Carotene and Vitamin A Content of Butter. J. Biol. Chem. 101, p. 547 (1933).
26. Thom, C., and Fisk, W. W. The Book of Cheese. The MacMillan Co., New York (1918).
27. Ellenberger, H. B. Cold Storage of Cottage and Other Soft Cheeses. Vt. Bul. 213 (1919).
28. Mueller, W. S., and Ritchie, W. S. Nutritive Value of Chocolate Flavored Milk. J. D. Sci. 20, pp. 359-369 (1937).
29. Charley, V. L. S. Flavored Milks D. Sci. Abst. 19:267-276 (1957).
30. Watt, B. K., and Merrill, A. L. Composition of Foods — Raw, Processed, Prepared. USDA Agr. Handbook No. 8 (1950).

REVIEW QUESTIONS

1. Why is protein an important food nutrient?
2. What purposes do the nutrients, carbohydrates and fat serve in the human body?
3. What is meant by mineralized milk?
4. What is a vitamin?
5. Is milk a good source of vitamin A? Vitamin C? Vitamin D?
6. Which has the greater energy value per unit of weight, proteins or fats? Why?
7. How does the composition of cows' milk differ from that of human milk?
8. What is meant by a soft curd milk?
9. Is mastitis milk apt to produce a hard or soft curd? Why?
10. What are some objections that have been raised to the pasteurization of milk?
11. How would you answer these objections?
12. Why does butter have a high caloric value?
13. Is it possible for storage butter to have a higher content of vitamin A than freshly made butter? Explain.
14. Is cheese a good substitute for meat? Why?
15. What are some of the advantages of ice cream as a food?

CHAPTER XV

The Chemistry of Cleaning

The flavor, bacteriological quality, and storage life of milk and its products are greatly influenced by the sanitary care of the equipment surfaces which contact the product during production and processing. Consequently, the changing patterns of milk handling — which emphasize less frequent pickup of milk from farms, lengthy periods of raw milk transport and storage prior to pasteurization, and considerable additional elapsed time between milk processing and its consumption in the home — have created much interest in the sanitation of dairy equipment.

The knowledge and application of certain basic chemical principles are essential for the proper sanitary care of milk handling equipment in our modern technology. While these cannot be covered in detail in the limited space available, it is hoped that brief mention of some of these chemical relationships will assist the reader in understanding the essentials of a suitable cleaning system. This subject was covered in some detail[1] in the Laboratory section of the Milk Industry Foundation Convention in 1947.

The efficiency of cleaning procedures normally used for food handling equipment is influenced by three independent but interrelated chemical systems. Thus, cleaning chemists must consider (1) the nature of the soil which is to be removed, (2) the characteristics of the water supply being used, and (3) the chemical makeup of the cleaning compounds which are desired. Many other factors must be known when planning a suitable cleaning program. These would include such considerations as the size, shape, composition, and condition of the surface to be cleaned, the temperature at which the cleaning is to be done, the frictional forces which might be applied, and the ability of the personnel to understand and apply cleaning instructions.

Milk as a Soil

Since our main interest in this text concerns milk and its products, we will confine our discussions to this area. However, as we review the complexity of the dairy industry, it is apparent that it is impossible to imagine a cleaning procedure which would be equally effective against all dairy product residues. Thus, it becomes necessary to separate and classify the dairy soils into some more workable pattern.

We have seen in earlier chapters that milk contains certain constituents which are water soluble and others which are insoluble in water. Certain constituents, such as the proteins and salts in milk, vary in solubility according to their previous treatment. Thus, a system based on soil solubility would have limited use as a general recommendation.

There would be some advantage to considering each of the major food constituents in milk as a separate cleaning problem. Thus, we could consider some of the problems associated with the removal of fats, proteins, carbohydrates, and salts from equipment surfaces. Another method of classifying food residues was proposed by Ordal,[2] who suggested they might be classified as (1) cooked-on deposits, (2) dried-on deposits, (3) slime deposits, (4) fats and oils, and (5) mineral and organic deposits.

If we consider cleaning as a problem of removing the residue of the major milk constituents, we find that the fatty materials can often be removed by rinsing action. This is particularly true if the rinse water is maintained at a temperature above the melting point of butterfat. A residual film of fatty materials usually requires the use of an emulsifying agent and/or a strong alkali to saponify the fat. Protein films are generally troublesome since they are likely to be very adhesive in nature and require peptizing, dissolving and suspending action for removal. A fresh protein film may be removed easily but a dried film or one which has been "cooked-on" during processing is only removed with difficulty. Lactose, like other simple carbohydrates is removed with relative ease.

Residues of mineral salts, like the proteins, can be a serious cleaning problem. Accumulations of milk and water minerals, commonly known as milkstone, cause serious problems in heat transfer, metal corrosion and bacteriological control as well as intensifying cleaning difficulties. Removal of these mineral residues generally depends on dissolving the minerals or forming complexes with the minerals which can be rinsed easily from the surface.

The choice of cleaning agents and cleaning procedures to effect the removal of product films will depend then somewhat on the relative amount of each of these constituents in the film to be removed, the history of the film, and the nature of the surface to be cleaned. Thus, we would expect to use different materials and methods in cleaning a wooden churn or an ice cream freezer than we would use on glass milk bottles.

Emphasis must be placed on the fact that the history of the milk film greatly influences the physical and chemical nature of the material to be removed. This factor must be considered when planning the cleaning operation. A fresh, raw milk film on milking equipment at the farm is removed with ease. The same film which has been allowed to dry on for several hours before cleaning may require special attention. In the fresh milk, the proteins are still in their normal state and the fat is still emulsified. A little wetting action followed by a thorough rinsing with soft (or softened) water will effectively clean the surfaces. On the other hand, the dry film will contain largely denatured protein and the fat emulsion will be at least partially broken to permit a continuous fat phase which covers other milk particles. This obviously results in a greatly intensified cleaning problem.

Cleaning of milk equipment on farms should not be a serious cleaning problem. If the surfaces are cleaned soon after milk is removed, a general purpose cleaner containing suitable wetting agents and water-conditioning ingredients should do the job. It is only when the film has been allowed to dry or milkstone deposits are present that the procedures become more complex.

There are many chemical formulations recommended for farm cleaning operations and also many recommended cleaning procedures. Recent studies[3] have indicated that any of these products or methods are effective if used as intended. Even the newer systems of milk handling — such as bulk tanks and pipeline milkers — create no particularly new cleaning problems from a chemical standpoint. The milk film remains the same as in the older systems. However, the use of cleaned-in-place (CIP) methods permits the use of higher temperatures and greater strength of the alkaline solutions than would be possible under hand cleaning methods. Here velocity is used to create the frictional force normally associated with hand brushing.

When milk films are not completely removed, buildups occur and the cleaning problems become more serious. Likewise, dried films or the residues formed during heat treatment of milk products in the processing plant require special formulations for effective cleaning. In some cases, heavy duty alkaline cleaners are recommended. Alternate cleaning, first with an organic acid mixture and then with an alkaline solution may be effective. The newly formulated chlorinated alkaline cleaners have been shown to speed the removal of such films. These compounds appear to solubilize proteinaceous films and then prevent the redeposition of precipitates on equipment surfaces.[4] MacGregor and coworkers[5] have attributed this action to the fact that the degradation products of protein are far more soluble in alkaline solution than is the case of the original protein molecule.

Films which are incompletely removed during cleaning tend to accumulate and appear as visible layers on equipment surfaces. When dry, they appear as a white film on the affected parts. However, they are not apparent to the eye when the surface is wet. Water salts, sediment, and fresh milk residues combine with the deposited film to form milkstone. The composition of milkstone is shown in Table 48.

Table 48. Extremes in Composition of Milkstone*

Constituent	Minimum	Maximum	Cleaner
Moisture	2.66	2.79	
Fat	3.63	17.66	Alkaline
Protein	4.14	43.83	Alkaline
Ash	42.03	67.33	Acid
CaO	21.05	34.66	Acid
P_2O_5	17.63	26.93	Acid
MgO	1.71	8.12	Acid
Fe_2O_3	0.00	0.29	Acid
Na_2O	1.40	7.33	Acid

*From Klenzade Dairy Sanitation Handbook.

Several factors contribute to the problems which result in an incomplete removal of films during cleanup. Many farms lack sufficient quantities of hot water or use water which is too hot or too cold. It must be remembered that cleaning is essentially a chemical process and temperature is extremely important. Lack of hot water lowers the efficiency of the cleaning compounds and, equally important, leaves the fat in a semi-solid state which can be removed only with difficulty. Too hot water will cause the precipitation of some of the milk proteins and milk salts, thus creating an additional chore for the cleaning solution. Water hardness must be treated before use or it, too, will form films which contribute to cleaning difficulties. If water is very hard (>10 grains per gallon) it would probably be more economical to soften the supply rather than to purchase cleaners having sufficient concentration of water softening chemicals to prevent precipitation of water salts.

Levowitz[6,7] has emphasized that cleaning operations in the dairy industry have been made considerably more difficult by the almost universal recommendation that utensils and equipment be rinsed with clear cold water immediately following use. He points out[6] that the calcium and magnesium content in milk would, if present in water, be roughly equivalent to 88 grains (1500 ppm) of hardness. That these salts do not precipitate out of milk on storage is due to the fact that they are "sequestered" by the phosphates which are also in the milk. Cleaning becomes more difficult because "when the milk film is water rinsed, (a) the fully soluble lactose and phosphate are removed quite completely; (b) some of the fat, protein, calcium, and magnesium may be removed, mechanically, by the turbulence of the rinse; (c) the fat, protein, calcium, and magnesium (now that the hydrophilic lactose and phosphate are gone), absorb more strongly on the surface, and (d) the calcium and magnesium, liberated from the phosphate sequestration, begin to precipitate. The hydrophobic film, when thick enough to be visible, is called 'milkstone.' "[7] Once the film has started to form, it is not removed by the ordinary alkaline cleaners and thus serves as the foundation of additional buildups. These can be removed efficiently only through the use of acids whose calcium and magnesium salts are soluble. The original milkstone layer can be prevented if sufficient phosphate or wetting agent is added to the first rinse water to keep the calcium and magnesium sequestered as they were in the normal fresh milk film.

Sanitation Chemicals

The cleaning of dairy residues is a complex problem. Soaps are undesirable as cleaning agents because they form sticky, gummy deposits in the presence of hard water salts. These deposits are caused by a chemical reaction between the soap and the hard water salts. The sodium ion of the soap is replaced by the calcium or magnesium of hard water to form compounds which are completely insoluble in water. Thus the soaps create cleaning difficulties rather than solve them. Sullivan[8] has noted that one hundred gallons of 20 grain water will require three pounds of an 88% soap before it can begin to have any cleaning effect.

The development of the sanitation chemical industry has been a product of modern technology. Cleaning chemists are constantly searching for new products to add to cleaning compounds to make them more efficient or to protect equipment surfaces from the action of other components of the compound cleaners. Synthetic detergents, in 1959, accounted for about 72% of all cleaners produced.[4]

An effective cleaning system must accomplish three distinct actions. The soil or residue must be separated from the surface to be cleaned. The separated soil must be held in suspension during the cleaning operation. Finally, the suspended soil must be completely rinsed away at the end of the cleaning cycle and leave no residue.[9]

Detergents are compounded to perform the variety of actions necessary to clean dairy equipment. The major duties or requirements of a dairy cleaner would include:

(1). Emulsification — The splitting of fat and oil particles into smaller globules which may be dispersed throughout the medium. The emulsion formed should be stable enough to permit complete rinsing without redepositing of the fat or oil. This may be facilitated by the addition of a suitable emulsifying agent.

(2). Saponification — The chemical reaction between an alkali and a fat to yield a soap. The insoluble fat is thus converted to a soluble soap.

(3). Wetting ability — The action brought about when the surface tension of a cleaning medium is lowered to enable the solution to more easily contact the surface to be cleaned. This property permits the chemicals in a cleaning solution to penetrate into cracks and crevices in soil particles and effectively dislodge them.

(4). Dispersion — Aggregates of soil are broken down into

small particles which can be more readily rinsed from the surfaces.

(5). Suspension — The action which maintains soil particles in the cleaning solutions so they will not redeposite before they can be flushed away.

(6). Peptizing — The breakup of protein soils into colloidal particles which are soluble in the cleaning solution.

(7). Water softening or sequestering — Removing or inactivating hard water salts by the formation of soluble complexes which do not interfere with the cleaning operation. The term "chelation" is used to denote sequestering activity of certain organic water-softening agents.

(8). Rinsability — The property which enables the detergent solution, with its dissolved or suspended soils, to be easily and completely removed from equipment surfaces.

(9). Solubility — A sanitation chemical should quickly and completely dissolve in water at the temperature desired and without necessity for vigorous agitation.

(10). Non-corrosive — The formulation should be non-corrosive to equipment and non-irritating to workers at use concentration.

The common ingredients of compound cleaners are shown in Table 49 with their relative cleaning values in the several areas defined above.

Many speakers at the lengthy series of Klenzade Educational Seminars have discussed the function of ingredients going into cleaning compounds. Fernandez[9] has divided them into six groups, (1) Surface active agents, (2) Sodium phosphates, (3) Silicates, (4) Alkalies, (5) Organic and Mineral Acids, and (6) Miscellaneous additives.

The surface active agents include both the soaps and the synthetic detergents. However, soaps are seldom used in dairy sanitation for reasons already mentioned. The surface active agents are extremely important wetting agents. They also have good emulsifying properties and aid greatly in rinsing the cleaned surfaces. Their action in forming suds is highly desired in some applications but quite undesirable in others.

The phosphates and polyphosphates are used for their ability to soften hard water and as a source of alkalinity or acidity to assist in the removal of various types of soil. They also improve

Table 49. Common Detergent Ingredients*

KEY TO CHART
A — High Value
B — Medium Value
C — Low Value
D — Negative Value
1 — Via Precipitation
2 — Via Sequestration
3 — Also Stable to Heat

INGREDIENTS	Emulsification	Saponification	Wetting	Dispersion	Suspension	Peptizing	Water Softening	Mineral Deposit Control	Rinsability	Suds Formation	Non-Corrosive	Non-Irritating
Basic Alkalis:												
Caustic Soda	C	A	C	C	C	C	C	D	D	C	D	D
Sodium Metasilicate	B	B	C	B	C	C	C	C	B	C	B	D
Soda Ash	C	B	C	C	C	C	C	D	C	C	C	D
Tri-Sodium Phosphate	B	B	C	B	B	B	A^1	D	B	C	$C+$	$C-$
Complex Phosphates:												
Sodium Tetra-Phosphate	A	C	C	A	A	A	B^2	B	A	C	AA	A
Sodium Tri-Polyphosphate	A	C	C	A	A	A	A^2	B	A	C	AA	B
Sodium Hexametaphosphate	A	C	C	A	A	A	B^2	B	A	C	AA	A
Tetrasodium Pyrophosphate	B	B	C	B	B	B	A^2	B	A	C	AA	B
Organic Compounds:												
Chelating Agents	C	C	C	C	C	A	$AA^{2,3}$	A	A	C	AA	A
Wetting Agents	AA	C	AA	A	B	B	C	C	AA	AAA	A	A
Organic Acids	C	C	C	C	C	B	A^3	AA	B	C	A	A
Mineral Acids	C	C	C	C	C	C	A^3	AA	C	C	D	D

*From Klenzade Dairy Sanitation Handbook.

the rinsing properties of the cleaning solutions and, in general, have value in all phases of the cleaning operation.

Sodium metasilicate is the best known member of the silicates associated with cleaning. The alkalinity of the silicates assists in the saponification and emulsification of fatty residues. They are valuable for suspending solids which have been removed from equipment surfaces. Perhaps their most important function is to protect the metal surfaces from the corrosion caused by other alkaline constituents in the cleaning compound.

The alkalies are the original ingredients of dairy cleaners. They are inexpensive sources of the alkalinity needed to remove fatty deposits and other residues which are soluble at a high pH. Generally speaking, their activity is limited to the removal of soil from equipment surfaces although tri-sodium phosphate (TSP) has some value as a water softener.

The acid constituents are also primarily concerned with the removal of residues from soiled surfaces. While they effectively remove some soils, and especially mineral deposits, they are quite corrosive to metal surfaces and must be handled with care. The common members of this group which are used in dairy cleaning operations would include hydrochloric acid, phosphoric acid, hydroxyacetic acid, and gluconic acid.

Many new additives are being used each year to improve the speed, efficiency, or convenience of cleaning operations. Some of these are used to prevent the powdered mixture from caking, to increase or decrease foaming, to offset the corrosive effect of other ingredients, or to better sequester the minerals in hard water. Other additives appear to have a synergistic effect, that is, they are far more effective in combination than any of the individual components of the system would suggest.

It must be remembered that all of these products are chemicals and the cleaning process is essentially a chemical action. While abrasive action of brushing or streaming water have a definite place in the cleaning procedure, the chemical reactions require a certain amount of time to perform their intended action. Soaking is the ideal cleaning situation. Where soaking is not practiced, as in bulk cooling tank or storage tank cleanup, the cleaning solution may be splashed on the surfaces. Brushing should be delayed for a few minutes in order to get maximum action from the cleaning solution.

The Water Supply

Sanitarians are showing more interest in water supplies than ever before. Munroe[10] has pointed out that water consumption on farms and rural areas doubled between 1900 and 1950. This increased use has made it necessary to accept secondary water sources and these frequently produce water of poor quality.

The ideal water supply for dairy operations would be soft and clear. It would be free from pathogenic bacteria, off-flavors and odors, and corrosive substances. It would be in sufficient quantity at low cost. Unfortunately, such water is extremely rare. The solvent nature of water is so pronounced that nearly all natural water contains various dissolved or suspended materials. Such waters have a pronounced effect on the cleaning operations.

Rain water, falling to earth, will dissolve carbon dioxide from the air. Additional CO_2 will be added as the water contacts decomposing organic matter in the soil and thus change to, what is in effect, a weak solution of carbonic acid. This acid solution then readily dissolves the soil and rock constituents which it contacts.[11] The average water supply may contain salts of calcium, magnesium, iron, and silica. Traces of copper, lead, tin, sulfur dioxide, hydrogen sulfide, and chlorine are also to be found.[8]

The dissolved solids in water are measured in parts per million or grains per gallon (17.1 ppm = 1 gpg.). They are referred to as water "hardness" and cause serious problems in the food industries due to their behavior during cleaning and cooling and in hot water systems. Water supplies under 10 grains hardness are considered to be relatively "soft" while those of 10 grains or over are called "hard."

Water hardness is divided into two types, "temporary" and "permanent." Temporary hardness is due to the presence of bicarbonates of calcium or magnesium. It is a major concern since these hardness salts precipitate readily from solution in the presence of either alkali or heat. The soluble bicarbonates change to insoluble carbonates or hydroxides and these form a white film on equipment surfaces. Permanent hardness is the type associated with the sulfates or chlorides of calcium or magnesium. These salts will precipitate through the action of alkalies but not through heat alone.

Table 50. Reactions of Temporary Hardness*

(1)
Calcium bicarbonate + Heat = Calcium carbonate + Water + Carbon dioxide
$Ca(HCO_3)_2$ + Heat = $CaCO_3$ + H_2O + CO_2
(Insoluble) (Gas)

(2)
Magnesium bicarbonate + Heat = Magnesium carbonate + Carbon dioxide + Water = Magnesium hydroxide + Carbon dioxide
$Mg(HCO_3)_2$ + Heat = $MgCO_3$ + CO_2 + H_2O = $Mg(OH)_2$ + CO_2
(Slightly soluble) (Gas) (Insoluble) (Gas)

Reactions of Permanent Hardness

(1)
Calcium sulfate + Heat = No change
$CaSO_4$ + Heat = No change

(2)
Calcium chloride + Sodium carbonate = Calcium carbonate + Sodium chloride
$CaCl_2$ + Na_2CO_3 = $CaCO_3$ + $2NaCl$
(Insoluble) (Soluble)

(3)
Magnesium nitrate + Sodium hydroxide = Magnesium hydroxide + Sodium nitrate
$Mg(NO_3)_2$ + $2NaOH$ = $Mg(OH)_2$ + $2NaNO_3$
(Insoluble) (Soluble)

*From Klenzade Dairy Sanitation Handbook.

The minerals in water are a handicap in cleaning operations and must be removed. In plant operations, the volume of water used and the several problems associated with water hardness make it economical to install water softening equipment. The two basic methods employed for such water softening involve (1) the precipitation and removal of calcium and magnesium salts, or (2) replacing the calcium or magnesium ions present in hard water with soluble sodium ions. The salts may be precipitated with hydrated lime or soda ash and then removed by filtration. The second method involves the use of zeolite minerals to exchange the hard water salts, leaving them as dissolved or stable complexes. Water hardness can be reduced to less than one part per million by either method if properly done.[11]

Farm water supplies which have excessive calcium or magnesium hardness are commonly softened by incorporating sequestering agents or chelating agents in the cleaning compounds. The proper use of synthetic detergents, complex phosphates, and organic sequestering agents will handle most of the hardness problems in farm water supplies. The non-ionic detergents do not ionize in solution and thus cause no reaction with the metal ions in water, therefore no metal salts are precipitated. However, these do not affect film formation caused by "bicarbonate" or "temporary" hardness.

Waters with iron or sulfur contamination should be treated prior to use in farm or milk plant operations. Water having more than 0.5 mg. FeO or 0.03 mg. Mn per liter is considered unfit for cleaning purposes.[12] When these minerals are present, polyphosphates may be fed into the water supply to sequester them so they will not settle on equipment surfaces. Sulfur is treated with 5 ppm of chlorine for each part sulfur to oxidize the sulfur, followed by filtration through an activated carbon filter to remove the objectionable materials.[13]

Table 51. Methods of Controlling and Removing Water Hardness*

1. Control of Hardness with Metaphosphate (Polyphosphate)

Calcium bicarbonate	Sodium metaphosphate	Sequestered calcium	Sodium bicarbonate
$Ca(HCO_3)_2$ + Hardness	$(NaPO_3)_6$	= $Na_2(CaP_6O_{18})$ +	$4NaHCO_3$

2. Removal of Hardness with Zeolite Softener

Calcium bicarbonate	Sodium zeolite	Calcium zeolite	Sodium bicarbonate
$Ca(HCO_3)_2$ + Hardness	Na_2Z New mineral	= CaZ + Exhausted mineral	$2NaHCO_3$

3. Removal of Hardness with Lime-Soda Softener

A.
Calcium bicarbonate	Calcium hydroxide	Calcium carbonate	Water
$Ca(HCO_3)_2$ + Temporary hardness	$Ca(OH)_2$ Lime	= $2CaCO_3$ Insoluble +	$2H_2O$

B.
Calcium sulfate	Sodium carbonate	Calcium carbonate	Sodium sulfate
$CaSO_4$ + Permanent hardness	Na_2CO_3 Soda ash	= $CaCO_3$ Insoluble +	Na_2SO_4

4. Removal of Hardness by Boiling

Calcium bicarbonate		Calcium carbonate	Water	Carbon dioxide
$Ca(HCO_3)_2$ + Hardness	Heat	= $CaCO_3$ Insoluble	+ H_2O	+ CO_2

5. Removal of Hardness by Acidification

Calcium bicarbonate	Acid	Calcium salt	Water	Carbon dioxide
$Ca(HCO_3)_2$ + Hardness	$2\ CH_2OHCOOH$	= $(CH_2OHCOO)_2Ca$ Soluble salt	+ $2H_2O$	+ $2CO_2$

6. Removal of Hardness by Chelation

Calcium bicarbonate	Chelating agent	Chelated calcium	Sodium bicarbonate
$Ca(HCO_3)_2$ + Hardness	$Na_4C_{10}H_{12}O_8N_2$ +	= $CaNa_2C_{10}H_{12}O_8N_2$ Soluble complex	+ $2NaHCO_3$

*From Klenzade Dairy Sanitation Handbook.

Chemical Sanitization

In the early days of the dairy industry, hot water or steam were the common means of sanitizing equipment in the milkhouse.

The use of chemical sanitizers dates back to about 1930. L. A. Black[14] discussed chemical sterilization before the Western Canada Dairy Convention in 1929 and stated the now generally accepted concept that it is not necessary to obtain absolute sterility of dairy equipment. He observed that strict sterilization certainly did not occur when utensils were treated with steam or hot water. All pathogenic or harmful bacteria should be destroyed. The non-pathogenic organisms should be so reduced in number as to have no effect on milk passing over the treated surfaces.

Selection of suitable sanitizing agents depend on many things. Swartling[15] lists the following requirements:

(a). Actively germicidal to produce immediate and complete destruction of all microorganisms.

(b). Must be reasonably stable in the presence of organic matter and hard water salts.

(c). Must be readily storageable under practical conditions.

(d). Must not affect metals and other similar materials.

(e). Must not have an offensive odor.

(f). Must be non-toxic and non-irritating to the skin.

(g). Must have acceptable chemical and physical properties.

(h). Price must not be prohibitive.

While many chemicals have germicidal value, the list of compounds which are soluble, non-toxic at use concentration, effective at relatively low temperatures, and which generally satisfy the above requirements have drastically cut down the number of acceptable products for use in dairy farm sanitation. Swartling[15] lists the common active ingredients in dairy farm sanitizers, as follows:

(a). Chlorine and chlorine-liberating compounds.

(b). Iodine-bearing compounds.

(c). Quaternary ammonium compounds.

(d). Amphoteric long-chain molecules.

(e). Chlorophenols.

Chlorine and Chlorine-liberating Compounds. The chlorine-bearing compounds were the first chemical sanitizers to find widespread acceptance in the dairy industry. These compounds now include:

(a). Chlorine (Cl_2)
(b). Calcium hypochlorite [$Ca(OCl)_2$]
(c). Sodium hypochlorite (NaOCl)
(d). Dichlorodimethylhydantoin
(e). Trichloroisocyanuric acid
(f). Chloramine B
(g). Chloramine T
(h). Dichloramine T

The active groups in these last three compounds are:

$$-\underset{ONa}{\overset{O}{\underset{\|}{S}}}=N-Cl \quad \text{or} \quad -\underset{ONa}{\overset{O}{\underset{\|}{S}}}-N{<}^{Cl}_{Cl}$$

The chlorine compounds have a very high sanitizing efficiency for very little expense. They destroy a wide variety of microorganisms, spores, and bacteriophage.[16] They are not particularly affected by water hardness although calcium hypochlorite may leave films or residues when used in hard water. On the other hand, the hypochlorites were found to be unstable in the presence of organic matter and very irritating to the skin.

The chloramines, dichlorodimethylhydantoin, and trichloroisocyanuric acid are much more expensive than the hypochlorites but release the chlorine slowly. Thus they are more stable in the presence of organic matter and are not as irritating. However, the chloramines do not have the germicidal value of the other chlorine compounds. All of the chlorine-bearing compounds are corrosive to stainless steel and other dairy metals. For this reason, they must be rinsed from equipment surfaces immediately after use.

In recent years it has been shown[16] that the hypochlorites are more effective bactericidal agents when used at somewhat lower alkalinity than was common in early usage. It appears that the pH influences the amount of chlorine present as hypochlorous acid (HOCl) and the amount present as hypochlorous ion (OCl)$^-$. The hypochlorous acid has been found to have far greater killing power than is found with the hypochlorous ion which is present in relatively greater amounts as the pH increases. Workers believe that the hypochlorous acid, in undissociated form, is capable of penetrating the cell walls of bacteria and then interfering with certain enzyme systems of the bacterial cell causing its death. Other explanations[15] suggest that the cell protein is denatured or that amino acids formed in protein digestion may undergo a reaction with chlorine in which active chlorine replaces the hydrogen of the amino group, causing destruction of the cell.

Chlorine sanitizing solutions are usually used at a concentration of 200-500 ppm. of available chlorine. Concentration may be decreased as exposure time is increased. Since high concentrations of chlorine are quite corrosive, exposure time should not exceed 30 minutes. The corrosive effect can be minimized by lowering chlorine concentration, decreasing contact time, lowering the temperature, or by using higher pH levels. However, it must be remembered that the bactericidal efficiency of hypochlorites is lowered as the pH is increased.

Iodine-bearing Compounds. The dairy industry has shown considerable interest in recent years in the iodophors, a group of compounds in which iodine is combined in loose chemical combination with suitable non-ionic wetting agents. Some acidic component, commonly phosphoric acid, is incorporated to lower the pH of the solution. The resulting product releases iodine slowly and is a very effective sanitizer for farm use. Bactericidal activity is high, especially against non-spore-forming bacteria, but is relatively slow against bacteriophage.[17]

The value of the iodophors as sanitizing agents in the milkhouse is very high at pH 3 to 5 where they are normally used. They are fast acting germicides which are practically non-corrosive and non-irritating. Because of the low pH, they are not affected by hard water salts. Toxicity is low except in high concentrations. The presence of milk solids lowers their germicidal efficiency. Vermont studies[18] have shown that residues of iodophor sanitizing solutions cannot be used to lower bacteria counts in milk. The same studies have shown that they do not produce an iodine flavor when added to milk in quantities associated with accidental contamination.

The iodophors are readily soluble in water, giving it a typical iodine color. As the sanitizer is used in the operation, the color fades until it is clear. When this happens, the solution still has cleaning ability but is of no more value as a sanitizer and must be renewed. One advantage frequently claimed for the iodophors is the fact that they give milkstone a noticeable brown color, thus aiding in its detection.

The iodophors should not be used if the medium exceeds pH 5 or if the temperature is above 120° F. Excessive heat causes the iodine to escape, causing loss of germicidal power and irritation to the workers. The iodophors are used in concentrations ranging from 12.5 to 100 ppm. of available iodine. The mode of action in cell destruction is a direct action on the cell caused by the precipitation of cell proteins.[15]

Quaternary Ammonium Compounds. The quaternary ammonium compounds, as the name suggests, are substituted ammonium halides with the type formula:

$$\left[R_1 - \underset{\underset{R_4}{|}}{\overset{\overset{R_2}{|}}{N}} - R_3 \right]^+ X^-$$

The fact that the substituted ammonium portion, the positive ion, is the active portion of the molecule and that these compounds have wetting ability have resulted in these compounds being known as catonic, surface active agents. The quaternary ammonium compounds, more commonly known as "quats," are relatively recent additions to the sanitation field.

The quaternaries have a high germicidal activity and are non-corrosive and non-irritating to the skin. Like the iodophors, they are widely used as a pre-milking udder wash.

They are compatible with sequestering agents, non-ionic wetting agents, and many other cleaner ingredients. This has established them as valuable components of cleaner-sanitizer preparations.

Solutions of these compounds are very stable in a suitable water supply if organic material and certain other interfering substances are absent. Quaternary ammonium solutions leave a bacteriostatic film on treated surfaces which prevents growth of organisms on equipment during storage periods between operations.

Since these compounds are cationic, they readily react with anionic wetting agents, certain polyphosphates, and sodium metasilicate. This renders them ineffective as sanitizers. Thus, it is imperative that cleaning solutions containing these interfering compounds be rinsed from equipment surfaces before they are sanitized with quaternary ammonium solutions.[19]

Hard water salts also inactivate the quaternaries. New evidence indicates that hard water sensitivity is influenced by the length of the alkyl chains in the molecule.[4] The C_{12} compounds will normally tolerate less than 500 ppm. of water hardness while those with C_{14} chains will tolerate more than 600 ppm. of hardness. If proper polyphosphates are added to quaternary compounds, the effect of the hard water salts can be overcome. In addition, the presence of the polyphosphates in the sanitizing solution appears to cause some "potentiating" effect since the activity of the quaternary plus polyphosphate is far greater than that of quaternary alone.[20]

The quaternary ammonium compounds are less sensitive to organic material than is the case with chlorine compounds. They are effective over a wide pH range. While active against a wide variety of microorganisms, they appear to be more effective against certain psychrophilic species in acid solution and against coliforms under alkaline conditions.[21] They are not as effective as the hypochlorites against spores or bacteriophage.

One of the major problems in using quaternary sanitizing solutions has been the problem of determining solution strength since tests are quite specific for individual compounds. There has also been some difficulty in determining how much of the bactericidal activity of the solution has been lost through the effect of hard water salts. Residues of quaternary sanitizing solutions have been blamed in recent years of difficulties in cheese-making operations and for off-flavors in milk but extensive experimental evidence fails to justify these criticisms.

Quaternary ammonium solutions are commonly used in concentrations of 200-400 ppm. of active sanitizer. Evidence suggests that the adsorption of quaternary on the bacterial cell causes a leakage of cell constituents through the cell wall. Others suggest that the quaternaries inactivate enzyme systems involved in cell respiration. Another possibility suggested is that they disorganize cell membranes and then denature proteins needed for cell growth and metabolism.[17]

Other Sanitizing Agents. There is some use in Europe of a sanitizer in which the active agent is an amphoteric long-chain molecule.[15] The structure of this compound is

$$R_1 - NH - R_2 - COOH$$

in which R_1 is a long-chain radical.

Certain of the chlorophenols have found application in dairy sanitation. These have had some use in environmental sanitation procedures in hospitals. Development has been held back by the need for a thorough understanding of the chemistry of phenolic derivatives before proper formulation could be made.[4]

Another new group of sanitizing agents which is receiving attention by the dairy industry is the products known as acid sanitizers.[22] These are composed of a suitable acid, commonly phosphoric, in combination with anionic or non-ionic wetting agents. The pH of these solutions at use strength is between 2 and 3.5. They appear to be quite effective on non-spore forming bacteria and are valuable in controllng water hardness and residual film formation.

Cleaner-Sanitizers. Cleaner-sanitizers are made from combinations of ingredients already mentioned and involve no particularly new products. The value of these preparations is based on the ability for one product to act both as a detergent and a sanitizer. The cleaner-sanitizers are combinations of compatible ingredients formulated to be effective under a wide range of conditions and against a wide variety of microorganisms. Of course, their action depends primarily on the effectiveness of individual components.

Although the quaternary ammonium compounds and the iodophors are commonly used as detergent-sanitizers, most of the groups of sanitizers can be combined into this type of product. The recent introduction of trichloroisocyanuric acid and dichlorodimethylhydantoin have permitted the formulation of chlorine-bearing cleaner-sanitizers.[23] It is anticipated that additional formulations for general dairy farm sanitation programs or for specific uses will be introduced at frequent intervals.[24]

While it is not the intent to discuss cleaning methods in this chapter, it must be remembered that essentially the same cleaning routine must be followed with the single-product cleaner-sanitizers as is recommended when separate compounds are used for the cleaning and the sanitizing operations. This means that the dairy equipment must be rinsed, then cleaned with a solution made from either a suitable cleaner or cleaner-sanitizer under proper conditions of solution strength and temperature. The solution of cleaner or cleaner-sanitizer containing the suspended or dissolved soil particles must be thoroughly rinsed with clear water before sanitizing with a *fresh* solution of sanitizer or cleaner-sanitizer. This step is important regardless of which type product is used. Equipment surfaces must be completely free of soil deposits when they are sanitized since all of the products are inactivated to a more or less extent by organic matter.

Neglect of these basic considerations can make any cleaning program ineffective regardless of the sanitation chemicals used.

Methods of Analysis. Many tests have been introduced to determine the composition of sanitation chemicals and water supplies as well as the strength of sanitation chemical solutions. The suppliers of cleaning chemicals have developed a number of general methods for rapid determination of specific properties. These commonly consist of titration procedures with selected indicators. A standard sample is treated with chemical solutions of calculated strength to enable the ready conversion of the number of drops or milliliters of test solution required to produce color changes into strength of sample in p.p.m., percent, etc.

A collection of such test solutions is shown below. With this outfit, no less than eleven different analyses of interest in sanitation technology can be determined.

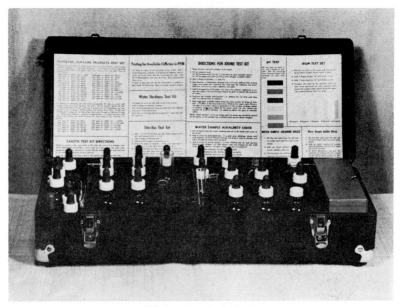

Water Analysis Kit
Courtesy Klenzade Products.

STANDARD ANALYSIS FOR WATER HARDNESS BY THE METHOD USING VERSENATE

For the determination of total hardness in water (calcium plus magnesium) this method is superior to the soap titration method in both accuracy and speed and is equal in accuracy to the time consuming soda-reagent method.[25]

Principle

An azo dye, Eriochrome Black T, has the ability to form a colored complex compound with magnesium. This complex is broken down by Versene to form the more stable magnesium chelate. The dye itself has a deep blue color in solution at a pH of 10.0.

In the presence of magnesium, however, a wine red complex is formed. When Versene is added to a magnesium solution containing small amounts of this dye, a sharp color change from red to blue occurs when sufficient Versene has been added to chelate all the magnesium. When calcium, which does not form a stable complex with the dye, is present, it is also titrated before the end point since Versene will first chelate the calcium ions.

Reagents

Buffer solution — Mix 1 part of 1M NH_4Cl and 5 parts of 1M NH_4OH, to obtain a solution with a pH just over 10.

Standard calcium chloride — Dissolve 1.000 g. of pure $CaCO_3$ in a minimum amount of dilute HCl. Neutralize with NH_4OH until slightly alkaline to litmus. Dilute to 1000 ml.

Or dissolve 1.109 g. of anhydrous reagent grade $CaCl_2$ in 1000 ml. of water. One ml. of this solution is equivalent to 1.00 mg. of calcium carbonate.

Indicator — A freshly prepared 0.2 per cent alcoholic solution of Eriochrome Black T.

Standard solution of Versene — Dissolve 4.00 g. of analytical reagent disodium dihydrogen ethylenediaminetetraacetic acid in water containing 0.5 g. of NaOH. Add to this solution 0.100 g. $MgCl_2 \cdot 6H_2O$ in water and dilute to 1 liter.

Titrate this solution according to the procedure given below using 50 ml. of the standard calcium solution. Add enough water to the standard solution of Versene so that 1 ml. of this solution is equivalent to 1.00 mg. of calcium carbonate.

Procedure

Pipette 50.0 ml. of the water to be tested into a 250 ml. Erlenmeyer flask. Add 10 ml. of the buffer solution and 3 drops of the indicator. Titrate with standard solution of Versene to a color change from wine red to blue.

Calculation

To obtain total hardness (calcium and magnesium) as ppm calcium carbonate, multiply the ml. of solution of Versene used by 20. Divide ppm by 17.1 to obtain grains per U.S. gallon.

TEST FOR QUATERNARY AMMONIUM COMPOUNDS

Work at Oregon State College under the direction of P. R. Elliker has contributed greatly to our knowledge of sanitation chemistry. One result is the test for quaternary ammonium compounds in water solutions and in milk,[26] which follows:

Reagents

Tetrachloroethane (technical grade).

Lactic acid (reagent grade), adjusted to 50% by weight with distilled water.

Eosin yellowish dye (Biological Color Commission Index 768) solution prepared at rate of 0.5 mg. dye per 1 ml. of distilled water.

4.0 Molar sodium hydroxide, analytical reagent.

Citric acid (monohydrate, analytical reagent) buffer prepared by adjusting a 25% solution of citric acid in distilled water to pH 4.5 with 50% sodium hydroxide (analytical reagent).

Anionic surface active solution prepared at rate of 0.1 mg. of 100% pure Aerosol OT (dioctyl sodium sulfosuccinate) per ml. of distilled water to provide final concentration of 0.01% of active Aerosol OT (Pure Aerosol OT obtained from American Cyanamid Co.).

Procedure for Milk

1. Place 5 ml. milk sample, 5 ml. tetrachlorethane, and 2 ml. 50% lactic acid in a test tube.
2. Cork tube and shake vigorously for one minute.
3. Add 2 ml. of 4.0 molar NaOH to tube and invert six times to mix contents. (Rubber stoppers are unsatisfactory.) Cork stopper or glass-stoppered tubes suitable for centrifuging are preferable.

4. Separate tetrachloroethane from milk solids and water by centrifugation. About 5 minutes are needed in a 10-inch centrifuge at 3200 r.p.m. It will take about 25 minutes in a Babcock centrifuge to complete separation.

   Three distinct layers should form. The top, aqueous layer should be clear; the middle, solid layer should consist chiefly of white, precipitated protein; and the lower layer should be clear tetrachloroethane containing any QAC originally in milk.

5. Remove top layer by decantation or with an aspirator.
6. Separate as much tetrachloroethane as possible from the curd.
7. Place a 2-ml. aliquot of the tetrachloroethane in a clean tube containing 0.5 of pH 4.5 buffer and 0.2 ml. of eosin solution.
8. Cork tube and shake vigorously. A pink to rose color in the tetrachloroethane indicates presence of QAC.
9. Titrate to a colorless end point with 0.01% Aerosol OT solution. Vigorous shaking is required during the titration.

Procedure for Determining QAC in Water

1. Place 2 ml. of the QAC solution to be analyzed in a test tube containing 2 ml. of tetrachloroethane, 0.5 ml. of pH 4.5 buffer, and 0.2 ml. of eosin solution.
2. Shake contents of tube vigorously. A pink to red color in the tetrachloroethane indicates the presence of 1 p.p.m. or more of QAC.
3. Titrate to a colorless end point with 0.01% Aerosol OT solution. Vigorous shaking is required during the titration.

Sensitivity of the direct water titration can be increased for low concentrations of QAC by employing 10 ml. of sample, 1 ml. of buffer and 0.2 ml. of eosin solution.

QAC in cleaner sanitizer solutions also can be titrated by the direct water method. More buffer may be required for highly acid or alkaline QAC solutions.

The quantity of anionic reagent required in the titration is directly proportional to the QAC concentration of the sample. The quantity of anionic reagent required agrees with that expected on the assumption of a reaction involving a 1:1 mole ratio of QAC to anionic surface active agent.

Table 52 — Quantity of Anionic Solution Required to Titrate Various Concentrations of Added QAC in Water and Milk*

Anionic solution required for following ppm. QAC contained in 2 ml. of sample.

QAC	Sample	1 ppm	2 ppm	5 ppm	10 ppm	25 ppm	50 ppm	100 ppm
A	Water	0.03	0.05	0.11	0.23	0.62	1.25	2.57
A	Milk	0.03	0.06	0.11	0.22	0.59	1.17	2.48
B	Water	0.03	0.06	0.15	0.27	0.70	1.39	2.81
B	Milk	0.03	0.05	0.13	0.26	0.69	1.34	2.70
C	Water	0.02	0.04	0.09	0.19	0.50	0.99	2.03
C	Milk	0.02	0.04	0.09	0.19	0.48	1.01	2.06
D	Water	0.02	0.04	0.09	0.19	0.56	1.20	2.53
D	Milk	0.02	0.04	0.08	0.19	0.39	0.78	1.61

A = Alkyl dimethyl ethyl benzyl ammonium chloride (M.W. = 374).
B = Alkyl dimethyl benzyl ammonium chloride (M.W. = 340).
C = Para di-isobutyl phenoxy ethoxy ethyl dimethyl benzyl ammonium chloride (M.W. = 448).
D = Methyl doecyl benzyl trimethyl ammonium chloride (M.W. = 368).
   Dioctyl sodium sulfosuccinate (Aerosol OT) (M.W. = 444).
*From data by Furlong and Elliker.[26]

BIBLIOGRAPHY

1. Trebler, H. A., and Harding, H. G. Dairy Cleaners — Their Constituents and Functions. Proc. Lab. Sect. Milk Ind. Found. Conv. (1947).
2. Ordal, Z. John. Soils Encountered in Food Processing and Their Detergent Requirements. Klenzade 18th Educ. Sem. (1955).
3. Atherton, H. V. A Field Study of the Sanitary Care of Milking Equipment on Dairy Farms. J. Milk and Food Tech. 22,9:273-278 (1959).
4. Barrett, R. B. New Aspects of Sanitation Chemistry. Klenzade 21st Educ. Sem. (1959).
5. MacGregor, D. R., et al. Effect of Added Hypochlorite on Detergent Activity of Alkaline Solutions in Recirculation Cleaning. J. Milk and Food Tech. 17:136-138 (1954).
6. Levowitz, David. Chemistry of Cleaning Farm Milking Equipment. The Milk Dealer, January 1950.
7. ————————————. Raw Milk of Low Total and Thermoduric Count. A Guide for Dairy Fieldmen. Ann. Conf. Dairy Fieldmen and Sanitarians. Raleigh, N. C. (1957).
8. Sullivan, Paul J. Significance of Water Composition in Selection of Detergents. Klenzade 19th Educ. Sem. (1957).
9. Fernandez, Louis. Definitions and Functions of Cleaner Ingredients. Klenzade 17th Educ. Sem. (1953).
10. Munroe, Henry F. Chlorination and Conditioning of Farm Water Supplies. Klenzade 19th Educ. Sem. (1957).
11. Livak, Charles W. Significance of Water Composition in Selection of Detergents. Klenzade 19th Educ. Sem. (1957).
12. Training Course. Production and Marketing of High Quality Milk. Denmark (1954).
13. Butcher, Leonard. Chlorination and Treatment of Farm Water Supplies. Klenzade 18th Educ. Sem. (1955).
14. Black, L. A. Chemical Sterilization. W. Can. D. Conv. (1929).
15. Swartling, P. The Influence of the Use of Detergents and Sanitizers on the Farm with Regard to the Quality of Milk and Milk Products. Review Article No. 75 D. Sci. Abst. 21,1:1-10 (1959).
16. Druckrey, I. C. New Developments in Bactericides. Klenzade 19th Educ. Sem. (1957).

17. Elliker, P. R. New Developments in Bactericides. Klenzade 18th Educ. Sem. (1955).
18. Atherton, H. V. Some Observations on Several Aspects of Milk Quality After Addition of Solutions of Dairy Sanitation Chemicals. Proc. Eastern Div. ADSA (1960).
19. Elliker, P. R. Quaternary Ammonium Compounds. Am. Milk Rev. (Oct. 1950).
20. —————————————. An Up-To-Date Evaluation of Sanitizers. Klenzade 17th Educ. Sem. (1953).
21. Schahfer, J. G. Use More Chemical Sanitizers. Am. Milk Rev. and Milk Pl. Mo. (Sept. 1958).
22. Barrett, R. B. New Developments in Sanitation Chemistry. Klenzade Southwestern Educ. Sem. (1958).
23. Elliker, P. R. Chemical Sanitizers Available to the Milk Industry. Proc. Lab. Sect. Milk Ind. Foundation Conv. (1957).
24. Fernandez, Louis. New Detergent Ingredients and Their Advantages. Klenzade 19th Educ. Sem. (1957).
25. Keys to Chelation© 1959. The Dow Chemical Company. Reprinted by permission.
26. Furlong, T. E., and Elliker, P. R. An Improved Method of Determining Concentration of Quaternary Ammonium Compounds in Water Solutions and in Milk. J. D. Sci. 36,3:225-234 (1953).

REVIEW QUESTIONS

1. What factors influence the efficiency of cleaning procedures?
2. Why is it necessary to remove milk films immediately after a container is emptied?
3. Why is milkstone a problem on dairy equipment?
4. Why are low levels of chlorine added to alkaline cleaning compounds?
5. Should utensils and equipment be rinsed with clear water after use? Why?
6. Why shouldn't soaps be used for milkhouse sanitation procedures?
7. The cleaning system must accomplish what three actions?
8. What are the requirements of a good dairy cleaner?
9. What ingredients go into cleaning compounds? What is the function of each?
10. Why is the water supply important in dairy farm sanitation?
11. What is the difference between "temporary" and "permanent" water hardness?
12. Name the main classes of chemical sanitizers.
13. What do we mean when we say a certain polyphosphate has a "potentiating" effect on quaternary ammonium compounds?
14. What is a "cleaner-sanitizer?"
15. Why is it important to rinse equipment after washing, regardless of the type of sanitizing agent to be used?

Appendix

LEGAL STANDARDS
UNITED STATES PUBLIC HEALTH SERVICE MILK CODE

Section 1. Definitions

The following definitions shall apply in the interpretation and enforcement of this ordinance:*

Caution — Any milk product defined herein, shipped or offered for shipment in interstate commerce, must conform with the applicable, detailed standards of identity under the Federal Food, Drug, and Cosmetic Act.

Milk — Milk is hereby defined to be the lacteal secretion, practically free from colostrum, obtained by the complete milking of one or more healthy cows, which contains not less than $8\frac{1}{4}$ percent milk solids-not-fat and not less than $3\frac{1}{4}$ percent milkfat.

Goat Milk — Goat Milk is the lacteal secretion, practically free from colostrum, obtained by the complete milking of healthy goats. The word "milk" shall be interpreted to include goat milk.

Milkfat — Milkfat, or butterfat, is the fat of milk.

Cream — Cream is a portion of milk which contains not less than 18 percent milkfat.

Half and Half — Half and half is a product consisting of a mixture of milk and cream which contains not less than 11.5 percent milkfat.

Whipped Cream — Whipped cream is cream to which a harmless gas has been added to cause whipping of the product. It may also contain sugar, other harmless flavoring, and a harmless stabilizer.

Concentrated Milk — Concentrated milk is a fluid product, unsterilized and unsweetened, resulting from the removal of a considerable portion of the water from milk. When recombined

---

* The milkfat and solids-not-fat standards in these definitions should be changed, if necessary, to conform with state laws.

with water, in accordance with instructions printed on the container, the resulting product conforms with the standards for milkfat and solids-not-fat of milk as defined above.

Buttermilk — Buttermilk is a fluid product resulting from the churning of milk or cream. It contains not less than $8\frac{1}{4}$ percent milk solids-not-fat.

Cultured Buttermilk — Cultured buttermilk is a fluid product resulting from the souring or treatment, by a lactic acid or other culture, of pasteurized skimmilk or pasteurized reconstituted skimmilk. It contains not less than $8\frac{1}{4}$ percent milk solids-not-fat.

Vitamin D Milk — Vitamin D milk is milk the vitamin D content of which has been increased by an approved method to at least 400 U.S.P. units per quart.

Cottage Cheese — Cottage cheese is the soft uncured cheese prepared from the curd obained by adding harmless, lactic-acid producing bacteria, with or without enzymatic action, to pasteurized skimmilk or pasteurized reconstituted skimmilk. It contains not more than 80 percent moisture.

Creamed Cottage Cheese — Creamed cottage cheese is the soft uncured cheese which is prepared by mixing cottage cheese with pasteurized cream, or with a pasteurized mixture of cream and milk or skimmilk, and which contains not less than 4 percent milkfat by weight, and not more than 80 percent moisture.

Homogenized Milk — Homogenized milk is milk which has been treated in such a manner as to insure break-up of the fat globules to such an extent that, after 48 hours of quiescent storage, no visible cream separation occurs on the milk, and the fat percentage of the top 100 milliliters of milk in a quart bottle, or of proportionate volumes in containers of other sizes, does not differ by more than 10 percent of itself from the fat percentage of the remaining milk as determined after thorough mixing. The word "milk" shall be interpreted to include homogenized milk.

Milk Products — Milk products shall be taken to mean and to include cream, sour cream, half and half, reconstituted half and half, whipped cream, concentrated milk, concentrated milk products, skimmilk, nonfat milk, flavored milk, flavored drink, flavored reconstituted milk, flavored reconstituted drink, buttermilk, cultured buttermilk, cultured milk, vitamin D milk, reconstituted or recombined milk, reconstituted cream, reconstituted skimmilk, cottage cheese, creamed cottage cheese, and any other product made by the addition of any substance to milk, or to any of these milk products, and used for similar purposes, and designated as a milk product by the health officer.

Pasteurization — The terms "pasteurization," "pasteurized," and similar terms shall be taken to refer to the process of heating every particle of milk or milk products to at least 143° F., and holding it at such temperature continuously for at least 30 minutes, or to at least 161° F., and holding it at such temperature continuously for at least 15 seconds, in approved and properly operated equipment: Provided, that nothing contained in this definition shall be construed as barring any other process which has been demonstrated to be equally efficient and which is approved by the State health authority.

Adulterated and Misbranded Milk and Milk Products — Any milk or cream to which water has been added, or any milk product which contains any unwholesome substance, or which if defined in this Ordinance does not conform with its definition, shall be deemed to be adulterated. Any milk or milk product which carries a grade label, unless such grade label has been awarded by the health officer and not revoked, or which fails to conform in any other respect with the statements on the label, shall be deemed to be misbranded.

DEFINITIONS AND STANDARDS OF CERTAIN DAIRY PRODUCTS

From the Federal Food, Drug, and Cosmetic Act
U. S. Department of Health, Education, and Welfare
Food and Drug Administration

Section 10.1 (c) of the general regulations relating to definitions and standards for food states: "No provision of any regulation prescribing a definition and standard of identity or standard of quality or fill of container under section 401 of the act shall be construed as in any way affecting the concurrent applicability of the general provisions of the act and the regulations thereunder relating to adulteration and misbranding. For example, all regulations under section 401 contemplate that the food and all articles used as components or ingredients thereof shall not be poisonous or deleterious and shall be clean, sound, and fit for food. A provision in such regulations for the use of coloring or flavoring does not authorize such use under circumstances or in a manner whereby damage or inferiority is concealed or whereby the food is made to appear better or of greater value than it is."

18.500 Cream class of food; identity. Cream is the class of food which is the sweet, fatty liquid or semi-liquid separated from milk, with or without the addition thereto and intimate admixture therewith of sweet milk or sweet skimmilk. It may be pasteurized and if it contains less than 30 percent of milk fat as determined by the method referred to in this section, it may be homogenized. It contains not less than 18 percent of milk fat, as determined by the method prescribed in "Official and Tentative Methods of Analysis of the Association of Official Agriculture Chemists," Fourth Edition, 1935, page 277 (Ed. note, 8th edition, 1955, p. 261, sec. 15.61) under "Fat, Roese-Gottlieb Method — Official." The word "milk" as used in this section means cow's milk.

18.501 Light cream, coffee cream, table cream; identity. Light cream, coffee cream, table cream, conforms to the definition and standard of identity prescribed for the cream class of food by 18.500, except that it contains less than 30 percent of milk fat, as determined by the method referred to in such section.

18.510 Whipping cream class of food; identity. Whipping cream is the class of food which conforms to the definition and standard of identity prescribed for the cream class of food by 18.500, except that it contains not less than 30 percent of milk fat, as determined by the method referred to in such section.

18.511 Light whipping cream; identity. Light whipping cream conforms to the definition and standard of identity prescribed for the whipping cream class of food by 18.510, except that it contains less than 36 percent of milk fat, as determined by the method referred to in 18.500.

18.515 Heavy cream, heavy whipping cream; identity. Heavy cream, heavy whipping cream, conforms to the definition and standard of identity prescribed for the whipping cream class of food by 18.510, except that it contains not less than 36 percent of milk fat, as determined by the method referred to in 18.500.

18.520 Evaporated milk; identity; label statement of optional ingredients. (a) Evaporated milk is the liquid food made by evaporating sweet milk to such point that it contains not less than 7.9 percent of milk fat and not less than 25.9 percent of total milk solids. It may contain one or both of the following optional ingredients:

(1) Disodium phosphate or sodium citrate or both, or calcium chloride, added in a total quantity of not more than 0.1 percent by weight of the finished evaporated milk.

(2) Vitamin D in such quantity as to increase the total vitamin D content to not less than 25 U.S.P. units per fluid ounce of the finished evaporated milk.

It may be homogenized. It is sealed in a container and so processed by heat as to prevent spoilage.

(b) When the optional ingredient specified in paragraph (a) (2) of this section is present, the label shall bear the statement "with increased vitamin D content" or "vitamin D content increased." Such statement shall immediately and conspicuously precede or follow the name "Evaporated Milk," without intervening written, printed, or graphic matter, wherever such name appears on the label so conspicuously as to be easily seen under customary conditions of purchase.

(c) For the purpose of this section:

(1) The word "milk" means cow's milk.

(2) Such milk may be adjusted, before or after evaporation, by the addition or abstraction of cream or sweet skimmilk, or by the addition of concentrated sweet skimmilk.

(3) The quantity of milk fat is determined by the method prescribed under "Fat—Official" on page 249 (Ed. note, 8th edition, 1955, p. 263, sec. 15.74) and the quantity of total milk solids is determined by the method prescribed under "Total Solids — Official" on page 248 (Ed. note, 8th edition, 1955, p. 263, sec. 15.72) of "Official Methods of Analysis of the Association of Official Agricultural Chemists," Seventh Edition, 1950.

(4) Vitamin D content may be increased by the application of radiant energy or by the addition of a concentrate of vitamin D (with any accompanying vitamin A when such vitamin D in such concentrate is obtained from natural sources) dissolved in a food oil; but if such oil is not milk fat the quantity thereof added is not more than 0.01 percent of the weight of the finished evaporated milk.

(5) The quantity of vitamin D is determined by the method prescribed in "Official Methods of Analysis of the Association of Official Agricultural Chemists," Seventh Edition, 1950, page 788 et seq., (Ed. note, 8th edition, 1955, p. 839 et seq., secs. 38.64, 38.75), under the heading "Vitamin D in Milk — Official."

18.525 Concentrated milk, plain condensed milk; identity; label statement of optional ingredients. Concentrated milk, plain condensed milk, conforms to the definition and standard of identity, and is subject to the requirements for label statement of optional ingredients, prescribed for evaporated milk by 18.520, except that;

(a) It is not processed by heat;

(b) Its container may be unsealed; and

(c) Optional ingredient 18.520 (a) (1) is not used.

18.530 Sweetened condensed milk; identity. (a) Sweetened condensed milk is the liquid or semi-liquid food made by evaporating a mixture of sweet milk and refined sugar (sucrose) or any combination of refined sugar (sucrose) and refined corn sugar (dextrose) to such point that the finished sweetened condensed milk contains not less than 28.0 percent of total milk solids and not less than 8.5 percent of milk fat. The quantity of refined sugar (sucrose) or combination of such sugar and refined corn sugar (dextrose) used is sufficient to prevent spoilage.

(b) For the purpose of this section:

(1) The word "milk" means cow's milk.

(2) Such milk may be adjusted, before or after evaporation, by the addition or abstraction of cream or sweet skimmilk, or the addition of concentrated sweet skimmilk.

(3) Milk fat is determined by the method prescribed in "Official and Tentative Methods of Analysis of the Association of Official Agricultural Chemists," Fourth Edition, 1935, page 281 (Ed. note, 8th edition, 1955, p. 264, sec. 15.86), under "Fat — Official."

18.535 Condensed milks which contain corn sirup; identity. (a) Condensed milks which contain corn sirup are the foods each of which conforms to the definition and standard of identity prescribed for sweetened condensed milk by 18.530 except that corn sirup or a mixture of corn sirup and sugar is used instead of sugar or a mixture of sugar and dextrose. For the purpose of this section the term "corn sirup" means a clarified and concentrated aqueous solution of the products obtained by the incomplete hydrolysis of corn-starch, and includes dried corn sirup; the solids of such corn sirup contain not less than 40 percent of weight of reducing sugars, calculated as anhydrous dextrose.

(b) The name of each such food is:

(1) "Corn sirup condensed milk," "condensed milk with corn sirup," or "condensed milk prepared with corn sirup," if corn sirup alone is used; or

(2) "...% Corn sirup solids ...% sugar condensed milk," "Condensed milk with ...% corn sirup solids ...% sugar," or "Condensed milk prepared with ...% corn sirup solids ...% sugar," if a mixture of corn sirup and sugar is used, the blanks being filled in with the whole numbers nearest the actual percentages of corn sirup solids and sugar in such food; alternately "...% sugar" may precede "...% corn sirup solids" in such names.

18.540 Dried skimmilk, powdered skimmilk, skimmilk powder; identity. Dried skimmilk, powdered skimmilk, skimmilk powder, is the food made by drying sweet skimmilk. It contains not more than 5 percent of moisture, as determined by the method prescribed in "Official and Tentative Methods of Analysis of the Association of Official Agricultural Chemists," Fourth Edition, 1935, page 282 (Ed. note, 8th edition, 1955, p. 265, sec. 15.93), under the caption "Moisture — Tentative." The term "skimmilk" as used in this section, means cow's milk from which the milk fat has been separated.

19.500 Cheddar cheese, cheese; identity; label statement of optional ingredients.

(a) Cheddar cheese, cheese, is the food prepared from milk and other ingredients specified in this section, by the procedure set forth in paragraph (b) of this section, or by another procedure which produces a finished cheese having the same physical and chemical properties as the cheese produced when the procedure set forth in paragraph (b) of this section is used. It contains not more than 39 percent of moisture and its solids contain not less than 50 percent of milk fat, as determined by the methods prescribed in paragraph (c) of this section. If the milk used is not pasteurized, the cheese so made is cured at a temperature of not less than 35° F. for not less than 60 days.

(b) Milk, which may be pasteurized or clarified or both, and which may be warmed, is subjected to the action of harmless lactic-acid-producing bacteria, present in such milk or added thereto. Harmless artificial coloring may be added. Sufficient rennet (with or without purified calcium chloride in a quantity not more than 0.02 percent, calculated as anhydrous calcium chloride, of the weight of the milk) is added to set the milk to a semi-solid mass. The mass is so cut, stirred, and heated with continued stirring, as to promote and regulate the separation of whey and curd. The whey is drained off, and the curd is matted into a cohesive mass. The mass is cut into slabs, which are so piled and handled as to promote the drainage of whey and the development of acidity. The slabs are then cut into pieces, which may be rinsed by sprinkling or pouring water over them, with free and continuous drainage; but the duration of such rinsing is so limited that only the whey on the surface of such pieces is removed. The curd is salted, stirred, further drained, and pressed into forms. A harmless preparation of enzymes of animal or plant origin capable of aiding in the curing or development of flavor of cheddar cheese may be added during the procedure, in such quantity that the weight of the solids of such preparation is not more than 0.1 percent of the weight of the milk used.

(c) Determine moisture by the method prescribed on page 262 (15.124) (Ed. note, 8th edition, 1955, p. 278, sec. 15.129), under "Moisture — Official," and milk fat by the method prescribed on page 263 (15.131) (Ed. note, 8th edition, 1955, p. 279, sec. 15.136), under "Fat — Official," of "Official Methods of Analysis of the Association of Official Agricultural Chemists," Seventh Edition, 1950. Subtract the percent of moisture found from 100; divide the remainder into the percent milk fat found. The quotient, multiplied by 100, shall be considered to be the percent of milk fat contained in the solids.

(d) Cheddar cheese in the form of slices or cuts in consumer-sized packages may contain not more than 0.2 percent by weight of sorbic acid.

(e) For the purposes of this section:

(1) The word "milk" means cow's milk, which may be adjusted by separating part of the fat therefrom or by adding thereto one or more of the following: Cream, skimmilk, concentrated skimmilk, nonfat dry milk, water in a quantity sufficient to reconstitute any concentrated skimmilk or nonfat dry milk used.

(2) Milk shall be deemed to have been pasteurized if it has been held at a temperature of not less than 143° F. for a period of not less than 30 minutes, or for a time and at a temperature equivalent thereto in phosphatase destruction. Cheddar cheese shall be deemed not to have been made from pasteurized milk if 0.25 gm. shows a phenol equivalent of more than 3 micrograms when tested by the method prescribed in paragraph (f) of this section.

Butter. The Act of March 4, 1923 (42 Stat. 1500), defines butter as "For the purposes of this chapter 'butter' shall be understood to mean the food product usually known as butter, and which is made exclusively from milk or cream, or both, with or without common salt, and with or without additional coloring matter, and containing not less than 80 per centum by weight of milk fat, all tolerances having been allowed for."

Nonfat Dry Milk. The Act of July 2, 1956 (70 Stat. 486), defines nonfat dry milk as follows: "* * * for the purposes of the Federal Food, Drug, and Cosmetic Act of June 25, sic, 1938 (ch. 675, sec. 1, 52 Stat. 1040), nonfat dry milk is the product resulting from the removal of fat and water from milk, and contains the lactose, milk proteins, and milk minerals in the same relative proportions as in the fresh milk from which made. It contains not over 5 per centum by weight of moisture. The fat content is not over 1½ per centum by weight unless otherwise indicated.

"The term 'milk,' when used herein, means sweet milk of cows."

Grade Names Used in U. S. Standards for Farm Products*

DAIRY PRODUCTS

Product	Grade Names			
Butter	U. S. Grade AA (U. S. 93 Score)	U. S. Grade A (U. S. 92 Score)	U. S. Grade B (U. S. 90 Score)	U. S. Grade C (U. S. 89 Score)
Cheddar Cheese	Grade AA	Grade A	Grade B	Grade C
Dry Buttermilk	Extra	Standard	—	—
Dry Whole Milk	Premium	Extra	Standard	—
Dry Whey	Extra	—	—	—
Nonfat Dry Milk	Extra	Standard	—	—
Swiss Cheese	Grade A	Grade B	Grade C	Grade D

*USDA Agricultural Handbook No. 157. Feb. 1960.

ATOMIC WEIGHTS
(1957)

Element	Symbol	Atom. No.	Atom. Wt.*	Element	Symbol	Atom. No.	Atom. Wt.*
Actinium	Ac	89	(227)	Mercury	Hg	80	200.61
Aluminum	Al	13	26.98	Molybdenum	Mo	42	95.95
Americium	Am	95	(243)	Neodymium	Nd	60	144.27
Antimony	Sb	51	121.76	Neon	Ne	10	20.183
Argon	Ar	18	39.944	Neptunium	Np	93	(237)
Arsenic	As	33	74.91	Nickel	Ni	28	58.71
Astatine	At	85	(210)	Niobium	Nb	41	92.91
Barium	Ba	56	137.36	Nitrogen	N	7	14.008
Berkelium	Bk	97	(249)	Nobelium	No	102	(254)
Beryllium	Be	4	9.013	Osmium	Os	76	190.2
Bismuth	Bi	83	209.00	Oxygen	O	8	16
Boron	B	5	10.82	Palladium	Pd	46	106.4
Bromine	Br	35	79.916	Phosphorus	P	15	30.975
Cadmium	Cd	48	112.41	Platinum	Pt	78	195.09
Calcium	Ca	20	40.08	Plutonium	Pu	94	(242)
Californium	Cf	98	(251)	Polonium	Po	84	(210)
Carbon	C	6	12.011	Potassium	K	19	39.100
Cerium	Ce	58	140.13	Praseodymium	Pr	59	140.92
Cesium	Cs	55	132.91	Promethium	Pm	61	(147)
Chlorine	Cl	17	35.457	Protactinium	Pa	91	(231)
Chromium	Cr	24	52.01	Radium	Ra	88	(226)
Cobalt	Co	27	58.94	Radon	Rn	86	(222)
Copper	Cu	29	63.54	Rhenium	Re	75	186.22
Curium	Cm	96	(247)	Rhodium	Rh	45	102.91
Dysprosium	Dy	66	162.51	Rubidium	Rb	37	85.48
Einsteinium	Es	99	(254)	Ruthenium	Ru	44	101.1
Erbium	Er	68	167.27	Samarium	Sm	62	150.35
Europium	Eu	63	152.0	Scandium	Sc	21	44.96
Fermium	Fm	100	(255)	Selenium	Se	34	78.96
Fluorine	F	9	19.00	Silicon	Si	14	28.09
Francium	Fr	87	(223)	Silver	Ag	47	107.880
Gadolinium	Gd	64	157.26	Sodium	Na	11	22.991
Gallium	Ga	31	69.72	Strontium	Sr	38	87.63
Germanium	Ge	32	72.60	Sulfur	S	16	32.066†
Gold	Au	79	197.0	Tantalum	Ta	73	180.95
Hafnium	Hf	72	178.50	Technetium	Tc	43	(99)
Helium	He	2	4.003	Tellurium	Te	52	127.61
Holmium	Ho	67	164.94	Terbium	Tb	65	158.93
Hydrogen	H	1	1.0080	Thallium	Tl	81	204.39
Indium	In	49	114.82	Thorium	Th	90	232.05
Iodine	I	53	126.91	Thulium	Tm	69	168.94
Iridium	Ir	77	192.2	Tin	Sn	50	118.70
Iron	Fe	26	55.85	Titanium	Ti	22	47.90
Krypton	Kr	36	83.80	Tungsten	W	74	183.86
Lanthanum	La	57	138.92	Uranium	U	92	238.07
Lead	Pb	82	207.21	Vanadium	V	23	50.95
Lithium	Li	3	6.940	Xenon	Xe	54	131.30
Lutetium	Lu	71	174.99	Ytterbium	Yb	70	173.04
Magnesium	Mg	12	24.32	Yttrium	Y	39	88.92
Manganese	Mn	25	54.94	Zinc	Zn	30	65.38
Mendelevium	Md	101	(256)	Zirconium	Zr	40	91.22

*A value given in brackets denotes the mass number of the most stable known isotope.

†The Atomic Weights Commission recommends that a range of ±0.003 be attached to the official value of 32.066.

Average Composition of Milk and Its Products

	Water	Fat	Casein and Albumen	Lactose	Ash	Sucrose	Authority
Cow's milk	87.00	4.00	3.30	5.00	0.70		
Cream 40%	54.70	40.00	1.94	2.95	0.41		Calculated[3]
Cream 20%	72.92	20.00	2.59	3.94	0.55		Calculated[3]
Skim milk (centrifuged)	90.30	0.10	3.55	5.25	0.80		Van Slyke
Buttermilk	90.65	0.50	3.60	4.50	0.75		
Whey	93.40	0.35	0.85	4.80	0.60		Van Slyke
Evaporated milk	73.00	8.30	7.50	9.70	1.50		Hunziker
Sweetened condensed (wh.)	27.47	9.28	7.42	13.35	1.88	40.60	Hunziker (Aver. 18 brands)
Sweetened condensed (sk.)	29.00	0.06	10.32	15.60	2.25	42.27	Vt. Exp. Sta.
Whole milk powder	1.40	29.20	26.92	36.48	6.00		Hunziker
Butter, salted	15.00	81.50	0.60		2.90[1]		
Cheddar cheese	36.80	33.75	23.75[2]		5.70[1]		Van Slyke

[1]Salt and ash. [2]Paracasein. [3]Mojonnier formula $F = 1.097 \times T.S - 9.70$.

Standard Units of Measure and Weight

Unit of Length

10 Angstrom units	= 1 millimicron (mμ)
1000 millimicrons	= 1 micron (μ)
1000 microns	= 1 millimeter (mm)
10 millimeters	= 1 centimeter (cm)
10 centimeters	= 1 decimeter (dm)
10 decimeters	= 1 meter (m)
1 centimeter	= 0.3937 inch (in.)
1 inch	= 2.54 centimeters
1 meter	= 39.37 inches or 1.0936 yards (yds.)
1 yard	= 0.9144 meter
1 mile	= 1.6093 kilometers (km)

Unit of Weight

1000 micrograms (μg)	= 1 milligram (mg)
10 milligrams	= 1 centigram (cg)
10 centigrams	= 1 decigram (dg)
10 decigrams	= 1 gram (g)
1000 grams	= 1 kilogram (kg)
1 kilogram	= 2.2046 pounds (lbs.)
1 ounce (avoir.)	= 28.3495 grams
1 pound (avoir.)	= 453.5924 grams

Unit of Capacity

10 milliliters (ml)	= 1 centiliter (cl)
10 centiliters	= 1 deciliter (dl)
10 deciliters	= 1 liter (l.)
1000 milliliters	= 1 liter
1 liter	= 1.05671 liquid quarts
1 liter	= 1000.027 cubic centimeters at 40°C. and 760 mm. pressure
1 quart	= 0.94633 liter

COMPARATIVE TEMPERATURE SCALES OF CENTIGRADE, REAUMUR AND FAHRENHEIT.

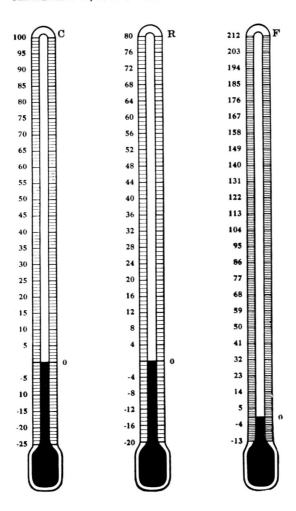

How to calculate the changes without the scale when only one temperature is stated:

It will be noted in comparing the three scales that for every 5 degrees Centigrade the Reaumur scale varies only 4 degrees, while the Fahrenheit scale varies 9 degrees.

Thus to reduce a Centigrade temperature to Reaumur multiply the known Centigrade reading by 4/5 and you have the equivalent on the Reaumur scale.

$$50° \text{ C.} \times 4/5 = 40° \text{ R.}$$
$$-20° \text{ C.} \times 4/5 = -16° \text{ R.}$$

To reduce Reaumur to Centigrade multiply the known Reaumur scale by 5/4.

Example:

$$40° \text{ R.} \times 5/4 = 50° \text{ C.}$$
$$-16° \text{ R.} \times 5/4 = -20° \text{ C.}$$

A little different procedure is necessary in reducing the Centigrade scale to Fahrenheit and vice versa. As noted above, every 5 degrees on the Centigrade scale equals 9 on the Fahrenheit scale. But we also must note that 0° F. starts 32 degrees below 0° on the Centigrade scale.

Thus we arrive at the result by multiplying the known Centigrade scale by 9/5 and adding 32.

Example:

$$50° \text{ C.} \times 9/5 = 90 + 32 = 122° \text{ F.}$$
$$-20° \text{ C.} \times 9/5 = -36 + 32 = -4° \text{ F.}$$

Reverse the operation for reducing Fahrenheit to Centigrade.

Example:

$$122° \text{ F.} - 32 = 90 \times 5/9 = 50° \text{ C.}$$
$$-4° \text{ F.} - 32 = -36 \times 5/9 = -20° \text{ C.}$$

To reduce the known Reaumur degree to the Fahrenheit scale the same procedure as Centigrade to Fahrenheit is employed, only that 9/4 and 4/9 is used instead of 9/5 and 5/9 as in the Centigrade Reduction.

Specific Gravity and Weight per Gallon of Water, Milk and Cream

Product	Fat %	Solids Not Fat %	Total Solids %	Specific Gravity (60° F.)	Weight Per Gal. Lbs.
Water	0.00	0.00	0.00	1.0000	8.34
Skim milk	.03	9.00	9.03	1.0360	8.64
Whole milk	3.00	8.33	11.33	1.0309	8.60
	3.25	8.49	11.74	1.0314	8.60
	3.50	8.60	12.10	1.0316	8.60
	3.75	8.70	12.45	1.0318	8.60
	4.00	8.79	12.79	1.0320	8.61
	4.25	8.88	13.13	1.0321	8.61
	4.50	8.95	13.45	1.0322	8.61
	4.75	9.02	13.77	1.0323	8.61
	5.00	9.10	14.10	1.0324	8.61
Cream	10.00	8.10	18.10	1.023	8.53
	12.00	7.92	19.92	1.020	8.51
	14.00	7.74	21.74	1.017	8.48
	16.00	7.58	23.58	1.015	8.47
	18.00	7.40	25.40	1.013	8.45
	20.00	7.20	27.20	1.011	8.43
	22.00	7.02	29.02	1.009	8.42
	24.00	6.84	30.84	1.007	8.40
	26.00	6.66	32.66	1.006	8.39
	28.00	6.48	34.48	1.004	8.37
	30.00	6.30	36.30	1.002	8.36
	32.00	6.12	38.12	1.000	8.34
	34.00	5.94	39.94	0.998	8.32
	36.00	5.76	41.76	0.997	8.31
	38.00	5.58	43.58	0.995	8.30
	40.00	5.40	45.40	0.993	8.28

Composition of Milk of Different Species of Animals

Species	Fat %	Protein %	Lactose %	Ash %	Total Solids %
Human[1]	3.75	1.63	6.98	0.21	12.57
Cow[1]	4.00	3.30	5.00	0.70	13.00
Goat[1]	3.92	3.32	4.62	0.84	12.70
Mare[2]	1.07	2.63	7.12	0.48	11.30
Ewe[3]	7.72	4.06	4.71	0.83	17.32
Sow[4]	5.19	6.75	3.89	1.03	16.86

[1]Average of various sources.
[2]Exp. Sta. Rec. IX, p. 685, 1897.
[3]Exp. Sta. Rec. XLI, p. 177, 1919.
[4]Vt. Exp. Sta. Bul. 195, 1916.

Composition of Milk of Different Breeds of Cattle*

Breed	Fat %	Protein %	Lactose %	Ash %	Total Solids %
Jersey	5.28	3.83	4.96	.74	14.81
Guernsey	5.00	3.83	4.96	.74	14.53
Ayrshire	3.91	3.44	4.90	.67	12.92
Holstein	3.41	3.23	4.76	.68	12.08
Brown Swiss	4.00	3.56	4.90	.73	13.19

*Average of various sources.

GLOSSARY OF TERMS

Agglutinate. To clump together.
Aldehyde. An oxidation product of an alcohol.
Amphoteric. Having the nature of both an acid and an alkali.
Anhydride. An oxide that becomes an acid by the addition of water, as sulphur dioxide and water forms sulphurous acid.
Burette. A graduated tube for measuring solutions.
Calibration. Checking glassware for accuracy.
Calorimeter. An apparatus for measuring heat or calories in a substance.
Carboy. A large bottle for holding such corrosive liquids as acids.
Caramelize. To darken sugar by excess heat.
Centigrade. A thermometer used mainly in scientific work rather than the Fahrenheit.
Cholesterol. One of the higher alcohols associated with fats.
Chromogens. Bacteria that produce color.
Colloid. A very small particle such as casein in milk.
Colostrum. The first milk after calving.
Conductivity. The power of transmitting heat.
Cryoscope. An instrument for determining freezing point of milk.
Cytoplasm. The protoplasmic substance of a cell less the nucleus.
Density. Weight of a substance per unit of volume.
Desiccator. A container in which samples may be kept dry.
Emulsion. A mixture of one liquid form in another as fat in milk.
Enzyme. An organic compound that brings about changes in other organic compounds without itself being changed.
Ester. A compound formed by the action of an acid with an alcohol.
Formaldehyde. A preservative which is a gas at ordinary temperatures.
Formalin. A 40 percent water solution of formaldehyde.
Geometric. A progression in the order of 2, 4, 8, 16, etc.
Glutathione. Contains the amino acids cysteine, glutamic acid and glycine.
Gram-negative. Organisms that do not take a certain stain.
Gravimetric. Measurement by weight as gram or pound.
Hemoglobin. The coloring matter of blood.
Hexose sugar. A sugar having six carbon atoms.
Hormone. A substance secreted by certain organs in the body which acts as stimulants to other organs.
Hydrate. To combine with water.
Ions. Electrically charged atoms or groups of atoms.
Isoelectric point. That hydrogen ion concentration at which proteins combine with neither acids nor bases.
Ketone. A compound in which CO is joined with two alkyl groups.
Lecithin. A phospholipid in milk.
Leucocytes. White blood corpuscles seen in milk.
Mastitis. Inflammation of the udder.
Meniscus. Curve in the fat column.
Neutralize. To make neither acidic nor basic.
Phenol. A derivative of coal tar as carbolic acid.

APPENDIX

Phospholipid. Belonging to the fats but contains a phosphoric acid and choline radicle.
Plasma. In milk all constituents except the fat.
Protoplasm. The living material of a cell.
Radical. That part of a compound that remains after one or more elements have been removed.
Rider. A small movable weight on a balance.
Saponify. To make into a soap.
Serum. In milk all constituents except the fat and casein.
Solvent. The liquid used to dissolve a substance.
Solute. The dissolved substance.
Specific gravity. The number times heavier a substance is than the standard, usually water.
Spore. A resistant cell fitted to reproduce under unfavorable conditions for respective species.
Titratable acidity. That acidity as measured by titration as against the use of a potentiometer.
Turbid. Solution not clear, riled.
Viscosity. Resistance to flow, as cream.
Vitamin. Chemical substance necessary for growth, development, etc.
Volatile. May pass off as a gas.
Volumetric. Measurement by volume as milliliter or quart.

Abbreviations

b.p.   boiling point
B. of H.   Board of Health lactometer
B. T. U.   British Thermal Unit
C.   Centigrade, calorie
F.   Fahrenheit
g.   gram
L.   Lactometer
mg.   milligram
ml.   milliliter
M   Molar
N   Normal
p.p.m.   parts per million
Q.   Quevenne lactometer
r.p.m.   revolutions per minute
S.N.F.   Solids Not Fat
Sp. gr.   Specific gravity
T. S.   Total Solids
var.   variety (in relation to bacteria)

Chemical Symbols

$AgCl$ Silver chloride
$AgNO_3$ Silver Nitrate
$Ca(OH)_2$ Calcium hydroxide
CH_2O Formaldehyde
$FeCl_3$ Ferric chloride
HCl Hydrochloric acid
HNO_3 Nitric acid
H_2SO_4 Sulphuric acid
H_3PO_4 Phosphoric acid
K_2CrO_4 Potassium chromate
$K_2Cr_2O_7$ Potassium bichromate
KOH Potassium hydroxide
$NaCl$ Sodium chloride
Na_2CO_3 Sodium carbonate
$NaOH$ Sodium hydroxide

FEDERAL AND STATE STANDARDS FOR THE COMPOSITION OF MILK PRODUCTS

	Whole milk			Skimmilk		
	Milkfat	Milk solids not fat	Total milk solids	Milkfat	Milk solids not fat	Total milk solids
	Min. %	Min. %	Min. %	Max. %	Min. %	Min. %
Federal	....	....	....	—	8.5	—
Alabama	3.25	8.5	11.75	—	8.5	—
Alaska	3.25	8.25	11.5	—3.25	8.25	—
Arizona	3.25	8.25	—	—3.25	8.25	—
Arkansas	3.25	8.0	—	—3.25	8.0	—
California	3.5	8.15	—	0.25	8.5	—
Colorado	3.2	8.25	—	—3.2	8.25	—
Connecticut	3.25	8.25	11.5	0.5	—	—
Delaware	3.5	8.5	—	—3.5	8.5	—
District of Columbia	3.5	—	11.5	—3.25	—	9.0
Florida	3.25	8.5	—	1.25	8.5	—
Georgia	3.25	8.5	11.75	—3.25	—	9.25
Hawaii	3.0	8.25	—	—3.0	8.25	—
Idaho	3.2	8.0	11.0	—3.2	—	9.3
Illinois	3.0	8.5	—	—3.0	—	9.25
Indiana	3.25	8.0	—	—3.25	—	8.5
Iowa	3.25	—	11.5	—3.25	—	—11.5
Kansas	3.25	8.25	—	—3.25	8.25	—
Kentucky	3.25	8.25	—	—3.25	—	—
Louisiana	3.8	8.5	12.3	—3.8	—	—
Maine	3.25	—	11.75	—3.25	—	—
Maryland	3.5	—	12.0	—3.5	—	—
Massachusetts	3.35	—	12.0	—3.35	—	—
Michigan	3.5	8.5	—	0.5	8.5	—
Minnesota	3.25	8.25	—	—3.25	8.25	—
Mississippi	3.5	8.25	—	—3.5	8.25	—
Missouri	3.25	8.0	—	—3.25	8.0	—
Montana	3.25	8.25	11.5	—3.25	8.25	—
Nebraska	3.0	8.25	—	—3.0	—	9.25
Nevada	3.25	8.5	—	—3.25	8.5	—
New Hampshire	3.35	—	11.85	—3.35	8.5	—
New Jersey	3.0	—	11.5	—3.0	—	—
New Mexico	3.25	8.25	—	—3.25	8.25	—
New York	3.0	—	11.5	—3.0	—	8.5
North Carolina	3.25	8.25	—	1.0	—	—
North Dakota	3.25	—	11.5	—3.25	8.5	—
Ohio	3.0	—	12.0	—3.0	—	—
Oklahoma	3.25	8.0	—	—3.25	—	—
Oregon	3.2	8.5	11.7	—3.2	8.5	—
Pennsylvania	3.25	—	12.0	0.5	—	—
Puerto Rico	3.0	9.0	12.0	—	—	—
Rhode Island	3.25	8.25	11.5	—3.25	8.25	—
South Carolina	3.8	8.25	—	—3.8	8.25	—
South Dakota	3.25	8.25	—	—3.25	—	9.25
Tennessee	3.5	8.5	—	—3.5	8.5	—
Texas	3.25	8.0	—	—3.25	8.0	—
Utah	3.2	8.3	11.5	—3.2	8.5	—
Vermont	3.5	8.5	11.75	—3.5	—	—
Virginia	3.25	8.5	11.75	—3.25	8.5	—
Washington	3.5	8.25	—	—3.5	—	—
West Virginia	3.0	8.5	11.5	—3.0	—	9.0
Wisconsin	3.0	8.25	—	—3.0	8.25	8.5
Wyoming	3.25	8.25	—	—3.25	—	8.25

APPENDIX

FEDERAL AND STATE STANDARDS FOR THE COMPOSITION OF MILK PRODUCTS

	Cream		Evaporated milk		Sweetened cond. (whole) milk	
	Light Milk-fat	Whipping Milk-fat	Milk-fat	Total milk solids	Milk-fat	Total milk solids
	Min. %	Min. %	Min. %	Min. %	Min. %	Min. %
Federal	18.0	30.0	7.9	25.9	8.5	28.0
Alabama	18.0	30.0	7.8	25.5	8.0	28.0
Alaska	18.0	30.0	(*)	(*)	(*)	(*)
Arizona	18.0	30.0	7.9	25.9	8.5	28.0
Arkansas	18.0	(*)	7.9	25.5	8.0	28.0
California	20.0	35.0	7.9	25.9	8.5	28.0
Colorado	18.0	30.0	7.7	—	7.7	28.0
Connecticut	16.0	36.0	7.9	25.9	8.5	28.0
Delaware	18.0	—	—	—	—	—
District of Columbia	20.0	—	(*)	(*)	(*)	(*)
Florida	18.0	—	7.8	25.5	8.0	28.0
Georgia	18.0	35.0	(*)	(*)	(*)	(*)
Hawaii	18.0	30.0	(*)	(*)	(*)	(*)
Idaho	18.0	—	(*)	(*)	(*)	(*)
Illinois	18.0	30.0	7.9	25.9	8.5	28.0
Indiana	18.0	30.0	7.9	25.9	8.5	28.0
Iowa	16.0	—	—	—	—	—
Kansas	18.0	30.0	7.8	25.5	8.0	28.0
Kentucky	18.0	30.0	7.9	25.9	8.5	28.0
Louisiana	18.0	30.0	7.9	25.5	8.0	28.0
Maine	18.0	—	7.9	25.9	8.5	28.0
Maryland	18.0	—	(*)	(*)	(*)	(*)
Massachusetts	16.0	—	7.8	—	8.0	—
Michigan	18.0	30.0				
Minnesota	20.0	30.0	7.9	25.9	8.5	28.0
Mississippi	18.0	—	7.8	25.5	8.0	28.0
Missouri	18.0	30.0	7.9	25.9	8.0	28.0
Montana	18.0	30.0	7.8	25.5	8.0	28.0
Nebraska	18.0	30.0	(*)	(*)	(*)	(*)
Nevada	22.0	36.0	7.9	25.9	8.5	28.0
New Hampshire	18.0	—	7.9	25.9	8.5	28.0
New Jersey	16.0	—	7.8	25.5	8.0	28.0
New Mexico	18.0	30.0	7.8	25.5	8.0	28.0
New York	18.0	—	7.8	25.5	8.0	28.0
North Carolina	18.0	36.0	7.8	25.5	8.0	28.0
North Dakota	20.0	30.0	7.9	25.9	8.5	28.0
Ohio	18.0	30.0	7.9	25.9	8.5	28.0
Oklahoma	18.0	30.0	7.9	25.9	8.5	28.0
Oregon	18.0	30.0	7.9	25.9	—	—
Pennsylvania	18.0	36.0	7.8	25.5	8.0	28.0
Puerto Rico	18.0	—	—	—	—	—
Rhode Island	18.0	30.0	7.8	25.5	8.0	28.0
South Carolina	18.0	30.0	—	—	—	—
South Dakota	18.0	30.0	(*)	(*)	(*)	(*)
Tennessee	18.0	30.0	7.8	25.5	8.0	28.0
Texas	18.0	30.0	7.9	25.9	8.5	28.0
Utah	18.0	30.0	7.8	25.5	(*)	(*)
Vermont	18.0	—	—	—	—	—
Virginia	18.0	(*)	(*)	(*)	(*)	(*)
Washington	20.0	30.0	7.9	25.9	—	28.0
West Virginia	18.0	30.0	7.8	25.5	7.8	28.0
Wisconsin	18.0	30.0	7.9	25.9	8.0	28.0
Wyoming	18.0	30.0	7.9	25.9	7.9	25.9

(*) Follow Federal Food and Drug Standards

356   CHEMISTRY AND TESTING OF DAIRY PRODUCTS

FEDERAL AND STATE STANDARDS FOR THE COMPOSITION OF MILK PRODUCTS

	Plain ice cream			Fruit, nut or chocolate ice cream		
	Milk-fat	Total milk solids	Wt. per gal.	Milk-fat	Total milk solids	Wt. per gal.
	Min. %	Min. %	Min. lbs.	Min. %	Min. %	Min. lbs.
Federal						
Alabama	10.0	18.0	—	8.0	18.0	—
Alaska	10.0	18.0	—	8.0	14.0	—
Arizona	10.0	—	4.5	8.0	—	4.5
Arkansas	10.0	18.0	—	8.0	14.0	—
California	10.0	—	—	8.0	—	—
Colorado	12.0	—	—	10.0	—	—
Connecticut	10.0	—	4.5	8.0	—	4.5
Delaware	12.0	+5.0	—	8.0	+5.0	—
District of Columbia	8.0	—	—	8.0	—	—
Florida	10.0	18.0	—	8.0	14.0	—
Georgia	10.0	—	4.5	8.0	—	4.5
Hawaii	12.0	18.0	4.5	10.0	15.0	4.5
Idaho	12.0	18.0	4.25	10.0	14.0	4.25
Illinois	12.0	—	—	10.0	—	—
Indiana	10.0	18.0	4.25	8.0	14.0	4.25
Iowa	12.0	20.0	4.5	10.0	16.0	4.5
Kansas	10.0	20.0	4.5	10.0	20.0	4.5
Kentucky	10.0	18.0	—	10.0	18.0	—
Louisiana	10.0	20.0	4.5	8.0	16.0	4.5
Maine	11.0	—	4.5	9.0	—	4.5
Maryland	12.0	20.0	4.5	8.0	15.0	4.5
Massachusetts	10.0	18.5	—	8.0	16.5	—
Michigan	12.0	—	4.5	10.0	—	4.5
Minnesota	12.0	20.0	4.5	10.0	16.0	4.5
Mississippi	10.0	—	4.5	8.0	—	4.5
Missouri	10.0	20.0	4.5	8.0	16.0	4.5
Montana	10.0	20.0	4.25	9.0	16.0	4.25
Nebraska	12.0	20.0	4.5	10.0	18.0	4.5
Nevada	14.0	—	4.5	14.0	—	4.5
New Hampshire	14.0	—	4.5	12.0	—	4.5
New Jersey	10.0	+5.0	—	8.0	+5.0	—
New Mexico	12.0	20.0	—	10.0	—	—
New York	10.0	18.0	—	8.0	14.0	—
North Carolina	10.0	—	4.5	8.0	—	4.5
North Dakota	12.0	—	4.5	10.0	—	4.5
Ohio	10.0	18.0	4.25	8.0	14.0	4.25
Oklahoma	10.0	18.0	4.5	8.0	18.0	4.5
Oregon	10.0	18.0	4.5	10.0	18.0	4.5
Pennsylvania	10.0	—	4.75	8.0	—	4.75
Puerto Rico	—	—	—	—	—	—
Rhode Island	10.0	20.0	4.5	8.0	16.0	4.5
South Carolina	10.0	18.0	4.25	8.0	16.0	4.25
South Dakota	12.0	—	4.5	10.0	—	4.5
Tennessee	10.0	18.0	—	8.0	14.0	—
Texas	8.0	—	4.5	6.0	—	4.5
Utah	12.0	20.0	4.2	10.0	16.0	4.2
Vermont	10.0	20.0	4.5	8.0	17.0	4.5
Virginia	10.0	20.0	4.5	8.0	16.0	4.5
Washington	10.0	20.0	—	10.0	20.0	—
West Virginia	8.0	—	—	8.0	—	—
Wisconsin	13.0	—	4.5	11.0	—	4.5
Wyoming	10.0	—	4.25	10.0	—	4.25

FEDERAL AND STATE STANDARDS FOR THE COMPOSITION OF MILK PRODUCTS

	Ice milk		
	Milkfat		Total milk solids
	Min. %	Max. %	Min. %
Federal			
Alabama	2.5	5.0	14.0
Alaska	3.0	8.0	14.0
Arizona	4.0	8.0	—
Arkansas	3.0	10.0	14.0
California	4.0	—	—
Colorado	—	—12.0	—
Connecticut	3.25	6.0	—
Delaware	4.0	—8.0	—
District of Columbia	3.5	—	—
Florida	3.0	10.0	14.0
Georgia	—	—	—
Hawaii	3.0	5.0	—
Idaho	4.0	—	14.0
Illinois	3.0	—	—
Indiana	Illegal		
Iowa	3.25	6.0	11.0
Kansas	—	—10.0	—
Kentucky	2.0	3.5	11.0
Louisiana	4.0	—	10.0
Maine	3.25	—	—
Maryland	Illegal		
Massachusetts	3.35	10.0	14.0
Michigan	Classified as sherbet		
Minnesota	2.0	12.0	14.0
Mississippi	3.0	5.0	—
Missouri	3.25	6.0	14.0
Montana	2.0	4.99	11.0
Nebraska	3.0	8.0	12.0
Nevada	4.0	—	—
New Hampshire	Illegal		
New Jersey	3.0	—10.0	14.0
New Mexico	—	—	—
New York			
North Carolina	4.0	—10.0	—
North Dakota	4.0	6.0	—
Ohio	3.5	—	16.0
Oklahoma	3.25	—	11.0
Oregon	3.2	10.0	14.0
Pennsylvania	—10.0	—	14.0
Puerto Rico	—	—	—
Rhode Island	3.25	6.0	11.0
South Carolina	4.0	—10.0	12.0
South Dakota	Illegal		
Tennessee	3.0	10.0	14.0
Texas	—	—	—
Utah	4.0	—	—
Vermont	3.5	6.0	—
Virginia	3.25	4.25	11.0
Washington	3.25	—	—
West Virginia	3.0	6.0	—
Wisconsin	3.0	—13.0	—
Wyoming	3.25	5.5	—

FEDERAL AND STATE STANDARDS FOR THE COMPOSITION OF MILK PRODUCTS

	Cottage cheese		
	Plain	Creamed	
	Moisture	Milkfat	Moisture
	Max. %	Min. %	Max. %
Federal	80.0	4.0	80.0
Alabama	—	—	—
Alaska	(*)	(*)	(*)
Arizona	—	4.0	—
Arkansas	(*)	(*)	(*)
California	80.0	4.0	80.0
Colorado	80.0	4.0	80.0
Connecticut	(*)	(*)	(*)
Delaware	80.0	—	—
District of Columbia	(*)	(*)	(*)
Florida	—	—	—
Georgia	(*)	(*)	(*)
Hawaii	(*)	(*)	(*)
Idaho	(*)	(*)	(*)
Illinois	80.0	4.0	80.0
Indiana	80.0	4.0	80.0
Iowa	—	—	—
Kansas	80.0	4.0	80.0
Kentucky	80.0	4.0	80.0
Louisiana	80.0	4.0	80.0
Maine	80.0	4.0	80.0
Maryland	80.0	4.0	80.0
Massachusetts	(*)	(*)	(*)
Michigan	80.0	4.0	80.0
Minnesota	80.0	4.0	80.0
Mississippi	80.0	4.0	80.0
Missouri	—	—	—
Montana	80.0	4.0	80.0
Nebraska	(*)	(*)	(*)
Nevada	—	4.0	—
New Hampshire	80.0	4.0	80.0
New Jersey	—	—	—
New Mexico	(*)	(*)	(*)
New York	(*)	(*)	(*)
North Carolina	80.0	4.0	80.0
North Dakota	—	—	—
Ohio	—	4.0	—
Oklahoma	80.0	4.0	80.0
Oregon	80.0	4.0	80.0
Pennsylvania	—	—	—
Puerto Rico	—	—	—
Rhode Island	(*)	(*)	(*)
South Carolina	80.0	4.0	80.0
South Dakota	(*)	(*)	(*)
Tennessee	—	—	—
Texas	80.0	4.0	80.0
Utah	(*)	4.0	(*)
Vermont	(*)	(*)	(*)
Virginia	(*)	(*)	(*)
Washington	—	4.0	—
West Virginia	(*)	(*)	(*)
Wisconsin	80.0	4.0	80.0
Wyoming	—	4.0	—

(*) Follow Federal Food and Drug Standards

Index

Acetic series 6
Acetylmethylcarbinol in
    starters 279
Acid measures 72
Acid phosphates in milk and
    amphoteric reaction ... 14
Acidity of
    butter 240
    buttermilk 239
    cheese 241
    chocolate milk 240
    cream 239
    evaporated milk 239
    milk 239
        apparent acidity 235
        real acidity 236
            factors affecting .244-246
    skimmilk 239
    whey 239
Acidity, relationship in different products 244
Acidity test, chemistry of .237, 241
Acidometer 79
Acid value 10
Adulterations of milk
    foreign residues 198
    skimming 158
    skimming and watering .. 160
    substitution of vegetable
        fats 193
    watering 159
Alanine, amino acid 16
Albumin in milk 20
    analysis of 22
Alcaligines viscosus, cause of
    ropy milk 282
Alcohol, amyl, use in Gerber
    test 98
    butyl, use in buttermilk
        test 116
Aldose sugar 25
Alpha lactose 24
Alpha tocopherol, active form
    of vitamin E 29
American Association test for
    buttermilk, skimmilk
    and whole milk 116
American Cheddar cheese 225, 310
Amino acids in milk proteins. 16
    essential, definition of ..15, 16
    non-essential, definition of 16

formulae of 17, 18, 19
A.O.A.C., test for solids in
    milk and cream 132
Amphoteric reaction 14, 236
Apparent acidity 235
Arginine, amino acid 16
Ascorbic acid, vitamin C ..30, 300
Ash in milk, minerals in ... 27
Aspartic acid, amino acid .. 16
Atomic weights 347
Autoclaving milk, browning
    effect 51
Babcock test, accuracy of .. 93
    apparatus required 70
    introduction 67
Babcock test for
    buttermilk 116
    churned milk 92
    cream 86
    frozen milk 93
    normal milk 70
    skimmilk 91
    sour milk 93
    whey 91
Babcock test bottles, specifications
    milk 70
    cream 86
    skimmilk 91, 92
Bacteria in milk, growth ... 263
    methylene blue, reduction
        test for 274
    microscopic count 270
    number 265
    plating for 267
    reproduction 263
    resazurin test for 277
    size 263
    sources 265
Base-exchange process, soft
    curd milk 306
Beimling test 69
Beta-lactoglobulin, amino
    acids in 16
Beta lactose 24
Bichromate, potassium, preservative 61
Biotin, vitamin 30, 302
Board of Health lactometer . 147
Boiled milk test 196
Boiling point of milk 40
Boric acid, test for in milk . 196

Bound water in milk, cream 46, 47
Breed method, counting bacteria in milk 270
British thermal unit 41
Brom cresol green, indicator 250
Brom cresol purple, indicator 250
Brom phenol blue, indicator. 250
Brom thymol blue, indicator. 250
Brom thymol blue test, for mastitis 284
Browning of milk 51
B.t.u., definition of 41
Buffers in milk 249
Butter, acid test 240
  Babcock test for 118
  composition 207
  food value 308
  free water in, test for 224
  Kohman method of analysis 220
  moisture test,
    factory method 211
    official method 209, 214
  preparation of samples ... 209
  salt test 215
  sampling methods 209
  sediment test 221, 223
  yeasts and molds in 289
Butterfat (milk fat) mixture of fats 5
Buttermilk, acid test 239
  fat tests, American Association 116
    Gerber 100
  composition 348
  food value 311
Butylene glycol, in starters . 279
Butyric acid, in milk fat ... 4, 5
Butyrometer 97
Calciferol, active form vitamin D 28
Calcium caseinate 14
Calibration of glassware ... 85
Calomel electrode 251
Calorie, definition of 41
Capric, fatty acid 4
Caproic, fatty acid 4
Caprylic, fatty acid 4
Carbonates, test for 196
Carotene in milk 28
Casein in
  butter 207, 348
  cheese 225
  milk 1, 14
Casein, amphoteric reaction in milk 14
  chemical analyses of 22
  specific gravity of 14
Catalase, enzyme 31
Cenco, solids test 138
Centigrade vs. Fahrenheit scale 349

Centipoise, definition of ... 45
Centrifuge, Babcock 72
Centrifuge, Gerber 100
  speed of, Babcock 81
Cephalin, phospholipid 12
Charred tests, cause of 79
Cheddar cheese, acid test .. 241
  composition 225
  fat test, Babcock 92
  fat test, Pennsylvania method 118
  food value 310
  moisture tests
    Karl Fischer titration method 227
    Olive Oil method 227
    Troy method 226
  salt tests, A.D.S.A., Marquardt, mercurimetric 230, 231
  sampling methods 225
  sediment test 228
Cheese, cottage, cream, Cheddar, Neufchatel, Roquefort, Swiss, composition and food value of 310
Chloramins, test for 201
Chloride test for mastitis .. 285
Chlor phenol red, indicator . 250
Chocolate milk
  fat test 96, 116
  food value 312
  formula for 312
Cholesterol, sterol in milk .. 13
Choline 13, 30, 303
Churned milk, fat test for .. 93
Citric acid, in milk 40
Citrulline, amino acid 16
Cleaning, chemistry of 316
  sanitizers — chlorine, iodophors, quaternary ammonium compounds .328-331
Cleaning glassware 83
Clumping of fat 49
Cobalt in vitamin B_{12} 299
Color of milk, factors affecting 35, 36
Colorimetric determination of pH 250
Colostrum, pH of 252
Composite samples 58
  accuracy of 62, 63
  care of 60
  fat test of
    cream 64
    milk 61
  methods of taking 58, 59
  preservatives for 61
Condensed milk, evaporated, defined 343

INDEX

sweetened condensed, defined 344
   fat tests for 117, 118
Conductivity 50
Cornell phosphatase test 167
Corrosive sublimate, preservative 61
Cottage cheese, fat test 118
   food value 310
Cream
   acid test 239
   composition 348
   definition 340
   fat tests 86, 88, 108
   sampling procedure 64
   sediment test 221
   surface tension 44
Cream cheese, composition and food value 310
   fat test 118
Cream line, factors affecting 39
Cream test bottle, specifications 86
Cresol red, indicator 250
Cresolphthalein, indicator .. 250
Cryoscopes, detection of watering
   Hortvet 36, 189
   thermistors 190
Curdled milk, acid test for . 239
Curdy tests, cause of 79
Cystine, amino acid 16
De Laval's lactocrite 68
Density, effect of temperature 148
Desiccator 214
Detergent tests for fat 122
   1. D.P.S. test, reagents .. 124
      cream, procedure ... 125
      ice cream, procedure . 126
      milk, procedure 124
   2. Schain test 122
   3. TeSa Reagent method . 126
      chocolate milk 128
      whole milks — composite, homogenized, pasteurized and raw 126
Diacetyl, in butter cultures . 279
Diameter of wheel, relation to speed 81
Diastase, enzyme 32
Dietert, total solids in milk products 135
Digestibility of nutrients in milk 304
Direct microscopic method .. 270
Dividers, use of 73
Dried milk, sampling 64
   test for fat 118
Drip sample 63
Dyne, definition of 43
Electrical conductivity 50

Electrodes, calomel, glass, quinhydrone 251
Emulsion 4
Energy value of milk 303
Enzyme activity, effect on acidity of milk 246
Enzymes in milk 31
Ergosterol 28
Escherichia-Aerobacter organisms 280
Essential amino acids 15, 16
Euglobulin 21
Evaporated milk, defined .. 343
   sampling and testing for fat 64
Fahrenheit vs. Centrigrade and Reaumur thermometers 349
Failyer and Willard's test .. 69
Falling drop, measure of surface tension 43
Fat in milk 1, 4
   clumping 49
   composition of 4, 5, 6
   expansion of 83
   factors affecting 8
   properties 8
   relation to specific gravity of milk 42
   relationship to solids-not-fat 42, 155
   solubility 9
   specific gravity of 9
Fat tests, see Babcock, Detergent, Gerber and Mojonnier
Fatty acids in milk fat 4
Feathering in cream 175
   test for 176
Fehling's solution 25
Fishy flavor 13
Foam in cream, milk and skimmilk 47
   factors affecting 48
Foamy fat tests, cause of .. 81
Folic acid 30, 302
Folin-Ciocalteau reagent for phosphatase test 165
Food value of
   butter 308
   buttermilk 311
   cheese 310
   chocolate milk 312
   ice cream 310
   milk 294
   skimmilk 311
Foreign residues in milk ... 198
Formaldehyde in milk, test for 195
Formalin, preservative 61

Freezing point of cream, of milk, of skimmilk ... 36, 37
Frozen dairy products, sampling 65
Frozen milk, composition of 38, 39
  effect on creaming 39
  fat test for 93
Fructose 23
Galactase, enzyme 31
Galactose, simple sugar in lactose 22, 23
Garrett-Overman test for ice cream 121
Gases in milk 32, 33
Gerber test for fat in dairy products
  accuracy of 101, 102
  buttermilk 100
  cream 100
  frozen desserts 101
  whole milk 96
  skimmilk 100
Germicidal period 264
Glass electrode 251
Glassware, cleaning 83
Globule, fat, size of 4
Globulin, analysis of 21, 22
Glossary of terms 352
Glucose 23
Glutamic acid, amino acid .. 16
Glycerol, component of fats. 4
Glycine, amino acid 16
Glymol 90
Heat, effect on milk 51
High quality milk, production of 290
Histidine, amino acid 16
Homogenized milk
  test for efficiency 183, 184
  test for fat, see whole milk
Hortvet cryoscope 189
Hotis test for mastitis 285
Human milk, composition .. 304
Hydrogen ion concentration . 247
Hydrometer 146
Hydroxyproline, amino acid . 16
Hypochlorites, test for 201
Ice cream, food value 311
Ice cream tests
  acetic-sulphuric acid 118
  Garrett-Overman 121
  Gerber 101
  Minnesota 119
  Mojonnier 108
  Nebraska 120
  sampling 65
Immersion refractometer ... 189
Indicators, table of 250
Inositol 30, 302
Invert sugar 23
Iodine number 9

Iodophors 330
Isoelectric point 15
Isoleucine, amino acid 16
Kay-Graham test for pasteurization 164
Ketose sugar 25
Kohman method, analyzing butter 220
Laboratory method, analyzing butter 211
Lacatlbumin 1, 20
Lactase enzyme 32
Lactation, effect on acidity of milk 245
Lactic acid, importance of .. 237
Lactobacillus, acidophilus, bulgaricus 280
Lactobutyrometer 67
Lactocrite test 68
Lactoglobulin 21
Lactometers 147
  lactometer readings, high vs. low fat testing milks 156
Lactose in milk
  alpha and beta 24
  composition of 22, 23
  food value 23
  lactic acid from 23
  reducing sugar 25
  relative sweetness 23
  specific gravity 40
  structural formula 26
Lauric, fatty acid 4
Lecithin, phospholipid in milk 12
Leffman and Beam's test ... 69
Leucine, amino acid 16
Liberman's test 68
Lignoceric, fatty acid 5
Linkage, single and double . 6
Linoleic acid 4
Linoleic series 6
Lipase, enzyme 31
Lipids, compound, derived . 11, 13
Lovibond tintometer 165
Lysine, amino acid 16
Malted milk, sampling 64
Maltose 23
Marchand's volumetric test . 67
Mastitis, defined 283
  efficiency of different tests 286
  effect on acidity of milk . 246
  tests for, brom thymol blue, chloride, Hotis, microscopic and Whiteside. 284, 285
McCay sampler 59
Melting point of fatty acids . 6
Meniscus defined 77
Mercuric chloride, preservative 61

INDEX

Meta cresol purple, indicator. 250
Methionine, amino acid 16
Methylene blue reduction test 274
Methyl red, indicator 250
Mho, definition of 50
Microscopic method of counting bacteria 270
Microscopic test for mastitis 284
Milk
   Babcock test for fat 70
   boiling point 40
   composition 1, 348
   composition of various milks 2
   definition 1
   energy value, formula for 13
   Gerber test for fat 96
   high quality, production of. 290
   human, composition of ... 304
   Mojonnier test for fat ... 106
   pasteurized 342
   sediment test 171
   soft curds 305
   specific gravity 40, 41
   specific heat 41
   surface tension 43, 44
   viscosity 45
Milk fat, mixture of fats 5
Milk plasma, definition of .. 1
Milk serum, definition of ... 1
Milk stone, cause of 317, 319
Mineral balance, effect on milk 52
Mineralized milk 298
Minerals in milk 27, 297
Minnesota test for ice cream. 119
Moisture and solids problems 258
Moisture test for
   butter 209
   cheese 225
Mojonnier test for fat and solids 103, 133
Molar solutions 256
Molds 288
   effect on fat tests 63
Myristic, fatty acid 4
Nebraska test for ice cream . 120
Neufchatel cheese, food value 310
Niacin, vitamin 29, 301
Non-reducing sugar 25, 26
Non-volatile fatty acids 7
Normal solution, definition of 255
Ohm, definition of 50
Oil of vitriol 72
Oiling-off test for cream ... 178
Oleic, fatty acid 4
Oleic series 6
Olive oil test, cheese moisture 227
Optical method, fat determination 129
Overrun in butter, churn, composition, factory ... 208

Oxalic acid, use in standardizing 257
Palmitic, fatty acid 4
Pantothenic acid 30, 302
Paracasein 14
Parson's test 68
Pasteurization, methods ... 342
Pasteurization, test for ..164, 167
Pasteurized milk vs. raw milk 308
Pathogenic bacteria 286
Penicillin, drug 23
   test for in milk 201
Pennsylvania test for fat in
   butter 118
   cheese 118
   chocolate milk 117
   dried whole milk 118
   evaporated milk 118
   ice cream 117
   sweetened condensed 117
Pepsin 14
Percentage solutions 256
Peroxidase, enzyme 32
pH, definition 247
   colorimetric determination of 250
   of milk 252
   potentiometric determination of 250
Phenol red, indicator 250
Phenolphthalein, indicator ..
   ....238, 250
Phenylalanine, amino acid .. 16
Phosphatase, enzyme 32
Phosphatase test for pasteurization of milk
   Cornell method 162
   factors affecting 170
   Scharer method 164
Phospholipids 11
Photometric method of determining fat in skimmilk . 130
Pipette for composite sampling 59
Plasma, milk, definition of . 1
Plasma solids, specific gravity of 40
Plate method of counting bacteria 267
Platinum ring, measure of surface tension 44
Poise, definition of 45
Potassium bichromate, preservative 61
Potassium chromate, indicator in salt test 215
Potentiometer, description of 250
Powdered milk, test for fat . 118
Preservatives 61, 195
Proline, amino acid 16
Protease, enzyme 31

Protein
  digestibility 304
  specific gravity 40
  stability, test for, Storrs .. 179
Pseudoglobulin 21
Psychrophiles 281
Pycnometer 151
Pyridoxine 30, 302
Quaternary ammonium compounds in milk or water,
  test for 336, 337
Quevenne lactometer 146
Quinhydrone electrode 251
Rancidity in milk, test for .. 181
Rapid acid test 242
Real acidity, cause of 236
Reaumur thermometer 349
Recknagle effect, on specific
  gravity 41
Reducing sugar 25, 26
Reductase, enzyme 32
Refractometer, immersion .. 189
Reichert-Meissl number 9
Remade milk 304
Rennet 14
Rennin, enzyme 14
Resazurin color test 277
Riboflavin, vitamin 20, 301
Ropy milk, cause of 282
Roquefort cheese 310
Salt bridge 251
Salt, sediment test of 221
Salt test for butter 215
Sampling
  butter 65
  cheese 65
  composite samples .. 58, 59, 63
  cream 64
  evaporated milk 64
  malted or dried milks ... 64
  sweetened condensed 64
  whole milk 55
Sampling tubes, McCay, Scovell 59
Saponification number 10
Saturated fatty acids 6
Scales for cream 86
Schardinger's enzyme 32
Scharer field test for pasteurization 165
Schmidt's test 68
Scovell sampler 59
Sediment test of
  butter 221
  cheese 228
  cream 221
  milk 171
Sepascope 130
Serine, amino acid 16
Serum, milk, definition of .. 1
Sharp's and Hart's formula
  for total solids 153

Short's test 68
Skimmilk
  acid test 239
  fat tests 91, 116
  food value 311
  surface tension 44
Soap 7
Sodium stearate 8
Soft curd milk 305
Solids in milk, formulae for .
  .... 152, 153, 154
Solids not fat, formulae for . 153
Solutions
  molar 256
  normal 255
  percentage 256
  problems 255, 256
  standardizing 257
Sour cream, sediment test .. 222
Specific gravity of
  milk and constituents ....
  .... 40, 41, 146
  milk fat 9
Specific heat of milk and its
  products 41, 43
Speed of centrifuge 81
Sphingomyelin 12
Standardization problems .. 259
Standardizing acid and alkali
  solutions 257
Standards, fat and solids 156, 340
Stearic, fatty acid 4
Stearyl olyl palmitin 5
Sterols 13
Storr's test for protein stability 179
Streptococcus citrovorus and
  paracitrovorus in starters 215
Streptococcus lactis 236, 278
Sucrose 23, 26
Sugar, milk 22, 23, 26
Sulfhydryls, cooked flavor .. 51
Sulphuric acid
  advantages 80
  care 79
  specific gravity 72
Surface tension of milk, measurements — falling drop,
  platinum ring, tensiometer 43, 44
Sweet cream, test for sediment 221
Sweet curdling 282
Sweetened condensed milk,
  .. fat test 117, 118
Swiss cheese, food value ... 310
Taste of milk, lactose-chloride relationship 35
Thiamin, vitamin 29, 300
Threonine, amino acid 16
Thymol blue, indicator 250

INDEX

Titratable acidity, factors affecting 244, 246
Toluene distillation for solids 139
Total solids, defined 1
Triglyceride 4
Trimethyl amine 13
Troy's formula for solids not fat 153
Tryptophane, amino acid ... 16
Unsaturated fatty acids 6
Valine, amino acid 16
Vegetable fats, substitutes .. 193
Viosterol 28
Viscosity of milk
  factors affecting 45
  measurements
    Borden flow meter 174
    pipette method 173
    viscosimeter 45
Vitamin A potency of butter . 309
Vitamins in milk 28-31, 300

Volatile fatty acids 7
Water bath 73, 76
Water hardness, test for ... 335
Water in milk 1, 4
Watering milk 159
  detection by
    freezing point, cryoscope 189
    lactometer 159
    refractometer 189
Water supply 325
Westphal balance 149
Whey
  acid test 239
  fat test 91, 116, 117
Whiteside test for mastitis .. 285
Whole milk powder, fat test . 118
Yeasts, torulae 23, 288
Zeiss, immersion refractometer 189
Zeolite 306
Zinc, in milk 298